Picturing Knowledge

Historical and Philosophical Problems Concerning the Use of Art in Science

The traditional concept of scientific knowledge places a premium on thinking, not visualizing. Scientific illustrations are still generally regarded as devices that serve as heuristic aids when reasoning breaks down. When scientific illustration is not used in this disparaging sense as a linguistic aid, it is most often employed as a metaphor with no special visual content. What distinguishes pictorial devices as resources for scientific study, and the special problems that are raised by the mere presence of visual elements in scientific treatises, tends to be overlooked.

The contributors to this volume examine the historical and philosophical issues concerning the role that scientific illustration plays in the creation of scientific knowledge. They regard both text and picture as resources that scientists employ in their practical activities, their value as scientific resources deriving from their ability to convey information.

(Toronto Studies in Philosophy)

BRIAN S. BAIGRIE is an associate professor in the Institute for History and Philosophy of Science and Technology at the University of Toronto.

Picturing Knowledge

Historical and Philosophical Problems Concerning the Use of Art in Science

Edited by
BRIAN S. BAIGRIE

UNIVERSITY OF TORONTO PRESS
Toronto Buffalo London

Printed in Canada

ISBN 0-8020-2985-X
ISBN 0-8020-7439-1

Printed on acid-free paper

Canadian Cataloguing in Publication Data

Main entry under title:

Picturing knowledge : historical and philosophical problems concerning the use of art in science

(Toronto studies in philosophy)
Includes bibliographical references and index.
ISBN 0-8020-2985-X (bound) ISBN 0-8020-7439-1 (pbk.)

1. Scientific illustration – History.
2. Scientific illustration – Philosophy.
3. Art and science – History. 4. Art and science – Philosophy. I. Baigrie, Brian Scott. II. Series.

Q222.P53 1996 502.2'2 C95-932643-X

University of Toronto Press acknowledges the financial assistance to its publishing program of the Canada Council and the Ontario Arts Council.

Contents

Illustrations

Introduction

This volume engages a cluster of historical and philosophical problems concerning the use of art in science. These issues criss-cross the borders of a number of disciplines – history of art, science, and technology; philosophy of art, science, and technology; cultural studies; medieval studies; anthropology; and the sociology of science. Though the contributors are drawn from a wide range of academic disciplines, each paper focuses on one particular issue that stamps these various problems with significance and connects them one to another – viz., the role that scientific illustration plays in the creation of scientific knowledge.

An appreciation of the role that illustration plays in the creation of scientific knowledge can only help to correct the privilege given to theory in the image of science inherited from logical empiricism. As Ronald Giere and Michael Ruse note, scientific knowledge has traditionally been taken to be encapsulated in theories which, in turn, are interpreted by the logical empiricist as axiomatic systems. The primary mode of scientific representation for the logical empiricist is linguistic. Thinking – and not visualizing – is held to be conducive to this activity. It is the rules of logic that legitimate inferences in science, and not the sorts of conventions that figure in the interpretation of works of art.

The logical empiricist image of science attaches no epistemic weight to the other cognitive and material resources that scientists routinely employ in their more practical activities. For example, in the wake of logical empiricism, experiment came to be seen as an activity that is performed in the service of theory; the familiar characterization of experiment as a 'litmus test' of theoretically derived results is a reflection of the cognitive authority that the logical empiricist conferred on linguistic representations. This cognitive authority resonates in standard portraits of the relationship between technology and science – technology, many scholars still believe, has only one life and it is given in the service of theory. Of course, this image is undergoing wholesale revision insofar as experiment and technology are concerned (see Buchwald 1995). Recent research has indicated, for example, that experiment is often most intensive and innovative when theoretical results are not at issue. Further to this, scholars have recently come to appreciate the extent to which technological interests dominate experimentation, the goal being a piece of equipment with wide industrial applications that performs more reliably.

Despite the publication of a number of important historical studies on the role of pictorial devices in particular sciences during the last quarter century (see Rudwick 1976; and Ferguson 1977), the textual deposits of scientists and scientific illustration are still generally regarded by science studies scholars as fundamentally different modes of expression – as different realms of language and imagery. Coupled with the deep-seated conviction that human thinking takes place in words, the supposition persists that the pictures in science are psychological devices that serve as heuristic aids when reasoning breaks down. When the term 'picture' is not used in this disparaging sense as a linguistic aid, it is most often employed in humanistic studies of science as a metaphor with no special visual content, as in Eduard Dijksterhuis's *The Mechanization of the World Picture* (1969) and countless scholarly articles with such titles as 'A Picture of Victorian Science.' A few scholarly anthologies (e.g., Lynch and Woolgar 1990) have raised important questions about the cognitive authority granted to linguistic representation but, if visual depictions are considered at all, they are typically subsumed under the general philosophical concept of 'representation' – a concept that speaks to a wide variety of scientific activities. What distinguishes pictorial devices as resources for doing science, and the special problems that are raised by the mere presence of visual elements in scientific treatises, tend to be eclipsed by philosophical worries about the nature of representation.

In this volume, the illustrations in science are understood in a circumscribed way – as images that are meant to be seen with associated text (see Knight 1985, and Topper, below). Although scientific illustrations are to be understood in the context of associated text, this does not mean that they exist solely to shed light on the text. Both text and picture are resources that scientists employ in their practical activities, their value as scientific resources presumably deriving from their ability to convey information. It is not surprising, therefore, that often it is the pictures that do the bulk of the scientific work. All the papers in this volume address some aspect of this thesis, but three seem especially relevant. Robert O'Hara's paper is the only contribution that has been published previously. He takes as his example the concept of a 'natural system' and explores the various ways that elements of this concept were visually depicted in the literature on the science of systematics in the nineteenth century. O'Hara's paper is reprinted here because it powerfully illustrates the thesis that some scientific concepts are best presented diagramatically. Michael Ruse takes as his example the influential 'adaptive landscapes' of the population geneticist Sewall Wright. Ruse insists that Wright's pictorial devices were indispensable, not only for the run-of-the-mill evolutionist, but for eminent evolutionists like T. Dobzhansky and G.L. Stebbins, who were unable to follow the mathematics of population genetics. And, finally, James Robert Brown argues that, in special cases, we can correctly infer theories from visualizable situations. This thesis is initially argued in the context of visualization in mathematics. However, Brown then extends his argument to the construction of phenomena out of data, noting that in the physical sciences there are different types of visual reasoning, one involved in the construction of phenomena (which Brown takes to be patterns or natural kinds that we can picture) and another involved in the creation of diagrams (e.g., Feynman diagrams) that bear no resemblance to the data that inform them.

The two features of scientific illustration that have been identified – namely, that they are to be seen with associated text and they impart information – are closely connected. This relationship is best appreciated if we classify scientific illustrations according to the degree to which they are integrated with textual and other kinds of symbolic elements. A raw image, say, a photograph of the lunar surface or a nesting bird, is most easily understood by the uninitiated – that is, people who are unfamiliar with cameras have no trouble seeing photos as pictures. By the same token, photos are the least informative: they contain so much raw information that they have little scientific utility. Visual depictions, whether mathematical proportions like the epicycles in standard works in Renaissance astronomy (see fig. 2.11) or Darwin's visualization of hypothetical divergence and extinction over time (fig. 7.9), give us tangible images of bits and pieces of nature, but they do not aim at recording what is visible. The difference between the naturalistic illustrations, say, in Andreas Vesalius's *De humani corporis fabrica* (see figs. 2.6 and 2.8), and a photograph of a human body is that Vesalius's pictures allow us to see things that strictly speaking are invisible in the photograph. As Martin Kemp notes, pictures make things visible – a purpose that applies just as well to the electron-scanning microscope as it does to depictions of epicycles (see Gombrich 1974, p. 184).

Though there are exceptions, of course, the identification of regions and individual objects, and the removal of shading, colour, and other distractions, promotes this end. In caricature, for example, the elimination of shading and real colour helps the illustrator highlight features that otherwise tend to blend in with a mass of detail. Line drawing – which is the simplest form of caricature in scientific illustration – lets the illustrator control exactly what the user sees. However, these devices are only useful for the initiated, who can still see the caricature as a picture of a particular figure (see Larkin and Simon 1987). The uninitiated may recognize that the picture is meant to caricature but not know what it is meant to portray.

When symbols are integrated into the picture to label objects that are depicted, the line drawing is transformed into a powerful, epistemological vehicle – the simple caricature is changed into a map or a plan. The integration of symbols can serve a number of distinct ends. For example, symbols can indicate that the depicted objects are meant to preserve the spatial relations (or some elements of the spatial order) between the objects symbolized. In the case of diagrams, in which there is near complete integration of the image with linguistic and other symbolic elements, spatial ordering may be set aside altogether so as to convey essential meanings. Bert Hall notes that a circuit diagram, for instance, bears little resemblance to the guts of a computer, but, as a plan of the salient features of computing machines, it helps the computer technician to identify sources of glitches in ways that are denied to raw photographs. Extraneous information is eliminated; the computer technician already knows, for example, how the different parts of the circuit board fit together, and so there is not need to include this in circuit diagrams.

For this reason, diagrams typically contain few hints of naturalism. Hall submits that they gain their epistemic purchase from the ability of the user to make the necessary inferences; that is, every diagram is a kind of encoding that demands a set of conventions that are shared by the illustration and the user. If the user is unfamiliar with the conventions at work, this compromises their utility. The development of a set of conventions or 'visual language' for a particular scientific community is chronicled in David Knight's paper on nineteenth-century chemistry. Along the way, the reader is taken on a tour of the many processes involved in the production of chemical illustrations – copperplate engravings featured in Lavoisier's *Elements of Chemistry: In a New Systematic Order* (1789), comparatively cheap lithographs that were introduced in the early nineteenth century, wood-engravings, and very inexpensive photographs that finally came into their own for work on spectra.

The suggestion that scientific illustration is a kind of encoding that rests on a set of conventions is a striking rebuttal of standard accounts of the rise of so-called naturalistic representation during the Renaissance, which contend that the use of such devices as linear perspective mitigate against lying (see Crombie 1985). In opposition, Hall and Knight submit that communication in the common language is all that matters – the only relevant factor is whether the users have a set of conventions within which discourse can occur. What's at issue here is the deeply rooted conviction that the epistemic purchase of scientific illustrations is directly tied to their faithfully representing physical objects. With reference to the mechanical contrivances illustrated in Agostino Ramelli's *Le Diverse et Artificiose Machine* (1588), Hall suggests that the success of Ramelli's illustrations as knowledge depended on their ability to persuade the user that they depict machines that actually work; that is, their cognitive authority as illustrations was tied to their ability to create conviction.

Hall's analysis is restricted to technological illustrations in the Renaissance. Brian Baigrie turns to the illustrations incorporated into the scientific treatises of the seventeenth-century natural philosopher René Descartes. This case is especially intriguing, in light of Descartes's contention that visualization is a principal source of error in the sciences. Baigrie argues that Descartes's use of pictorial devices needs to be situated in terms of his mechanical philosophy, which seeks to model the world as a system of interconnected machines. His illustrations are designed to help his reader conceive natural phenomena as constrained mechanical systems, since the lesson Descartes and his fellow mechanists extracted from the science of machinery was that insensible causes of things can be 'seen,' in a manner of speaking, and rendered intelligible by reconceptualizing natural phenomena in terms of systems of rigid parts that collaborate in the production of mechanical effects. The persuasive power of Descartes's illustrations is that they gave the would-be mechanical philosopher reason to believe that the world had been fabricated by God as a solution to an engineering problem. Whether they faithfully represent natural phenomena is simply beside the point.

One assumption that scholars make all too often is that there is a single unifying theme in illustration – that is, there is a single way of reasoning with diagrams. Martin Kemp challenges this assumption in the most dramatic way by juxtaposing the two seminal scientific treatises that were published in 1543 – Vesalius's *De humani corporis fabrica* and Copernicus's *De revolutionibus orbium caelestium.* Kemp submits that, while Renaissance developments in anatomy and astronomy involved similar factors of realism, rhetoric, and aesthetics, Copernicus's means of visual representation display no signs of the direct veridical representation that runs through Vesalius's work. Such new forms of representation, Kemp contends, became effective in astronomy only when celestial bodies could be observed as individual objects, with discernible features. With the invention of the telescope, Galileo, who was familiar with the science of perspective and the artist's understanding of cast shadows, was in a position to claim that the changing patterns of light and shadow, disclosed by telescopic studies of the lunar surface, were cast by the moon's topographical features.

Stephanie Moser's paper strongly reinforces Kemp's thesis that there is no single story to be told about illustration in the sciences. Moser turns to a science that is explicitly visual – prehistoric archaeology, particularly depictions of the daily life of our hominid ancestors – and details the ways that these representations argue in a distinctly visual manner, in ways, that is, that are denied to verbal text. The goal of research into human origins has been to define the boundary between humans and apes – namely, to establish the point at which our ape-like ancestors acquired human-like behaviour. What's intriguing about Moser's paper is her contention that visual reconstructions of prehistoric life enjoy great persuasive power because they use a range of icons and symbols that draw on our own experience. They differ from other types of scientific illustration, including those used in archaeology, in that they exploit naturalistic and familiar forms of representation. For example, they are full of gratuitous details that serve only to make the illustration seem more like a photograph of our hominid ancestors. Although Moser's conclusion appears to make concessions to naturalism, her claim is that depictions of prehistoric life rely on a battery of conventions. It is just that here the conventions are drawn from ordinary experience.

In addition to furnishing the reader with a welcome overview of the issues, David Topper raises a question that is at the heart of the marginalization of illustrations in historical and philosophical studies of science: namely, the question of the demarcation between art and science. Philosophers of science have wrestled with variations on this problem for some time, but their interest in demarcation problems has been almost exclusively restricted to the demarcation between science and pseudo-science or systems of belief that pass themselves off as genuine science. Topper's paper addresses a much more intriguing question, granted that art is often called into the service of science: what are our respective criteria of demarcation for art and science? The articles by O'Hara, Moser, Brown, and Ruse testify that visual depictions bring about conviction, having much the same effect as arguments. However, each of these authors deals with pictorial devices that were created to serve scientific ends. Topper reminds us, however, that pictures are often used as science even when they were originally created to serve aesthetic and artistic ends. Sometimes works of art are transformed into works of science. To further muddy the line between science and pseudo-science, Topper points out that sometimes art is transformed into pseudo-science, as in the case of Petrus Camper's depictions relating to physical anthropology (see fig. 7.6). In order to make sense of the plasticity of visual images, Topper submits that any visual scribble (e.g., field drawings and geometrical drawings in laboratory notebooks) is a potential scientific illustration and, conversely, that scientific artifacts are potential works of art, as evidenced by Max Ernst's reliance on scientific illustrations for motifs on which to base his surrealist and abstract paintings. It's all a question of media, style, and context.

Ronald N. Giere tackles one of the thorny philosophical issues that runs through the many contributions to this volume. The logical empiricist places a premium on linguistic representation, generating an enormous chasm between the scientific text and the visual image as resources for doing science. Giere seeks to close the gap between the abstract theoretical models that form the core of any scientific theory and visual presentations of data by drawing on the work of the geophysicist Alfred Wegener, who argued that the positions of the continents and oceans have shifted in geological time. On Giere's view, scientific theories are to be understood as a family of idealized models or prototypes of material things that are judged to be of that type. Whereas many historians of science and art (e.g., Arnheim 1969; Rudwick 1976; and Knight 1985) have attempted to bridge the gap between word and image by suggesting that scientific illustration involves a kind of visual language, Giere leaves us with the provocative suggestion that a scientific theory is more like a picture than vice versa – essentially a non-linguistic entity.

ACKNOWLEDGMENTS

This collection is far from complete – there is little on contemporary science and nothing on the use of the up-to-date computer technology that has recently produced such dramatic results in astrophysics and in biogenetics research. Needless to say, scientific illustration is much too grand and dynamic a field to be squeezed into one volume, and so these studies will have to wait for another volume. As editor, my guiding principle has been to include handcrafted studies that speak powerfully to the theme of this volume – the capacity of illustrations to impart information. A number of these papers focus on science and technology in the Renaissance and in the early modern period. Perhaps this emphasis on early modern science is appropriate given that it was at this time that scientific illustration came to prominence.

The volume has taken quite a bit of time to come to the press. Though I am enormously grateful to those who have agreed to let us reproduce illustrations for this volume, negotiating for copyright material can be tedious. I have done my best to give credit where it is due and offer an apology for any oversight. A number of people deserve some recognition. Thanks to Beth Rainey and the staff of the Rare Books Collections in Durham University Library for arranging the photography of the illustrations for David Knight's article. Thanks are due to my research assistants, Julian Smith and Gordon Baker, for work on the illustrations and the bibliography; Ron Schoeffel and University of Toronto Press for patience and encouragement; and to my partner, Patricia Kazan, for unconditional support. This book was published with the generous assistance of the Office of the President of Victoria University.

Picturing Knowledge

1. The Didactic and the Elegant: Some Thoughts on Scientific and Technological Illustrations in the Middle Ages and Renaissance

BERT S. HALL

Now because oftentimes more may be expressed in a small Picture of a thing, than can be done by a Description of the same thing in as many words as will fill a Sheet; it will be often necessary to add the Pictures of those Observables that will not otherwise be so fully and sensibly exprest by Verbal Description: But in the doing of this, as a great Art and Circumspection is to be used in the Delineation, so ought there to be very much Judgment and Caution in the use of it. For the Pictures of things which only serve for Ornament or Pleasure, or the Explication of such things as can be better describ'd by words is rather noxious than useful, and serves to divert and disturb the Minds, and sways it with a kind of Partiality or Respect.

Robert Hooke, *Of the True Method of Building a Solid Philosophy or of a Philosophical Algebra*

1. INTRODUCTION

The problems of early illustrations portraying scientific and technological topics remain among the more intractable aspects of the history of science and technology.[1] This is true despite some useful and suggestive publications dealing with early scientific and technical illustrations (Topper 1990a, 1990b; Ashworth 1984, 1987). We still lack clear-cut conceptual maps where illustrations are concerned. We don't really quite know what we are looking *for* as we look *at* the pictures, sketches, diagrams, and prints that make up the raw materials for discussion. For the most part, images remain an unusual subject, peripheral to the mainstream of investigations and still largely unexplicated in respect to their broader cultural significance. 'In the beginning was the Word ...' says the Bible, and true to our monastic origins, we scholars remain convinced of the primacy of words as the conveyors of meaning. Images rest in shadows, in the attic, always there and amusing to look at from time to time, but probably not worth any serious effort to understand.

Not surprisingly, many agendas can be imposed on such a body of materials. One overarching problem for any attempt to analyse early illustrations stems from the world of art history and the grand patterns of expectations that are generated when we use the word 'Renaissance' in an art historian's sense. It is a commonplace that images came to occupy a position of great importance in Renaissance culture, and we are naturally led to expect such a 'revolution' to have something to do with the 'Scientific Revolution' of the seventeenth century. A critical look at scientific and technical illustrations needs some background in art history, to be sure, but it also needs a different framework of interpretation. It is helpful to ask how images came to have authority in science, how they came to be accepted as bearers of authentic information (Ashworth 1991). Such a *Fragestellung* will keep us from barking up some very wrong trees.

2. PRINTING AND 'PLINY'S PROBLEM'

The most widely read views concerning our subject are those imbedded in a much broader discussion of the impact of printing on Western thought and culture. Of course, the leading proponent, indeed the apostle, of a thesis that sees print as an epoch-making event in Western history is Elizabeth Eisenstein (1969, 1970, 1979, 1983). Eisenstein is not only interested in the influence of printed texts in science, but also in the way that 'exactly repeatable pictorial statements' made themselves felt in the development of scientific and technical discourse, and this places her work squarely within the field we have to consider. Her views on printed illustrations seem to depend heavily on the work of William Ivins, whom we may safely label as an older defender of the value of naturalistic illustrations for scientific endeavours.

It is not easy to summarize Eisenstein's views within a brief span but, in general, she argues that the conditions of textual transmission before the advent of print mitigated quite strongly against the possibility of retaining accurate illustrations in any body of commentaries or treatises. The problem is vividly set forth in some remarks made by Pliny the Elder in his *Natural History*:

In addition ... there are some Greek writers who have treated [botany]. Among these ... Crateuas, Dionysius and Metrodorus adopted a very attractive method of description, though one which has done little more than prove the remarkable difficulties which attended it. It was their plan to delineate the various plants in colours, and then to add in writing a description of the properties which they possessed. Pictures, however, are very apt to mislead, and more particularly where such a number of tints is required for the imitation of nature with any success; in addition to which, the diversity of copyists from the original paintings, and their comparative degrees of skill, add very considerably to the chances of losing the necessary degree of resemblance to the originals. Hence it is that other writers have confined themselves to a verbal description of the plants; indeed, some of them have not so much as described them even, but have contented themselves for the most part with a bare recital of their names, considering it sufficient if they pointed out their virtues and properties to such as might feel inclined to make further inquiries into the subject. (Book 25, chapters 4–5, cited by Ivins 1953, p. 14)

Pliny describes what we might call a failed research program in classical antiquity, and at the same time presents for our consideration a reasonably complete inventory of the problems that would have plagued any attempt at making a pictorial record. Nature is difficult to 'imitate' in paintings, and the chances of retaining the 'necessary degree of resemblance' to the original over generations of hand-made copies made the effort simply not worthwhile. For a science seeming to require precise descriptions and depictions, such as botany, the weight of these difficulties is overwhelming. All specialists in the field can do is to record the names and properties, and perhaps a scant verbal description, hoping to inform someone who presumably already knows the plants in question through firsthand experience.

From this beginning, it is easy to assume that nothing significant could happen to change the situation until the twin revolutions of fifteenth-century Europe – linear perspective and the block print. The former, understood as a catch-phrase meant to stand for all forms of 'representational realism' in Renaissance art, enabled and encouraged artists to 'follow' or 'imitate' nature more closely than their predecessors, while the latter preserved and transmitted their successful efforts for the edification of others. The Eisenstein-Ivins thesis attributes to the combination of naturalism and printing most of the important characteristics of the world of learning in the centuries that followed the fifteenth, at least insofar as that world relied on pictures as sources of information. Descriptive sciences, such as botany, are held to have been virtually the creations of the Renaissance, inasmuch as only in the fifteenth century did conditions crystallize that would enable such sciences to be practised cumulatively and progressively over long periods of time.

This 'not until ...' argument neatly imbeds the issue within a matrix of larger art historical and technological changes, and it guides us away from any evidence that would contradict its strong dichotomy between 'manuscript culture' and 'print culture.' Yet Pliny's words speak of a much more complex classical attitude towards the role of images and of experience in the formation of knowledge. 'Pictures are very apt to mislead ...' not merely because they are difficult to copy, but because they are themselves untrustworthy representatives of the immediate experience contained in observation. Note also that Pliny's botanists do not seek to sketch just the shape of the plant, a compromise that might have enhanced their chances. Their program called for coloured drawings, despite the fact that colours fade or turn depending on the state of the papyrus and its conditions of storage. Why, one asks, did the best become the enemy of the good in this case? The answer probably lies in the primordial Greek philosophical distinction between superficial appearance and underlying levels of truth. What was of principal interest to the botanist was not the transient appearance of the plant, which could vary from day to day as the specimen grew, aged, and decayed away, but the underlying, persistent 'properties' and 'virtues' the plant possesses. Knowledge of these constituted the real 'science' of botany, not mere acquaintance with how plants might look today or tomorrow.

The tension between the specific and the generic, between the individual (with all its accidental features) and the typical (where variations have all disappeared in favour of insight into underlying structures), represents one of the enduring and characteristic problems of natural philosophy and of modern science. It shapes and is in turned shaped by the problem of authority in science. The medical student learns a variety of subjects in textbooks illustrated by drawings chosen in preference to photographs. Drawings can often serve a didactic function better than photos simply because they represent the state of a cell or a tissue as the authoritative doctrines of medical science claim that state ought to be, while omitting anything that, from the same viewpoint, is merely circumstantial, accidental, or simply 'ought not to be there.' The specific, individualistic, and idiosyncratic is suppressed in favour of the generic and the typical. The same medical student performing her mandatory dissection is trained to see through the 'superficial' layer of 'accidental' details to the underlying anatomical structures that the instructor regards as significant. The student's view of matters has no authority compared with that of the professor or demonstrator. Indeed, only in the case of gross malformations – old surgical lesions, for example – will a student's comments on 'accidental' features of the cadaver be permitted, and then only as evidence of the student's 'eye for detail' and in the assumption that training in spotting anomalies augers well for the future doctor's diagnostic acumen. Similar anecdotes can be found in almost any branch of science where observation plays a role.

Yet what we might tentatively label 'Pliny's Problem' did not completely eliminate the use of illustrative material from scientific discourse before the advent of the printing press and the woodblock. Some knowledge of the pictorial legacy of medieval scientific manuscripts can be gained by viewing the collection of images in the volume of the Album of Science Series devoted to antiquity and the Middle Ages (Murdoch 1984). It seems apparent that medieval attitudes towards illustrations were ambivalent. On the one hand, it is undoubtedly true that illustrations are quite rare in much of the mainstream of classical science that passed to the medieval period; Aristotle and Galen especially seem barren of pictures (Murdoch 1984, p. x). On the other hand, medieval absorption of classical attitudes seems always to have been rather uneven and haphazard, and there are many cases where pictures were employed as bearers of important information.

One case in point comes from the very field that Pliny addressed – botany. Several late ancient and early medieval manuscripts include plant descriptions and illustrations. In figure 1.1 we see the dracontea or cuckoopint plant (*Aurum maculatum*) as described by Dioscorides and as depicted in the Ancia Juliana text of Dioscorides done in A.D. 512. Cuckoopint's ivy-like leaves, upright stalk, and bulbous root are all plain. The second illustration is of the same plant from the *Herbarius* of Pseudo-Apuleius in a seventh-century copy; the leaves look quite different, but the bulbous root is accurate and the grape-cluster seeds described by Dioscorides and omitted from the sixth-century copy are here included. A later hand has added in German the words 'Schlangen Kraut' next to the drawing, and it was indeed as a specific prophylactic against snakebite that Dioscorides recommended dracontea. In the last drawing (from an anonymous Beneventan manuscript of the ninth or tenth century), Dioscorides' words appear in a Latin trans-

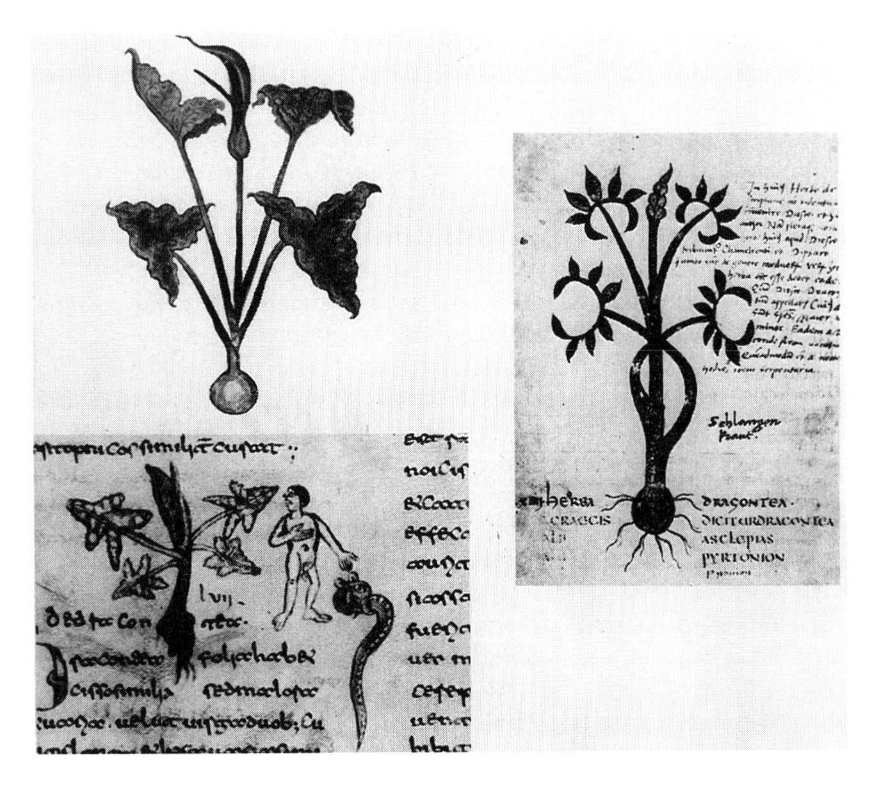

1.1 Three depictions of dracontea.

lation, next to a drawing of a plant that looks at first glance only somewhat like the sixth-century original. Yet the white spots on the ivy-like leaves that Dioscorides describes are plainly visible in this manuscript. The most serious transformation in this illustration is the prolongation of the bulbous root into the stem, the whole structure having a somewhat snake-like appearance. And indeed, next to the plant is a miniature figure of a man and a snake, illuminating perhaps the most important specific property of cuckoopint, the protection against snakebite that it gives.

The question of whether these are 'good' representations of *Aurum maculatum* is moot; from the Renaissance point of view with its emphasis on naturalistic representation, none of them is very prepossessing. Yet each drawing emphasizes certain physical features of the plant and omits others, and the selection of features to be emphasized seems largely to have been guided by the artists' differing assumptions about what is significant about dracontea. Is there much conceptual distance between these illustrations selectively emphasizing certain features and the modern illustrations in gross anatomy textbooks, where a similar process of selective emphasis has taken place (informed, to be sure, by very different notions of what needs to be emphasized)?

3. DIAGRAMS AND NATURALISM

The easiest way around the discomfort such a comparison evokes in us is to distinguish sharply between a naturalistic illustration and a 'diagram.' Diagrams are usually simplified figures, caricatures in a way, intended to convey essential meanings. Diagrams by their very nature do not pretend to be naturalistic; they seek to represent whatever their author regards as the salient features of the subject. A circuit diagram, for example, looks very little like the actual insides of a computer or a television set, but, as a 'map' that represents the functioning of the complex whole, it serves to guide the installer or repair-person in a way that a photograph or naturalistic sketch never could. Diagrams usually function best where the viewer is able to make the necessary inferences to move from the image to the object. Computer repairers know only too well what circuit boards look like; the diagram tells them what they need to know about *this* circuit board. Only the most innocent tourist would expect the London underground to look like its famous diagrammatic map.

Medieval illustrations, we are prone to say, are really diagrams. They presuppose the ability to make inferences, to 'overcome their inaccuracies,' and they are lodged in contexts where their medieval readers supposedly had sufficient additional information to bridge the gap between image and object. All this is indisputably true, but saying it helps remind us that the difference between a diagram and a naturalistic illustration is more a difference of degree than a difference of kind. Diagrams simply represent one end of a spectrum of demands that images may make on their viewers. Diagrams, of course, lend themselves to the repetition of the familiar, the expected, the typical, and thus they are apt to become reifications of the metaphysical preconceptions of ancient and medieval science in favour of essentials and against accidentals. (The diagrammatic cuckoopint illustrations certainly follow this rule, even to the point of making the plant look like the beast whose bite it cures, a snake.)

Yet is this so very different from the way naturalistic illustrations actually function when seen as bearers of information? All images, we need to remember, are crowded with features arising from expectations that the image-maker brings to the subject; every portrayal is a type of encoding that demands, at a minimum, a certain set of conventions common to both the individual doing the act of representation and the viewer of the representation. Once such a common 'language' has been established, communication is possible, and like all communication, it is equally as easy to 'lie' as to 'tell the truth.' Carried to an extreme, this sort of relativistic position argues that the exact nature of the visual language is unimportant, that like verbal languages, the only important consideration is whether the users have an agreed-upon set of rules within which discourse can occur. Could we therefore de-privilege the naturalistic illustrations of the Renaissance, imagining a counter-factual history in which modern science grew up in a visual world of medieval diagrams?

Faced with this question, defenders of the importance we assign to Renaissance naturalism usually reply that, unlike verbal languages, certain visual conventions mitigate against 'lying' and facilitate 'telling the truth' more readily than do others. Naturalistic Renaissance art, one could say, is more prone to make 'truthful' representations of the natural world than is diagrammatic medieval art. On the face of things, one is inclined to concede something to this claim. The closer an artist's mode of representation to the way we 'actually' see something, the less training the viewer needs to decode the image on the page or canvas. Much has been written about the claims of artistic naturalism, and about the limitations such claims are subject to (Gombrich 1968). Certainly we can no longer regard the act of seeing as a simple or mechanical 'representation' within ourselves of the world 'as it really exists' outside. 'Seeing' not only involves many parts of the brain (Zeki 1992), it is inevitably heavily influenced by what the observer wants or expects to see. Claims in favour of naturalism are circumscribed by these insights, perhaps even fatally compromised. Historically at least, one need only argue that diagrams are even more predisposed to problems of interpretation than self-consciously naturalistic attempts at portrayal. One might argue that a commitment to naturalism represents a kind of discipline for the artist, a defence against imposing too many of his own preconceived notions on the object being represented (granted always that one could probably never reduce to absolute zero the tendency to impose such notions). This is the essence of the claim that naturalistic art makes it more difficult to 'tell lies' than other forms of representation.

For the historian the problem of naturalism *versus* diagrammatic representation cries out for some historical examples. One medieval case showing the capabilities of the diagram concerns a figure in a manuscript chronicle done in the 1120s and 1130s at the monastic house in Worcester (fig. 1.2). On 8 December 1128, John of Worcester tells us, 'two black balls' appeared on the sun's disk and remained there all day long. The upper was somewhat larger than the lower, and the two spots remained opposite each other 'as this sort of figure shows' (ad huiusmodi figuram). The figure is imbedded in the writing, and the text flows about the diagram (even compressing and stretching to accommodate the round figure). It is apparent from these facts that the figure was put on paper before the text and was meant to serve as the principal source of information on the sunspots. No hint of naturalism is visible – no coloration, none of the blurring at the sunspots' edges that any observer must have perceived. We have here something approaching a 'pure' diagram – a phenomenological diagram – one that shows only the essential, observable features of the phenomenon. What is all the more remarkable is that it could not have been informed in any way by a pre-existing doctrine telling the chronicler what a sunspot ought to look like, since sunspots are not supposed to occur in medieval cosmology. It is not merely that they are absent; they are effectively ruled out by Aristotelian notions of the perfection of the celestial realm. The Worcester manuscript serves to remind us that diagrams are not, therefore, utterly dependent on convention for their didactic force.

Some further thoughts about the role of naturalism are suggested by another discipline where representation is usually thought to be critically important. In figure 1.3 we see a medieval anatomical commonplace, three of a total of five drawings illustrating an anatomical treatise. The so-called 'bone man' is at the upper left, the 'muscle man' at the lower left, and the 'nerve man' at the upper right. The 'artery man' and the 'vein man' make up the balance of the set. The illustrations and the text were done in 1247 at the Bavarian monastery of Scheyern by a Brother Conrad. The next image (fig. 1.4) for comparison comes from the *Anatomy* of an Italian physician living in Paris at the royal court in the mid-fourteenth century. Guido da Vigevano was physician to the Queen, then the King, of France before being killed when plague swept through Paris in the winter of 1348–9.

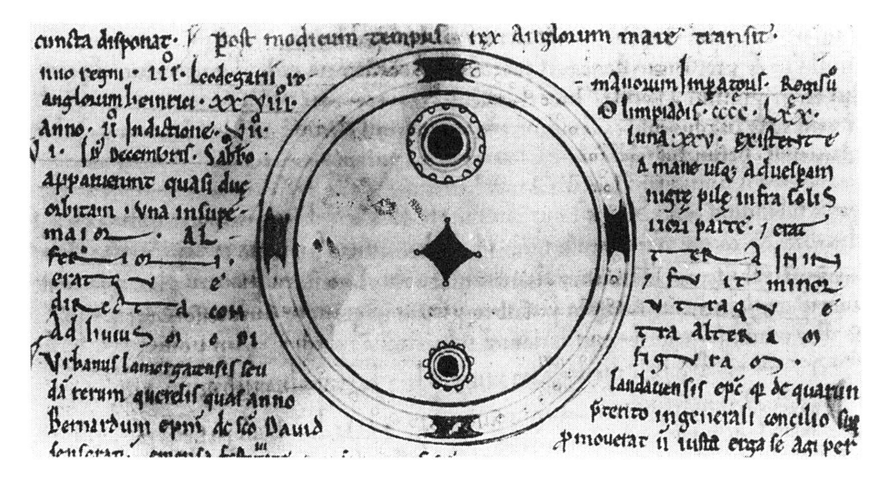

1.2 Sunspots, twelfth century.

Guido's *Anatomy*, which must date from about 1345, contains a novel way of representing the relationship between the human form and the organs contained within. Here we see the figure on the manuscript, but his rib cage is drawn on two flaps of parchment that can be folded back rather like French doors to reveal another layer of drawing below, showing the abdominal and thoracic organs *in situ*. Moving ahead some two centuries, we see (fig. 1.5) a female anatomized on a broadside published by Jacob Fröhlich in Strasbourg in 1544. Like the 'anatomy men,' she sits surrounded by text describing the individual organs, and these are shown *in situ* in a manner similar to Guido's depiction, by folding back the flap of paper on which the torso is drawn to reveal the abdominal cavity and part of the thorax. There is a strong resemblance between her posture and that of the 'anatomy men,' the only difference being in the elegance of her somewhat naturalistic pose.

How much difference ought we to see in these anatomical representations stretching over three centuries? We should note straight away that there is no serious question of 'progress' in anatomical knowledge during this period. Fröhlich published before the epoch-making work of Vesalius, and he knew scarcely more anatomy than was known at Prüfening in the twelfth century, indeed perhaps slightly less than Guido da Vigevano. Like the Prüfening monks and Guido, Fröhlich was deeply dependent on Galen. Fröhlich's figure of an anatomized body

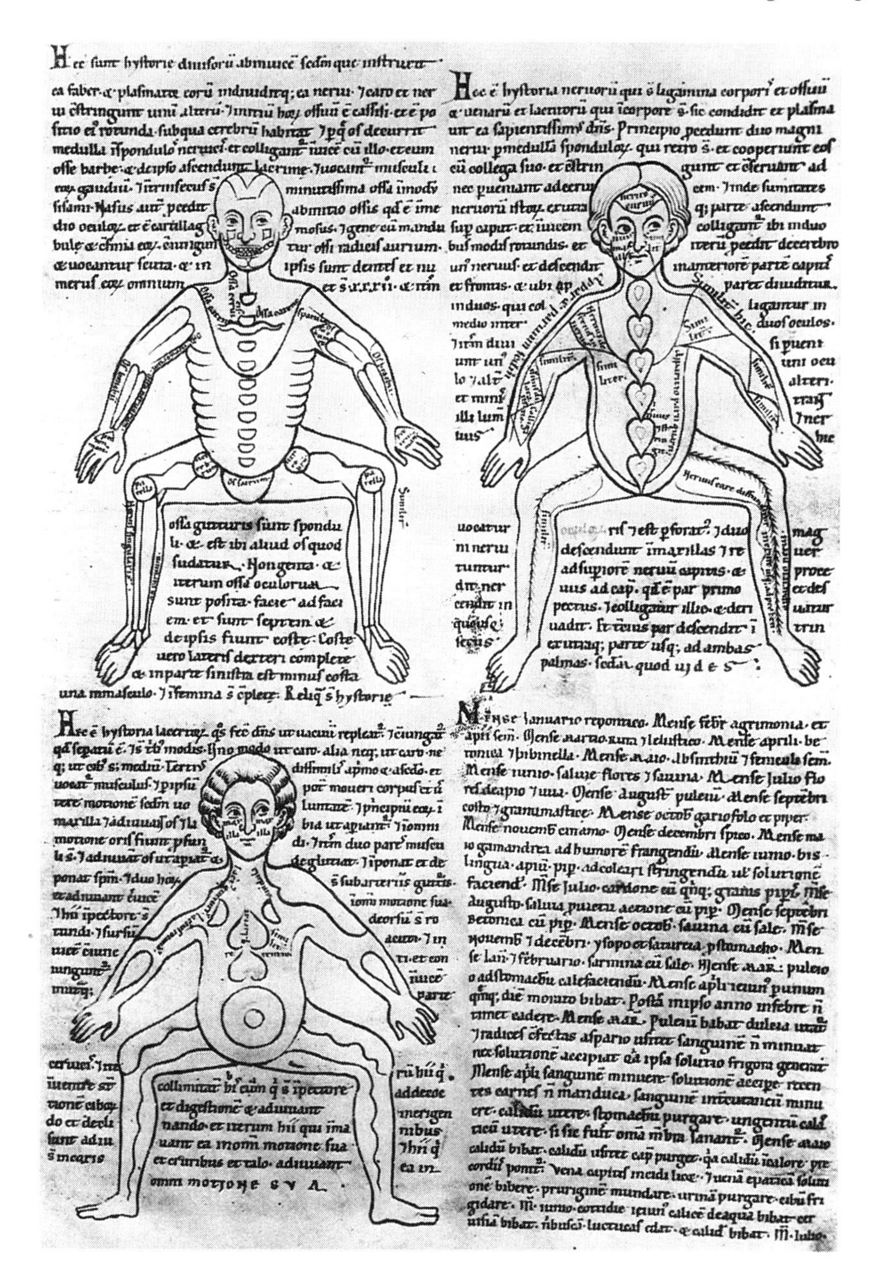

1.3 Bone man, etc.

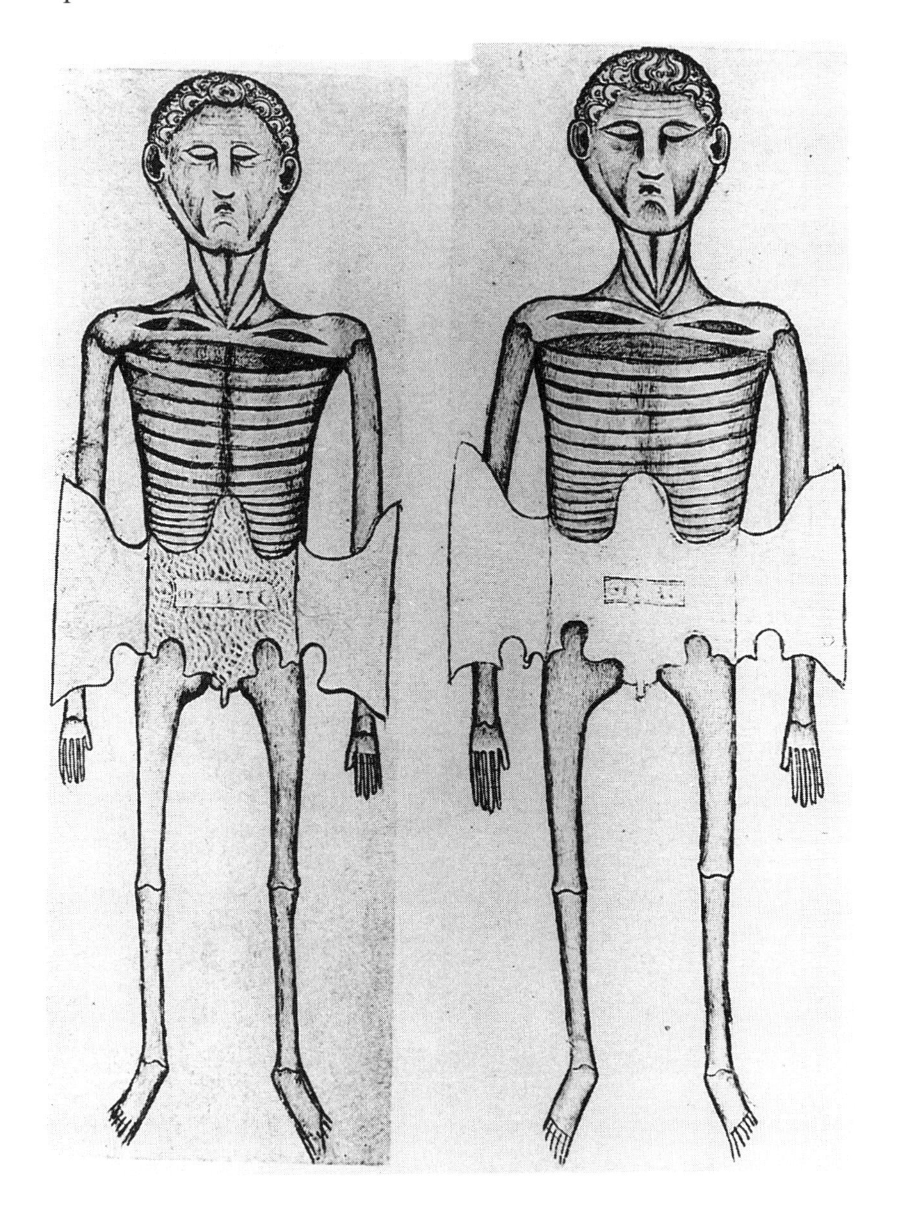

1.4 Anatomized cadaver. Guido da Vigevano.

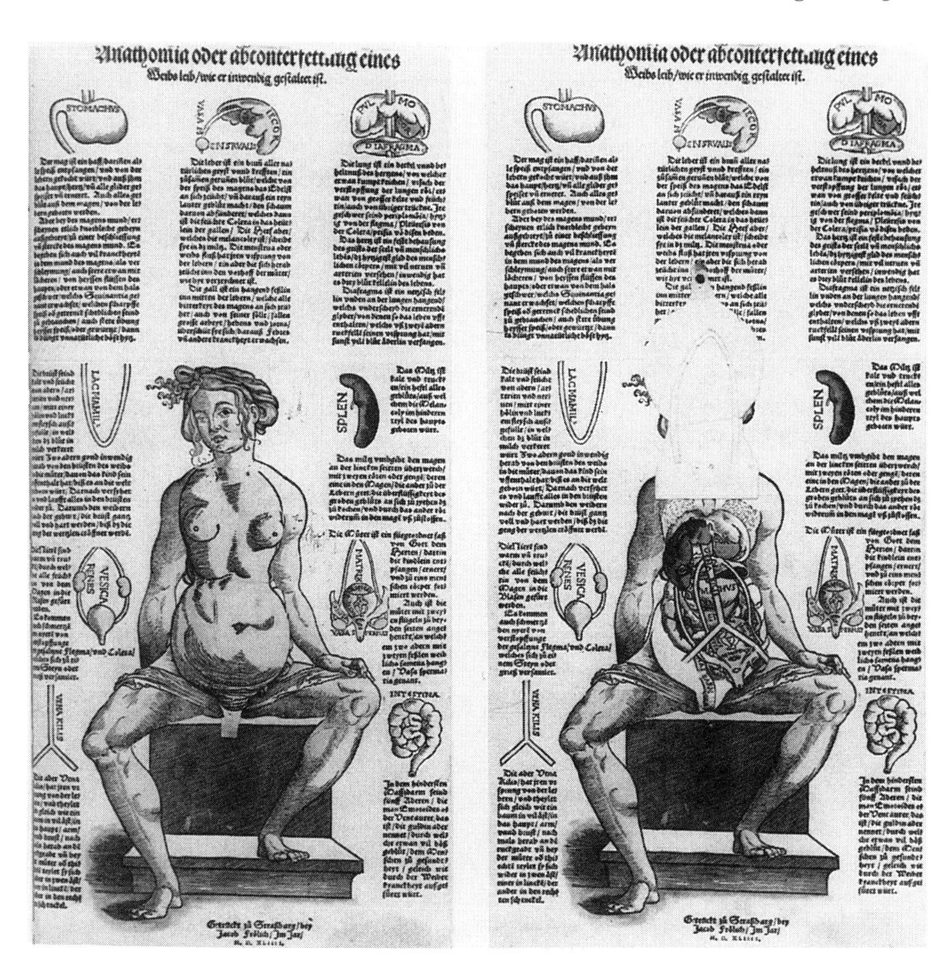

1.5 Anatomized woman. Fröhlich.

obviously represents a synthesis of the pictorial style of the medieval 'anatomy men' and the open-door technique of Guido. What strikes us, of course, is that the sixteenth-century drawing is done in a 'correct perspective,' that is to say, according to the rules of linear perspective Alberti had laid down about a century earlier. From the point of view of its elegance, the Fröhlich drawing is considerably more appealing to our sensibilities than its precursors, but is it better anatomy? We might look briefly at the small insert drawings – for example, of the stomach in the upper left – where the duodenum appears to be on the upper aspect of the stomach, not far from the pyloric sphincter. Were this the true configuration of the human stomach, we would be doomed to perpetual indigestion. Similar observations about simplifications or misrepresentations could be made about the mammaries (second on left), the kidneys and ureters (below the mammary), the liver and gall bladder (upper middle), and the uterus (second from bottom right). Likewise in the ensemble, the woman figure's uterus and vagina have been shown radically displaced in order to allow the anus to be made visible.

This series of anatomized figures poses a challenge to the common assumption that what appeals to the eye is necessarily more truthful. At first glance, most viewers would probably have awarded the prize for anatomy to Fröhlich, simply because his illustrations soothe our visual sensibilities. In the vocabulary I have chosen for the title of this essay, Fröhlich's drawings are 'elegant' in a way that Guido's or the anatomy men are not. Such is the power of elegance over the merely didactic that we routinely assume 'prettier' drawings contain more accurate information than 'uglier.' The problem that we need to address in respect to early illustrations is how to avoid permitting our own responses to the elegance of Renaissance perspective drawings from sweeping us along towards the assumption that they are invariably more informative. Quite simply, they are not, and we err in according naturalism a privileged position, as if it could create through its verisimilitude the authority that science claims to speak with.

4. IMAGES, SPECIMENS, AND EMPIRICISM

Even worse, we have to be on guard against the assumption that scientific illustrators actually saw their best – that is, their didactically superior – drawings published in the same way that we can enjoy their efforts today. Quite clearly some Renaissance authors who produced illustrated scientific treatises did indeed have in mind to employ perspectival drawings to didactic ends through the medium of print. The Swiss naturalist Conrad Gesner, for example, was not only a gifted observer, but someone who could sketch from nature with extraordinary facility, as in his sketch of pines and larches (fig. 1.6). Yet these drawings remained unpublished during his lifetime, despite the fact that Gesner prepared a collection of some fifteen hundred illustrations for a companion volume to his *Historia animalium* that would deal with plants. When Gesner died in 1565, his friend Kaspar Wolf made an attempt to publish the botanical *Historia*, but he was unable to do so. Finally, some of the images appeared in a decorative manner in an appendix to J. Simler's *Vita ... Conradi Gesneri* (1566). Other images from the collection were used to illustrate various works by Joachim Camerarius the Younger (Arber 1986, p. 111). In those works that Gesner did publish during his lifetime, most especially the *Historia animalium* (1558), he personally supervised the engraving of plates that rendered his sketches into living, accurate depictions of all that he had surveyed. Gesner was deeply aware of how pictures could supplement the printed word; his illustrations were intended 'so that students may more easily recognize objects that cannot be very easily described in words' (quoted in Cohen 1980, p. 150). So ingrained within us is the habit of attaching meanings like Gesner's to any scientific illustration that we have difficulty grasping how any other attitude towards them could even be possible. We attribute cases such as Gesner's ill-fated *History of Plants* to sheer bad luck, to the accidents of this life. We reserve our consideration for the potentialities of naturalistic drawings in the world of textual information, never recognizing that these possibilities were mediated by severe practical and attitudinal difficulties.

Sixteenth- and seventeenth-century printers were, for the most part, small businessmen interested in producing a product that would sell well on the market while minimizing their own costs. Illustrations seem to have been an expensive part of any book, probably far more costly on a page-by-page basis than text could ever be. This led, quite understandably, to the reuse of the engraved plates from which illustrations were printed. Engravers themselves often had no models to rely upon except earlier printed works, and in the absence of any laws or customs prohibiting such plagiarism, even those plates freshly cut for a new work sometimes simply repeated older published materials. This can result in some very scrambled relations between text and images. Awareness of these problems forces us to realize that attitudes towards illustrations varied considerably from one period to another.

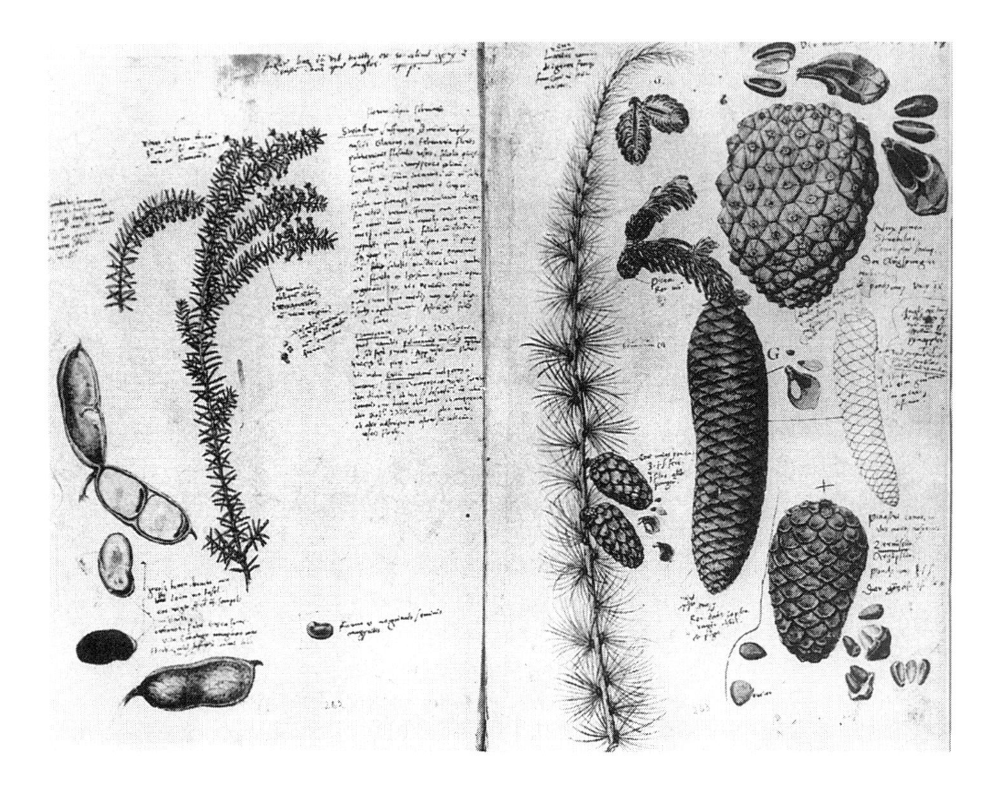

1.6 Pines and larches. Gesner.

Consider, for example, the matter of botanical illustrations (Ashworth 1991). Certainly the decades of the 1530s and 1540s must be considered a golden period in the publication of works concerning plants. Otto von Brunfels's *Herbarum vivae eicones* (1530) with its vivid illustrations by Hans Weidlitz (a pupil of Dürer's methods) was a fully realized and quite self-conscious attempt to fulfil the commitment manifest in the title ('Living Portraits of Plants'). Weidlitz even reproduced withered and insect-damaged leaves as they appeared on the specimens from which he worked (Arber 1986, pp. 206–9 and fig. 99). Brunfels was followed by two other pioneering botanists, the Swiss Leonhart Fuchs (*De historia stirpium* [1542]) and the Italian Peirandrea Mattioli (*De Pedacio Dioscoride Anazarbeo libri cinque* [1544]), whose works are notable for the freshness of their illustrations (Arber 1986, p. 64 [Fuchs] and p. 92 [Mattioli]). But by the middle of the sixteenth century, the urge to reuse images begins to manifest itself, as is apparent in Rembert Dodoens's very influential works published in the 1560s (Arber 1986, p. 82). By the end of the century, publishers such as the Frankfurt house of Nicholas Bassaeus had no inhibitions about publishing works like Tabernaemontanus's *Eicones plantarum* with blocks frankly copied directly from Fuchs and Mattioli (Arber 1986, pp. 76, 281).

This sort of borrowing was, of course, common enough in the sixteenth century, and it was probably harmless if sufficient editorial control was exercised over the 'fit' between images and words. Often, however, this was not the case. John Gerard's *The Herball, or Generall Historie of Plants* (1597) is a good example of bad practices. It began as an attempt, sponsored by the London publisher John Norton, to translate parts of Dodoens into English. A certain Dr Priest laboured on this project but died before it could be completed. Gerard, an ambitious barber-surgeon, completed the job but altered the arrangements of Dodoens and sought to publish the work under his own name, claiming that the Priest translation had been lost after the latter's death. Unfortunately, Gerard's knowledge was inadequate to the task of rearrangement, and Norton, his publisher, called on the aid of the French refugee botanist Lobelius (Mathias de l'Obel) to set matters right. Gerard, enraged by jealousy, successfully sabotaged this effort (which if successful might actually have enhanced his book's reputation) and somehow forced Norton to issue the work in mangled form. To add to the confusion, Norton illustrated the volume with plates taken from Tabernaemontanus's *Eicones plantarum*, which, as we have seen, were in turn borrowed from Fuchs and Mattioli (Arber 1986, p. 129).

Despite these defects, Gerard's *Herball* proved popular enough to warrant a new reworking in 1633 at the hands of Thomas Johnson, a London apothecary and botanist. Johnson was a scrupulous editor, anticipating modern methods of scholarship in his treatment of Gerard, but in his treatment of illustrations he remained a man of his age. He threw out the Tabernaemontanus illustrations, but in their place he could do no better than to borrow some 2,766 woodblocks that had previously been used by the Antwerp publisher Christophe Plantin to illustrate works by Lobelius and Clusius, among others (Arber 1986, p. 79). Indeed, in some cases Johnson went even farther afield in his search for images. Perhaps the most famous Renaissance manuscript 'find' in respect to botany was the Dioscorides manuscript known as the Ancia Juliana codex brought from Constantinople to the Hapsburg library in Vienna in 1569. *Codex Vindobonensis Med. Gr. 1* (as it is somewhat lamely known today) must date from around A.D. 512, but its illustrations seem descended from originals by that same Crateuas whom Pliny mentioned and who must date from the first century B.C. Johnson, of course, was aware of the Vienna Dioscorides, for Dodoens had included some prints based on the manuscript in his works. Johnson himself included the woodcuts based on Dioscorides in his edition of Gerard, despite their occasionally jarring discontinuity with the text (Ashworth 1991).

What does all this prove? One example, to be sure, does not make a case, but along with modern knowledge of perception, it does suggest that our prejudices in favour of naturalistic illustrations need closer examination. Indeed, in the present state of scholarship, it seems rash to conclude that scientific and technical illustrations were significantly aided by the development of naturalism or their newly acquired ability to be printed in as many 'exactly repeatable' examples as the printer saw fit to produce. Not only do we sense that naturalistic drawings may be approximately as theory-laden (or as theory-free) as 'diagrams,' there is simply no warrant for the assumption that artistic naturalism is accompanied by a deep commitment to what we may as well call 'empiricism' on the part of scientific authors. Empiricism, of course, can mean many things to many people, but we commonly imply when we use the term about a given author that illustrations and textual descriptions will usually match fairly closely. Indeed, the whole authority of images as didactic devices depends on our conviction that they somehow represent aspects of a larger natural world 'out there' that it is the scientist's job to describe as accurately and as carefully as possible.[2]

Images seem to us particularly appropriate to those sciences that depend on large numbers of specimens, some of them rare, many perhaps from exotic and inaccessible locations, to serve as a foundation for large generalizations, schemata, or taxonomies. These 'inventory sciences' (Zeller 1987), at least in the modern world, are critically dependent on recorded images, and we 'naturally' assume that the images we find in printed sources are – if not entirely free of theoretical overburdens – at least representations of specific specimens. Only in the very warts-and-all specificity of a particular representation might there lie residual or 'hidden information' whose 'correct' interpretation would force a modification of the classificatory scheme or any other high-order generalization. This epistemological necessity that underlies our use of images is of course valid for the modern period, but it should not mislead us into falsely interpreting the history of didactic illustrated materials. As Johnson's use of the Vienna Dioscorides images reveals, attitudes towards visual information varied – sometimes sharply – from our own during the first two centuries of printing, and any attempts at glib generalizations need to be tempered with the realization that pictures were both commodities and symbols during most of the Renaissance. It is not until the latter third of the seventeenth century that pictures took on the roles we are most familiar with today, a development that for the English-speaking world was ratified with the publication of Robert Hooke's *Micrographia* in 1665.[3]

5. THE EDGERTON-MAHONEY DEBATE

We have now reached a point where the discussion can be broadened somewhat. As a rule, we expect images that appear in the company of texts to explicate the material covered in the text in some fashion or other. Behind this lies the further assumption that both words and images are related to an external reality, a world 'out there' whose description is one of the principal tasks of the scientist-writer whose texts we are reading. Yet it should be obvious by now that drawings cannot play so seemingly simple a role without an elaborate structure of institutional authority and personal credibility on the part of the author, and that we have failed to assess the importance of these intangible structures in our histories of scientific illustrations. Drawings, however, can play other roles as well, roles that are not necessarily governed by the assumptions just mentioned. It has been suggested that Renaissance naturalism has far broader implications than merely providing a descriptive language; Samuel Y. Edgerton (1985) argues that the practice of linear perspective and chiaroscuro created essential preconditions for the success of the Scientific Revolution of the seventeenth century. This thesis, first put forth by Erwin Panofsky, seeks similarities between the reorientation of vision that we call by the code name linear perspective and the dramatic alteration in science known as the Scientific Revolution. It is not, in Edgerton's restatements, a simple cause-and-effect thesis, but instead one that suggests, in the phrasing of one of its critics, 'the new pictorial techniques were a prerequisite for the new science of mechanics and for the world-machine described by that mechanics' (Mahoney 1985, p. 199). For Edgerton, the critical moment in Western art comes with its acceptance of Euclidean space as the unyielding frame within which three-dimensional objects must be placed. The

Western artist from the Renaissance to the nineteenth century sees reality through a window-frame, as it were, with objects and persons related to each other within the picture-space just as they would be were a real observer to see them through the 'window' represented by the picture frame (Edgerton 1985, p. 169; and 1980). In a way, this is an almost literal restatement of Alexandre Koyré's view that the essence of the Scientific Revolution lay in the geometrization of nature. Edgerton's thesis does not exactly depend on a judgment of how closely or how 'accurately' the naturalistic illustration represents the world.[4] Rather, in this thesis the artist becomes a kind of 'quantifier,' someone whose very act of representing the world imposes a mathematical order on space, forcing it into the three-dimensional framework of Euclidean geometry. The parallels with the mathematization of physics during the course of the seventeenth century are obvious.

The favourite sons within Edgerton's thesis are the artist-engineers of the Italian and Northern Renaissance who produced a multitude of highly illustrated machine books, largely military in character, between the early fifteenth and the mid-seventeenth century: Mariano Taccola, Francesco di Giorgio Martini, Leonardo da Vinci, and Agostino Ramelli. Edgerton seeks to demonstrate his thesis by suggesting how 'unnatural' the geometrized way of looking at the world is, and how this particular social construction of reality becomes a defining characteristic of European culture. He notes the fate of the artist-engineers' machine books when they were taken to China as part of the Jesuits' effort to convert the Middle Kingdom starting in the late sixteenth century. Western technology was a calling card for the Jesuits, a means to attract the interest of Chinese who might then be susceptible to conversion. The Jesuit mission sponsored Chinese translations of excerpts from the European machine literature, including redrawings in the Chinese style of the printed copperplate engravings. Comparison of original and copy reveals that, even within a technologically and artistically sophisticated social order like China's, the illustrations were profoundly misunderstood. The seemingly simple pictures in fact involved a great many established conventions of representation that simply did not translate into a Chinese tradition that lacked linear perspective.

Edgerton is not arguing the now-outmoded Eurocentric view that China was simply 'backward.' There can be no question of 'backwardness' or 'primitiveness' here, given what we know about China's long and distinguished history of significant accomplishments in both science and technology (Needham 1954– and 1982). There are deep cultural differences between China and the West, however, and Edgerton argues that these differences include how pictorial space is to be treated. The published Chinese versions of Western machines are often completely unworkable, even unintelligible *as a result of the failure of Chinese illustrators to understand the conventions of European images.* In figure 1.7 we see one of Ramelli's Baroque conceits, a water-well windlass in the form of a crankshaft. The gears and the winding drum such a mechanism requires are carefully concealed in a covered pit next to the well shaft. But to clarify the operation, Ramelli's artistic device is to show the pit cover as if it were transparent. In addition, the winding drum and the spur gear with which it shares an axle are depicted as lying on the ground in front of the well, like spare parts.

Ramelli's work was taken into China by the Jesuits as part of their missionary effort to attract Chinese interest, and under Jesuit auspices a Chinese encyclopaedia of 'Diagrams and Explanations of Wonderful Machines' from European sources was produced. Wang Cheng, a Chinese convert to Christianity, and Fr Johann Schreck selected some fifty illustrations from Western machine books for inclusion in this *Ji Qi Tu Shuo*, printed in 1627. We see in figure 1.8 how Wang Cheng's Chinese draughtsman rendered the original (Edgerton 1985, p. 191). The cutaway or transparent view seems to have proven especially troublesome, and the artist has rendered it in a convention remarkably like that used to suggest magical or miraculous events in traditional Chinese iconography. The winding rope appears twice, once above ground, and again on the underground winding drum, but the connection between these two ends is not made at all clear. In a further entropic process that Pliny might well have appreciated, some of the 1627 encyclopaedia illustrations were recut as woodblock prints for another work that appeared ninety-nine years later, the *Tu Shu Ji Cheng.* Clearly the later illustrator had never seen Ramelli's original, nor had he any further insights into the conventions of European pictorial representation, for the picture is distorted to such a degree that no one could hope to build Ramelli's well-windlass from this picture alone. For Edgerton the early Ch'ing Dynasty's failure to grasp the meaning of Western technical illustrations forms a kind of reversing mirror in which we can apprehend the significance of Renaissance Europe's accomplishments in the treatment of scientific and technological images. The Renaissance artist, like the Renaissance scientist, is a 'quantifier' of reality, and thus the possessor of powerful new tools to describe the natural world.

1.7 Crankshaft well windlass (Ramelli 1588).

1.8 Windlass well. From *Ji Qi Tu Shuo.*

The response to Edgerton's thesis from professional historians of science is found in an article by Michael Mahoney that appeared in the same volume as Edgerton's most advanced and articulate version of his argument. Mahoney accepts the central position that Edgerton's thesis gives to mechanics in the Scientific Revolution, but he argues for a strong division between machine design in the seventeenth century and the science of mechanics. In particular, he notes

> it is difficult to see how more accurate depiction of the basic phenomena as physical objects could have conduced to their abstraction into general systems. For the defining terms of the systems lay in conceptual realms ever farther removed from the physical space the artists had become so adept at depicting. (Mahoney 1985, p. 200)

Mathematical treatment of the elements of mechanics in particular seems to move in directions quite opposite to what Edgerton needs to assume. That is, as the science of mechanics develops in the seventeenth century, it is assisted by (and some would say it in turn assists) the development of new forms of mathematics, the analysis of infinitesimals and symbolic algebra. What is important, Mahoney asserts, about the transformation of mechanics in the Scientific Revolution is precisely this new method of treating the properties of real systems, be they individual machines or celestial mechanics itself. The new analytical mathematics of the seventeenth century, though it grew out of the geometrical methods of the Greeks, was considerably more abstract, and thus considerably less susceptible to being conveyed through drawings or diagrams:

> Only by reaching into realms for which no physical correlates existed, for example the realm of imaginary numbers, could mathematicians achieve the theoretical generality they claimed for their subject. Mathematicians reached these realms not by looking at the world in new ways, but by looking beyond it altogether. To the extent that mechanicians followed suit, the science of mechanics that epitomizes the Scientific Revolution manifests modes of thought antithetical to Edgerton's inventive [artist-engineers]. (Mahoney 1985, p. 201)

Mahoney provides examples in support of his thesis with selected cases: Galileo's geometrical treatment of the Law of the Lever from Day Two of the *Discorsi*; Huygens's derivation of the mathematical curve for a pendulum to swing in true isochrony; and Newton's Theorem 1 of the *Principia*, treating a body moving under the influence of a centripetal force. There is a clear progression from less to more abstract modes of reasoning, from greater reliance on the physical appearance of things to greater demands on the mathematically trained inner eye of the mind. From drawing to diagram to pure abstraction, the course of mechanics as a branch of physics seems clear. Mahoney's criticism draws a definite line across the landscape, one that separates the core of the mathematical sciences from the realm of depictions of three-dimensional space. No equivocation on the word 'quantification,' no wordplay on 'seeing' or 'viewing,' is going to cross that boundary. So far as the very important central sciences, the mathematical sciences, are concerned, the value of any sort of illustrations must be held to have always been minimal and to have diminished still further as those sciences developed into their characteristic modern forms. As Mahoney concludes, 'to link in a directly causal manner new techniques for the accurate depiction of machines with the emergence of the science of mechanics is to ignore the line of thought that drove the diagram from dynamics' (Mahoney 1985, p. 217).

It would seem in light of Mahoney's criticism of Edgerton's thesis that we must be prepared to put aside any attempts to see linear perspective naturalism and printed images as crucial elements in the formation of the central enterprise of 'modern' forms of science; namely, mathematical mechanics. Yet illustrations did come to play an important role in the 'inventory sciences' and, it is now apparent, a very different role from their place in mathematical sciences. If we accept this distinction, we can take another step if we make another distinction, that between science *per se* and technology, and ask whether illustrations may have played a different role in the latter. Edgerton's error lay in assimilating under the banner of 'science' all forms of what we would plainly now call technology (machine design, for example) along with biological or 'life' sciences and such studies as astronomy, mathematics, and physics. This is a commonplace of modern discourse, but it is usually not useful to the historian to regard science and technology as a single, unitary enterprise, all branches of which grow at the same speed and in the same direction. The separation of science from technology is intellectually easy for some and perhaps more difficult for others. One might also want to admit that a strong distinction between science and technology is not characteristic of the Renaissance. Yet for heuristic reasons, if for no others, it is helpful to try to analyse their development separately. For all its anachronism, we may be able to use this distinction as part of an effort to suggest how illustrations served different purposes in the period.

6. ELEGANCE AND THE 'MIND'S EYE'

Another distinction needs to be made, that of my title, between the didactic and the elegant. By this I mean to suggest that illustrations may be considered from two different and at times contrasting angles of vision. On the one hand, illustrations may be used to clarify a problem, to indicate to a viewer how something looks, be that something a specimen, a piece of apparatus, or the path of a moving body. It does not matter in such a case whether the drawing be crude or finished, skilfully done or barely intelligible (provided only that it be intelligible to the viewer in respect to the details being conveyed); the didactic drawing is purely instrumental, merely a convenient means of conveying information to the viewer that otherwise could only be put across using masses of words – if then. It is almost exactly what we mean when we use the hackneyed phrase 'a picture is worth a thousand words.' On the other hand, 'elegant,' a word more often found in fashion magazines than in academic discourse, is meant to convey the other function illustrations usually serve, that of enhancing the appearance of the illustrated in the eye of the beholder. Elegance is an aesthetic quality, often identified with 'beauty,' and usually intended by the illustrator to appeal to some already existing sensibility in the viewer and to elicit a sympathetic response to the subject illustrated. In perhaps its most basic sense, we see something as 'elegant' when it 'just fits,' when it connects most fully with our experience. Extending this primitive sense, any number of people, objects, or ideas can be said to possess 'elegance,' mathematical equations or machines as well as women and men.

It should go without saying that an elegant drawing may be didactic, and a didactic drawing may also be elegant (in any one of several senses). J.J. Audubon's *Birds of America* leaps to mind as a work filled with illustrations both didactic and extremely elegant at the same time. But if we consider the problem somewhat further, it should also strike us that the didactic and the elegant dimensions may lead in quite different directions. In one possible conflict between them, the elegance of an illustration may seduce the viewer into accepting as true something that is not. It is just this possibility that the ancients seem to have found so threatening, and that led them to have concluded that 'pictures ... are very apt to mislead.' Most of us, I think, are less inclined than the Greeks and Romans to mistrust images, but in considering aesthetic and informative functions, we are quite likely to think that they have little to do with each other. A didactic drawing calls to mind a textbook illustration, plain and serviceable, but utterly lacking in charm. An elegant picture, by contrast, may affect us emotionally so strongly that we unconsciously employ words derived from witchcraft to describe our feelings – we are 'charmed,' 'beguiled,' or 'entranced.' This tension (loosely that between the intellectual and the emotional) has usually been viewed only from the above perspective, the problem of pictures' ability to seduce. Looked at from another angle of vision, cannot pictures acquire authority through their power to convince?

Examining the terms of our discussion, it seems that we put on a certain kind of perceptual blinkers when defining as our subject 'scientific illustrations.' In a manner of speaking, there were no scientific illustrations in any specific sense until the second half of the seventeenth century. But there was a very large-scale tendency throughout all aspects of European manuscript and book production towards more and more heavily illustrated works. This overall love of images in pages appears from the fourteenth century onward in lavishly illustrated devotional works and the cheap and humble block-print images with inscriptions. This same tendency carried over into – and was expanded considerably during – the early centuries of printing. It is still manifest in the High Baroque. The influence of this quite widespread shift in tastes is very pervasive, and it is particularly important for the development of technology.

For most of us, technology is fairly easily defined as the application of science to practical problems. This modern commonplace stresses the dependency of technical achievements on a scientific base, but it ignores the insights of those who work within the engineering disciplines. Those people tell us that *design* is the central element of the engineer's craft. Design is not easy to define with rigour, but it can be characterized as 'the purposive adaptation of means to reach a preconceived end, the very essence of engineering' (Layton 1976, p. 696). Eugene S. Ferguson (1992) has recently published a study that emphasizes how engineers, unlike scientists, are trained to create, to fashion from materials as diverse as reinforced concrete, thermoplastic resins, or lines of computer code, something that did not exist before. This design-centred vision of engineering is, admittedly, somewhat controversial; for one thing, it runs counter to a great deal of modern engineering pedagogy, which stresses ever more advanced modes of mathematical analysis.[5] It is also the case that design is notoriously difficult to teach, far more so than even the most demanding modes of analysis, and a design-centred vision of engineering verges on the mysterious, if not the mystical. Yet it remains true that the engineer is a manifestation of *homo faber* in a way that scientists evidently are not, and to this extent engineering design resembles an art disciplined by knowledge of properties of materials more than it does the sciences that describe the materials used in the design process.

Ferguson stresses how engineering design works by means of a process that employs visualization as a fundamental element. Engineering design, for Ferguson, involves imaginative strategies that envision an object while projecting its behaviour or properties under working conditions. This 'visual thinking' is for Ferguson (1992, pp. 47–54) a consistent aspect of engineering practice from Watt's steam engines to Whitcomb's 'area rule' for supersonic aircraft. (It is also an informal method that virtually all artisans use in the fabrication of simple tools or implements; indeed, it is so thoroughly intuitive and pre-verbal that it has left little trace in historical records other than the results of the artisan's activities, the artifacts themselves.) When we consider that pre-modern modes of teaching technical knowledge rested on a foundation of personal relationships and oral instruction, we need no special pleading to account for our lack of historical information about early design. With the development of differentiated forms of technical activity, or more simply, as the architect-designer-engineer is separated from the artisan-labourer, there emerges in our historical period a necessary medium of communication between these now independent participants in the activity of building, and this is why the Renaissance gives birth to the illustrated technical manuscript.

Ferguson's recasting of the engineer's role, making it into that of an artist, offers a new perspective on the role images play in the act of engineering design. Ferguson also offers us a model, a virtual template, for integrating the illustrated Renaissance technical manuscript into the longer history of technology and engineering. As the recipient of patronage through princely courts (and these must be understood to include their republican and municipal counterparts as well), the engineer was a participant in the same milieu that saw the birth and development of linear perspective drawings. This new mode of representation was for the engineer more than a new artistic style: it was a powerful new tool that could be adapted to the task of communicating technical ideas (Ferguson 1992, p. 77). Engineers (who were often artists themselves or *architetti* with artistic training) quickly adopted the perspective sketch as their principal means of communicating all manner of novelties, whether building plans and elevations, machine designs, or fountains and gardens. Ferguson, of course, is aware that modern engineers rarely use perspective sketches, at least in their formal presentations, preferring instead a specific and highly stylized form of drawing known as *orthographic projection.* Engineering drawings, in the modern sense of the term, allow designs on paper to be translated into three-dimensional objects that must be cast or milled to achieve their final form, and this imposes demands (for dimensional exactitude, for example) that cannot be met in linear perspective sketches. If modern engineers no longer speak the visual language of academic painting and magazine illustrations, it remains true that their professional ancestors once did, and that engineering drawing evolved in the eighteenth century from an ancestral form of depiction that was once the pictorial *lingua franca* of European culture (Booker 1963; Deforge 1981; Belofsky 1991).

7. EARLY TECHNICAL ILLUSTRATION

Ferguson's setting of the context in which engineers learned to draw, while correct in its essentials, is only the starting point for a reinterpretation of early technical drawings. Specifically, we need to try to grasp how this body of pictorial literature expressed its own aesthetic ideals, how it blended the didactic and the elegant, and how this blending shaped the Western perception and presentation of technology down to the present. These are tall demands, too tall, indeed, for a paper of this length, but we are now, at last, in a position to ask the right questions. From the fourteenth century onward, we have a body of textual material that is broadly interested in scientific or technological subjects, but in ways that defy neat categorization. These texts were not produced in the schools or the cloisters. They are part of the more urban, more secular, less traditional context that late medieval Europe could provide and that was epitomized by the princely court. 'Courtly' texts were aimed at a general audience, but one probably better instructed in the matters under discussion than today's common reader is likely to be, and were patently meant to be interesting to persons without deep 'professional' backgrounds in the subjects – regimens of health for patients, not diagnostic treatises for doctors, or books about technology for courtiers, not artisans. One characteristic of this growing literature is that it is quite often profusely illustrated. What this means, in terms of the simple dichotomy we have established here, is that these illustrations were meant to be elegant as well as didactic, to charm as well as to instruct. The rather undifferentiated environment in which such writers worked – where they might design fortifications, construction machinery, military engines, and stage sets and props, all while pursuing an 'official' career as a physician or a cleric – meant that they were driven to use drawings as didactic devices and to give their works an air of elegance. In time, with the advent of linear perspective, their mastery of the universal language of pictorial representation gave them the ability to communicate not only with artisans and workmen, but with patrons and distant readers in a common tongue.

Consider, for example, a page (fig. 1.9) from Giovanni de'Dondi's treatise on the construction of the elaborate astronomical clock that he completed in 1364. Giovanni de'Dondi was neither an artist nor an engineer by profession, but a physician and the son of a physician (Barzon 1960; Bedini and Maddison 1966; White 1978). Giovanni's text that describes the device, *Tractatus astrarii*, gives us only a brief and inadequate description of the mechanical heart of the device, its escapement, much to the frustration of the technical historian. But the principal manuscript provides an elaborate coloured image of a verge and crown-wheel escapement through which the text twines. This is a good symbol for the marriage of the didactic and the elegant that characterizes much of later technical imagery; not only does the picture make good the text's defect in respect to technical information, it also enhances the appearance of the page in ways that are reminiscent of Gothic manuscript marginalia.[6] It is worth noting that without the drawing, Giovanni's *Tractatus* would be as uninformative as that of his predecessor in the making of elaborate astronomical clocks, Richard of Wallingford, whose *Tractatus horologii astronomici* (completed about 1330) is equally laconic about the escapement, but without any drawing of it (North 1976, vol. 1, pp. 473–83, and vol. 2, p. 328). The difference, of course, is that Richard, Abbot of St Albans, wrote in a severe monastic milieu that had different attitudes towards the role of pictures than those current in Italy nearly two generations later.

The example that one might expect at this point is the singular character of Leonardo da Vinci. Countless modern commentators have seen in Leonardo a blending of artistic and scientific elements that might make him a natural choice for an essay of this type. Yet, leaving aside the question of whether modern writers see in Leonardo nothing

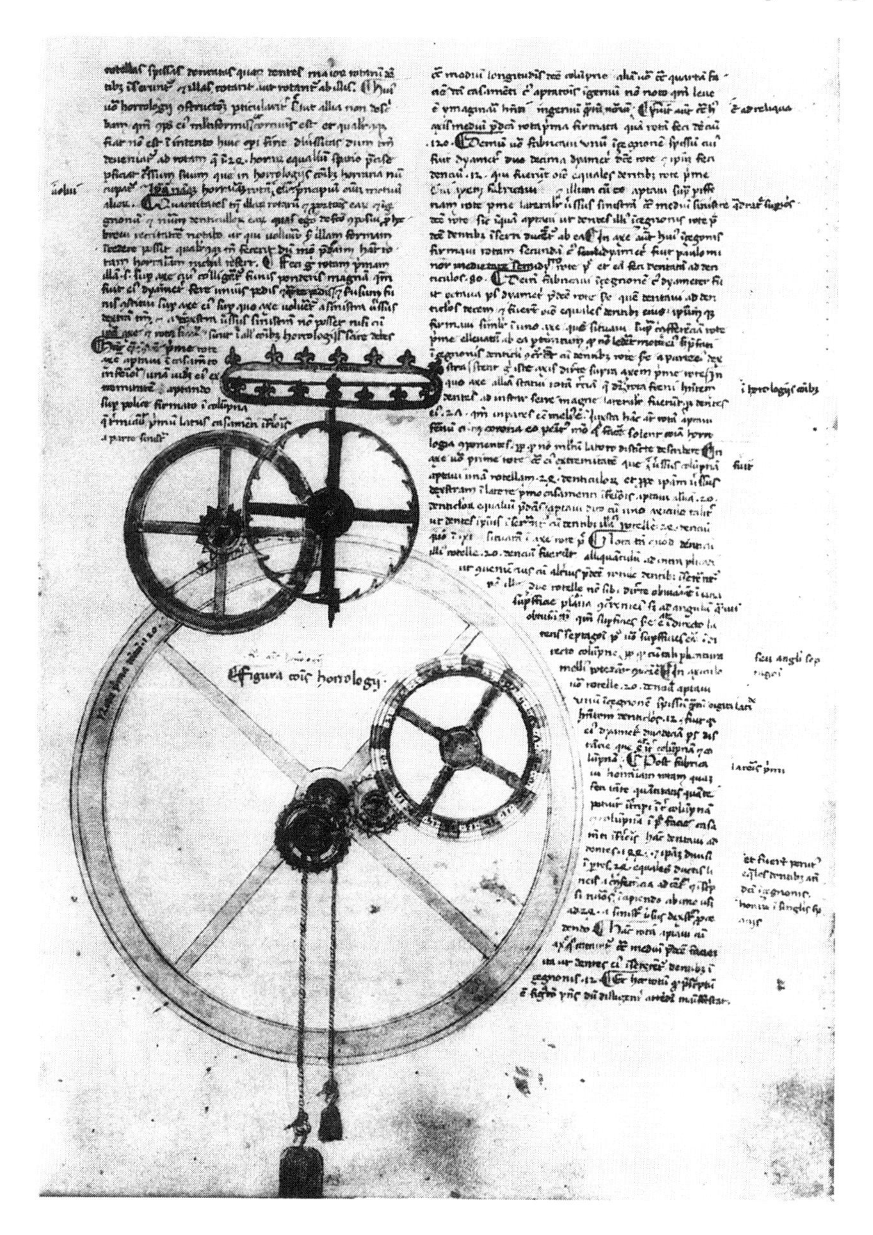

1.9 Verge and escapement for an elaborate astronomical clock. Giovanni de'Dondi.

but a creature of their own imaginings, it remains true that the great Florentine is not easily captured in a few sentences. He deserves a study to himself, one that would focus on how drawings formed part of a dialogue of invention (Hall and Bates 1976; Hall 1976; Galluzzi 1987 and 1990). But Leonardo is a singularly inappropriate specimen on which to base larger generalizations about the course of technical drawings in the Renaissance and later.

A far more appropriate example of how Renaissance technical 'writings' interweave the themes of thinking in pictures and elegant display may be seen in the 'Theatre of Machines' produced by Agostino Ramelli in 1588, *Le Diverse et Artificiose Machine.* Ramelli's is a virtual encyclopaedia of mechanical contrivances – cranes and hoists, pumps and earth movers, siege engines, and ancillary devices such as jacks (useful for breaking into barred entrances). Ramelli was a military engineer who served the Catholic side in the Wars of Religion, and he was a participant in the siege of La Rochelle. Ramelli's machines – those that seem to have puzzled the Chinese copyist so deeply – are all shown in their customary settings – pumps in gardens, cranes at dockside – where they are attended by workers. The 'naturalism' of the setting is contradicted by some distinctly unusual pictorial conventions; for example, where it is necessary to show the inner workings of a machine, Ramelli simply leaves out a wall or part of the framework. Underground, buried elements are visible to the viewer through strategically placed holes in the pavement or pathways. Details of constructions, elements or sub-assemblies of the larger ensemble, are usually indicated by spare parts left lying about the picture, as if by some careless repairman, but always rather carefully arranged so as to show how the parts relate each to the other. The accompanying text is effectively just a gloss on the illustration, explicating such fine points as might not be apparent upon inspection. Following a convention pioneered by Leonardo, it contains letters keyed to the drawings, so that textual references to specific parts can be made unequivocal.

Ramelli's didactic intent is always framed within an effort at charming elegance. He wishes to display a catalogue of whole machines and to show how each works, but this is blended with the goal of showing these contrivances in a manner that pleases and delights while instructing. In his dedicatory epistle to his royal patron, Henri III, Ramelli expresses the hope that the book may 'bring ... you pleasure and not a little satisfaction,' and that the king might 'on occasion put ... them into operation,' all the while hoping that they may 'serve most usefully all your brave captains and soldiers' (Ramelli 1588, p. 44). It is critically important to grasp one element in technology that distinguishes it from any form of natural science. Science is always 'about' nature, in some sense or other; technology is not. Technology is 'about' itself; technology is inherently self-referencing. Ramelli implicitly points to this in his title, which presents certain difficulties in modern translation. *Artificiose machine,* 'artificial machines,' sounds pleonastic to the modern ear, and Ramelli's modern editors wisely substitute 'ingenious machines' in their title. Yet for Ramelli and his generation, an 'artificial machine' is, of course, distinct from the great machinery of nature that God has created. Modern words such as 'fabricate,' 'forge,' or 'counterfeit,' all of which have come to mean some manner of illicit creation – despite the fact that etymologically no such sense attaches to them *ab origine*[7] – serve to remind us of the radical difference between the natural and the artificial. For Ramelli, the overcoming of this disjuncture cannot be achieved in the modern manner by asserting some specious unity of science and technology, of nature and artifice, but must be achieved contextually, by placing the machine in a garden, or a park, or some other 'natural' setting. Leonardo had already very nearly perfected the pictorial device of isolating the machine or the machine element from its surroundings – as we moderns prefer – and thus Ramelli appears to today's critics to represent some form of retrogression back to a stage in which machinery had to be seen whole and in situ. We thus overlook the way his attempts at elegant presentation serve to integrate artifice and nature while preserving didactic intent and clarity of insight.

Ramelli is frequently the target of criticism from modern engineers, who invariably note that his machines are too elaborate, pay no attention to frictional losses, or are simply frivolous in their ultimate purpose. The usual formula is to dismiss the 'Theatre of Machines' as 'coffee table books.' I think much of this modern criticism misses the point. To my knowledge no one has ever discovered in Ramelli a machine that will not work kinematically, where gears or connecting rods are simply positioned wrongly. Ramelli's machines were meant to be statements in themselves, articulations of what it might be possible to do given the mechanical elements at Ramelli's disposal. These are elegant machines, not practical plans, and they represent what nearly three centuries of court-centred treatises could produce. Ramelli occupies a historical moment when the elegant and the didactic become inseparable, when presentation and substance are impossible to separate. This can be achieved in technology to a degree that cannot be achieved in science, simply because in technology there is no 'nature' against which comparison is possible. Rather than representing a biological specimen or a physical process, Ramelli's engravings depict something very close to 'pure' or 'absolute' machinery. Modern engineers, grounded in the natural sciences and oriented towards utilitarian modes of analysis, find it objectionable to think that machines can even be said to exist without some intimate connections to the world of nature and the world of human purpose. Yet it would seem that technology does have the potential for such a form of existence, and that historically, it has existed in such a manner.

8. THE AUTHORITY OF PICTURES

To return to the towering question no scholar in this field has yet answered: How did images acquire their power to persuade? What happened in the cultural shift from medieval to Renaissance modes of communication to enable pictures to be vested with intellectual authority? Perhaps some consideration of technological (as distinct from purely scientific) illustrations will provide a clue. Whatever it is that makes us believe we understand something better if we can create an image of it, that something also allows us to create things that do not yet exist. (In circles other than science and technology, this capacity is usually called fantasy,[8] but I have learned that engineers dislike, indeed detest, the notion that their art owes anything to fantasy, so that perhaps it would be better to see in Ramelli an extreme example of 'thinking with pictures.')

By admitting that it is possible to 'think' in images, we come closer to a Renaissance view of how pictures actually function, and thus to how pictures achieved their authoritative stature. For a thought picture is not necessarily either 'true' or 'false,' but instead occupies a realm of its own. It is a kind of mental construct that may (or may not) be subject to whatever passes for 'rational analysis' within the framework of the thinker's world. Such analysis must necessarily come *after* the imaginary entity has been 'created.' This business of thinking-in-pictures is clearly not 'science.' (Indeed, as Mahoney reminds us, it was the business of science to stand apart from these imaginative excurses and to provide increasingly sophisticated means for their rational analysis and criticism.) And yet it is not unrelated to science either. Both a well-developed capacity for imaginative representation of possibilities and the rational criticism of such representations seem to have been complementary and equally necessary features of the development of the twin siblings of science and technology as we know them today. In this process, technological illustrations helped create the context in which pictures could be seen as authoritative. Machines, designs, and plans that existed nowhere in the 'real' world but that existed 'on paper' are not just *represented* by pictures; their whole existence *is* in the image. Such pictures persuade entirely through their effect on the viewer. The image goes beyond mere fantasy, however, in that it seeks to reveal the hidden, inner workings of the machine to the viewer. The imagined machine will *seem* to work, and it will be *seen* to work by the viewer. In this way, the image draws the viewer into a process of verification that establishes the image's plausibility, and it creates its own authority thereby. Unlike the image of a plant from another climate or a New World animal, whose veracity was always subject to some question and whose authority ultimately depended on the credibility of the image's author, drawings of machines carried within themselves the crude means for their own verification. So long as the hidden workings of the gears, pulleys, and levers were exposed to the eye of the knowledgeable observer, by tracing the machine's motions, the viewer can see how the operation is supposed to proceed. Images usually serve as adjuncts to words; images depict what is described elsewhere. In technology, the priorities are reversed: the image is the primary object of the viewer/reader's attention, while words of explication serve to illuminate details that might not be apparent at first glance. By displaying how this thing works – and by extension how any similar thing might work – technical drawings persuade the viewer to accept whatever is being illustrated as a possible machine. In that sense, technological illustrations are *self-authenticating*.

In seeking to grasp the complex process by which images came to be vested with the ability to create conviction, we must see these self-authenticating technological drawings as pioneers. They emerge earlier – in the latter sixteenth century – than did credible scientific illustrations, which appear only in the seventeenth century. Keep in mind that scientific illustrations' credibility was plagued by those problems we sketched above: the printers' commercial motive to reuse plates and the editors' cavalier ways of mating text and image. Scientific illustrations had the added problem (noted above) of dealing with the tension between typical and specific views. In the end, scientific illustrations could only become fully credible by appealing to the contextual authority of institutionalized 'science' itself. Technology faced none of these obstacles. Technological illustrations in the High Renaissance and early Baroque era were widespread harbingers of new attitudes towards the didactic value of pictures. They paved the way for genuinely scientific illustrations in the seventeenth century, and our interpretation of the history of science will always be incomplete unless the role of technological publications is taken into account.

NOTES

1 I have written on these problems in the past. See Hall 1978, 1979a, 1979b, 1982a, 1982b, and 1982c.
2 Note that this implication of the word 'empirical' is different from a purely technological empiricism, in which the 'empirical' knowledge of the craftsman is rooted in experience of materials and formed by the urge to produce similar outcomes whenever technical processes are repeated (see Eamon 1980).
3 Ashworth (1991) discusses this trend especially as it becomes evident in 'official' methodologies of learned societies and specimen collections. Recent studies of Hooke underline the importance of *Micrographia*, whose subject matter and methodology, after all, excluded all witnesses save the sole observer at the eyepiece; illustrations had to take the place of *viva voce* testimony. (See Dennis 1989; and Harwood 1989.) Harwood comments that printed images were critical for Hooke and even quotes Pliny (pp. 135-6) without concerning himself with the historical distance that separates Hooke from Gutenberg.
4 Although Edgerton (1984) does defend the cause of naturalism in some form.
5 It also tends to give pride of place to 'engineering sciences,' such as mechanics of solids and fluids, thermodynamics, electrical theory, and properties of materials. Critics charge that this emphasis in the academic training of engineers impoverishes the cultivation of design abilities, which tend to be learned indirectly rather than at first hand. Significantly, many engineering schools hold design competitions for their students, requiring them to build working models of, for example, bridges or robots. The displays are often more reminiscent of sculpture competitions or architectural design contests than any other activity.

6 These were ways that would prove impossible to replicate on the printed page. Some attention needs to be given to how printing *restricted* relations between text and image as well as how it expanded them. (For a catalogue of Gothic marginalia, see Randall 1966.)

7 To be sure, there is also a concurrent theme in many treatments of the mechanical arts to regard them as somehow 'fraudulent,' producing only inferior replicas of things 'natural.' Martin of Laon (d. 875) went so far as to derive the Latin *machina* from the Greek *moichos* or 'adulterer,' and Hugh of St Victor regarded them as 'adulterine' or 'mixed.' These attitudes seem rooted in the Graeco-Roman notion of 'mechanical' arts as somehow inferior to 'liberal' pursuits (see Whitney 1990).

8 The word is, however, precisely correct in this historical context (see Kemp 1977).

2. Temples of the Body and Temples of the Cosmos: Vision and Visualization in the Vesalian and Copernican Revolutions

MARTIN KEMP

The words or the language, as they are written or spoken, do not seem to play any role in my mechanism of thought. The psychical entities which serve as elements in thought are certain signs and more or less clear images which can be 'voluntarily' reproduced or combined ... The above mentioned elements are, in my case, of visual and some of muscular type. Conventional words or other signs have to be sought for laboriously only in a secondary stage.

Einstein in Hadamard (1954), pp. 142–3

1. INTRODUCTION

The conjunction of the rise of the printed book as a prime means of transmitting information and the Renaissance reformulation of the means of visual representation was clearly an integral part of what we call the Scientific Revolution. On one level, it seems perfectly obvious that to be able to represent (say) a plant in a convincingly naturalistic manner in a printed botanical treatise would serve to provide straightforward instruction and to transmit checkable information to students of the natural world. Indeed, the polemic in favour of illustration by Leonhart Fuchs, introducing his great book of botanical science in 1542, provides early support for this view. He confronts those who 'will cite the most insipid authority of Galen that no one who wants to describe plants should try to make pictures of them' (Fuchs 1545, pp. x–xi). Fuchs asks rhetorically, 'who in his right mind would condemn pictures which can communicate information much more clearly than the words of even the most eloquent men?' In a similar manner, Leonardo, that most fervent advocate of visual communication, had already demanded, 'O Writer, with what words will you describe with such perfection the entire configuration which the drawing here does?' (Keele and Pedretti 1979). And, comparably if somewhat more unexpectedly, Michael Mästlin's referee's report on Kepler's *Mysterium cosmographicum* for Tübingen suggests that 'Kepler might provide a diagram and numerical tabulations [of the order and sizes of the spheres according to Copernicus], because the subject is absolutely incomprehensible without a diagram' (Kepler 1938–88, XIII, p. 85; trans. Rosen 1975, p. 325).

It has been claimed that the new techniques of systematic naturalism in the visual arts – above all the artists' new science of perspective – are inseparable from the 'search for truth' in Renaissance science. A nice formulation of this view is provided by Alistair Crombie:

> The conception of the *virtuoso*, the rational artist aiming at reasoned and examined control alike of his own thoughts and intentions and actions and of his surroundings, seems to me to be the essence of European morality, meaning both habits and ethics, out of which the European scientific movement was generated and engineered. In this context the rational artist and the rational experimental scientist appear as exemplary products of the same intellectual culture. (Crombie 1985, pp. 15–16; cf. Ackerman 1985 and Root-Bernstein 1985)

For present purposes, it matters not whether this intellectual culture is seen as triumphantly progressive or (as Foucauldians would have us believe) imperialistically oppressive. The complementarity of the cultural symptoms remains essentially the same. Indeed, to go even further in forging the conceptual alliance between art and science, perspectival representation has been seen in the Panofskian tradition as the 'symbolic form' of the Renaissance – as the conceptual model through which vision was radically redirected, the world was made to look different to the observer, and the transmission of knowledge was reformed (Panofsky 1927).[1] Specific incidents have been adduced – as discussed later in this study – to show that the interpretation of new visual phenomena, such as those revealed by the telescope, were most effectively conducted by observers who were literate in the painters' methods of three-dimensional design, particularly in the sciences of cast shadows and perspective (Edgerton 1984; Kemp 1991, pp. 94–6).

I find a sharp contrast between these big claims about visual representation and the levels of understanding we have achieved about the roles actually played by depiction at each stage in the processes that lie behind the making of an illustrated scientific text.[2] These processes potentially involve, in a complex and not necessarily sequential manner, variant combinations of observation, visualization, graphic modelling, publication, communication, and reception. Furthermore, the framework within which a particular combination of processes is realized will differ substantially over time and even within the same period. Our habit of assuming certain kinds of roles for representation in our various modern sciences may provide us with very misleading criteria when we approach the texts and images of past eras. I remain sufficient of an empiricist to believe that the characterization of the role of representation in science cannot be adequately achieved without a close study of how illustrations actually functioned in their particular historical environments. My contribution to this book on scientific illustration is designed to take *the* two sciences of 1543, anatomy and astronomy, sciences that apparently rely upon very different modes of visualization and representation, and to look at how illustrative material functioned in relation to the agendas of the scientists. By choosing such different sciences, we will also be able to broach if not to answer the question as to the extent to which the visual representations as realized on the page provide access to the conceptual models in the scientist's minds – the kinds of non-verbal models of which Einstein spoke.

My tactic in the sections that follow will be to ask to what extent were the innovatory features in the two sciences conveyed through visual representation. On one hand, we will need to consider whether acts of accurate representation in the new Renaissance manner were in their own right essential to the innovations or whether the main burden of the scientific content was transmitted through other means (textual and diagrammatic). In anatomy, Vesalius will dominate our considerations, since his great book essayed solutions to almost all the illustrative problems in descriptive anatomy before the eighteenth century. My thesis with respect to anatomy will be that Vesalius recognized the potential of veridical illustrations as a tool in a way that was integral to his reform of anatomical science, and that his use of illustrations was far more varied in type and function, both in their own right and with respect to his text, than has generally been recognized. By function, I do not just mean modes of anatomical description, but I also include the factors of communication to his audience through the particularly magnificent new medium of the large-scale printed volume. These factors embrace the aesthetics of what I will be calling the 'rhetoric of the real,' as well as more technical questions of what is actually possible in his chosen medium of illustration. When we turn to astronomy, I will be arguing that the new vision of the universe formulated by Copernicus involved not dissimilar factors of realism, rhetoric, and aesthetics, but that his means of argument and visual presentation remained untouched by new forms of representation. Perhaps there is nothing surprising in this absence of new presentational means, since the kind of astronomy that was practised in the era before the telescope was predominantly a matter of measurement and mathematics. However, a number of astronomers in the succession of Copernicus, most notably Tycho Brahe and Johannes Kepler, saw how new systems of representation in printed books could offer a powerful tool in the broadcasting of certain aspects of their science, above all its instrumentation.

2. THE TWO SCIENCES – SOME GENERALIZATIONS

Since I have asserted airily that anatomy and astronomy are very different in their visual characteristics, I think it is only fair that I give at this stage some general idea of what I mean – though this idea will necessarily depend upon some crunching generalizations. Anatomy is *par excellence* a descriptive science, at least in its modern sense, and its primary subjects of interest can be viewed to good effect with the naked eye, even if other techniques of examination (including microscopy) have in the post-Renaissance period amplified the scope of observation. Linked to the physical process of dissection, anatomical illustration lends itself to sequential, step-by-step exposition in which the visual presentation acts as a surrogate for the eye-witness experience or as a visual summation of many eye-witness experiences. In the hands of Vesalius and many of his successors, anatomical illustration lent itself to what I am calling the 'rhetoric of reality'; that is, the use of recognizable visual signals of uncompromising naturalism to convince the viewer that the forms are portrayed from life. These visual signals were frequently accompanied by texts or captions that emphasized the concrete situations and procedures by which the representations were generated, and by visual references to the act of dissection itself, through such devices as the display of tools.

In astronomy, by contrast, the plain description of the appearance of the heavens to an unaided eye at a single moment would serve little purpose, and even a series of sequential pictures would generate forecastable patterns and little else. The appearance of the heavens only becomes eloquent to the enquirer after structure when coupled with systematic measurements in which the eye serves as just one component in an instrumental system of controlled recording over a period of time. The translation of these measurements into coherent visual form involves the representation of things that cannot literally be seen, such as the orbs that enclose the paths of the planets, the points that mark the centres around which they turn, or the circles that map out the invisible spheres (crystalline or notional) which determine the motions of the celestial mechanism. The rhetoric in this case is very different. It is the 'rhetoric of irrefutable precision,' conveyed by tables of figures and flat geometrical diagrams. Yet it is this very translation of the visual phenomena into mathematical schemata remote from immediate sensory and physical experience that contributed to the vulnerability of the representations, since a particular geometrical diagram of the cosmos may be just one of a number of analogue models which can be contrived to fit the appearances. It was this long-recognized dilemma that gave Osiander his licence in the foreword to *De revolutionibus* to argue that the heliocentric theory was a fruitful new hypothesis rather than a representation of the physical actuality of the universe – a licence that could draw some partial support from Copernicus's argument that relative motions produce a 'reversible agreement,' though Copernicus casts his arguments in predominantly realist terms (Osiander in Copernicus 1543, pp. iv–vi, and *Works*, 1972, II, p. xvi).[3] However, as we will see, the new breed of astronomers found alternative ways to build the 'rhetoric of the real' into the visual presentation of their reformed science.

3. ANATOMY BEFORE VESALIUS

In looking at the sciences of 1543, it seems wise to begin with anatomy since it apparently presents the simpler case and anatomical illustration has been more widely discussed in the existing literature than the role of illustrations in astronomy. This is not to say, however, that extensive discussion necessarily results in adequate understanding. Even recent histories of anatomical illustration show a notable reluctance to discard the traditionally triumphalist view in which the central purpose of the historical narrative is to outline the inevitable progress in depictions of the body according to the procession of perfectible naturalism (Roberts and Tomlinson 1992). To my mind, this remains a valid narrative within its own limited frames of reference, but it casts aside all those factors which might explain the nature of the imagery in its broader social, intellectual, and aesthetic aspects. Even on its own terms, the narrative of perfected representation causes problems when set within the history of observational science, since the logical consequence of any insistence upon observing the real thing is that illustration is at best a limited substitute for the primary experience and at worst a dangerous evasion of the obligation to undertake firsthand observation. It should not come as too much of a surprise to find Vesalius, the second authentic hero of the standard story, asserting that

> I believe it is not only difficult but entirely futile and impossible to hope to obtain an understanding of the parts of the body or the use of simples from pictures or formulae alone, but no one will deny that they assist greatly in strengthening the memory in such matters. (Vesalius 1538, letter of dedication; and Saunders and O'Malley 1950, p. 233)

Vesalius's reference to memory is unlikely to have been casual, given the prominent emphasis upon the need to cultivate the art of memory in an era in which the continued cost and limited availability of books and manuscripts meant that much information had necessarily to be carried around in the mind.

The first authentic hero of the conventional story is, of course, Leonardo da Vinci, who would not have been inclined to accept Vesalius's qualification on the limits of the understanding of anatomy that could be gleaned from illustrations. Indeed, he emphasized that his drawings were superior to the witnessing of a single dissection, given the considerable practical problems of dissecting and the need to combine results from many dissections. However, just taking one of Leonardo's drawings – one of his most famous (see fig. 2.1) – we will be readily able to see how much more complicated are the visual and intellectual factors than his own claims for representation might lead us to assume. The study of a foetus in the womb, with related diagrams and notes, demonstrates all his skills as a draftsman in conveying the three-dimensional presence of objects and his extraordinary inventiveness in devising methods of demonstration – most notably in the upper diagrams of the interdigitations of the placenta and uterus wall (Keele and Pedretti 1979–80, no. 198r).[4] Yet underlying his personal rhetoric of reality – both in the drawings and in the discussions of dissections in the sets of related notes – are a series of complex dialogues with various kinds of tradition and meaning. Most obviously, as consistent with Galenism, he has incorporated features from animals, as in the cotyledonous placenta derived from his study of ungulates. One of his notes

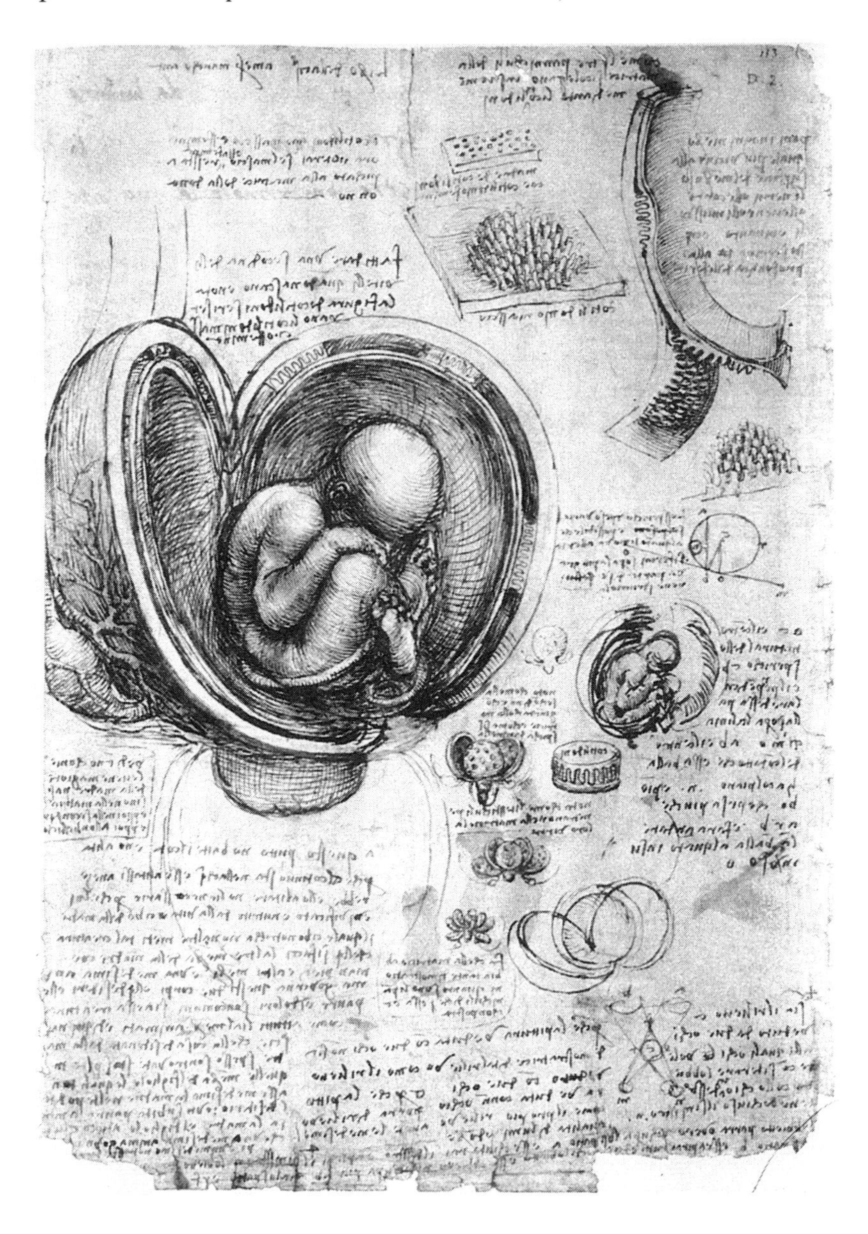

2.1 Study of the foetus and the womb, with optical and mechanical diagrams. Leonardo da Vinci.

speculates on the traditional question of the relationship of the souls of the mother and foetus, so that 'something desired by the mother is often found imprinted on the limbs of the infant' – a concept based on the notion of the soul as the 'form' (or form-generating agency) of the body. The whole set-up of the image, particularly as revealed in the small sketches of the enclosing coats of the womb, assumes its full effect in the context of his theory of the microcosm, in which the constituent parts of nature express the profound analogies within the whole. In this case, the parallel is between the womb and an opening bud or seed-case. At centre right is an entirely diagrammatic figure exploring the behaviour of a spherical body with a heavy weight at its periphery on an inclined plane, which may have been occasioned by his thinking about the orientation of the foetus with its heavy head in the womb. In the bottom right corner is an optical diagram and note which explains 'why a picture seen with one eye will not demonstrate such relief as the relief seen with both eyes' – which indicates that even for Leonardo the illusion of three dimensions on a two-dimensional surface possessed inherent limitations compared with the viewing of the real thing. However, the assertive language of objectivity spoken by the drawings is not such as to encourage the spectator to be openly aware of the limits and pitfalls of naturalistic representation.

The earliest published illustration that lays overt claims to be a true picture of an actual dissection makes a startlingly direct assertion of presenting the unvarnished truth. This is the print by Hans Wächtlin (or Wechtlin) of a dissection by Dr von Brackenau of a hanged man in Strasburg in 1517 (fig. 2.2), first published by Lorenz Fries a year later (Fries 1518). As befits an illustration by a printmaker in the succession of Dürer, who himself depicted plants and animals with uncompromising attention to their individual peculiarities and accidental damage, the criminal is portrayed with tortured face and savagely twisted right arm. The fact that the man was a vile criminal is underlined in the caption as a strategy to sanction the gory display – and to set it in a nexus of German imagery which would include Hans Baldung Grien's macabre iconography of death.[5] It is, I think, no coincidence that Wächtlin's rawly direct style should (like the successor images) have been used to illustrate a book in the German vernacular rather than learned Latin. The relatively unobtrusive labelling, which encroaches on the main image as little as possible, is designed to enhance the sense that we are looking at a true picture. Successive derivations of this much copied image show its translation into more schematized formats,

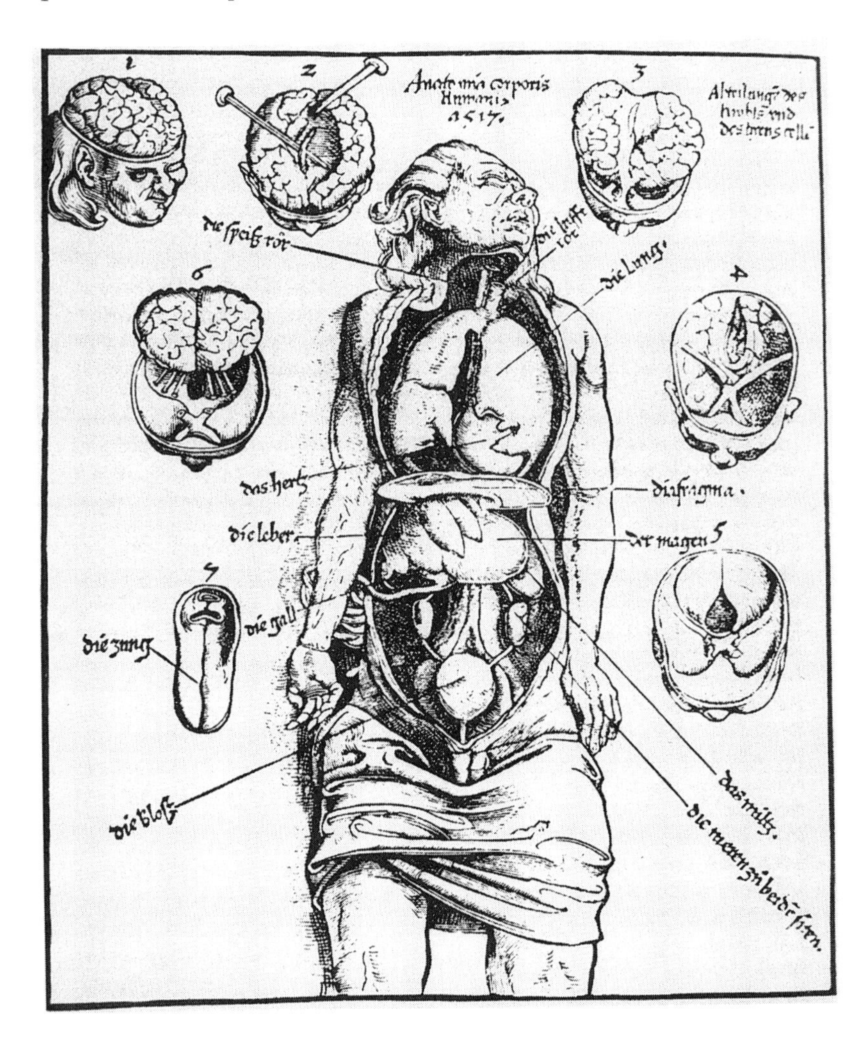

2.2 Dissection of the brain, thorax, and abdomen (Hans Wächtlin 1517).

as in Lorenz Fries's 1519 treatise, or adapted to serve a different function as a blood-letting figure in 1540 (Fries 1519 and von Gersdorf 1517).

Even Wächtlin's apparently direct image, however, raises problems about how the anatomical content entered the representation. To take just one feature, the lobed liver corresponds to stock accounts and representations (Hundt 1501). We must assume some kind of mechanism by which the schemata of traditional anatomy were available to the draftsman and provided a visual foundation for his representation of features. It should be remembered that in an inevitably messy dissection 'seeing' would certainly not have been readily translated into 'knowing.' The apparent naturalism does not mean that the image is necessarily to be more trusted than the earlier woodcut, but it does mean that it is making implicit and explicit *claims* to be trusted. The same point can be made by looking at one of Leonardo's drawings of the muscles of the abdomen, which, even on a small and summary scale, conveys something of the conviction of his draughtsmanship (Keele and Pedretti 1979–80, no. 111r). For all its air of objective directness, the diagonally criss-cross muscles depend closely upon Pietro d'Abano's *Conciliator* of 1496 (cxcix) and upon his desire to emphasize graphically that 'every muscle uses its force along the line of its length.'

Viewed in the light of such complications, the traditional reservations about illustrations in anatomical texts appear more understandable. Thus Berengario da Carpi, whose *Commentaria* of 1521 and *Isagogae breves* of 1522 are the first anatomical books in which illustrations make a really substantial impact on the tone of the whole production, warned the reader his figure of the vertebrae, for example, 'does not exhibit their true likeness ... [and] their actual form is better seen in dried vertebrae in cemeteries' (Berengario 1522, trans. Lind 1959, p. 160; see French 1985 and Kemp 1993). His much admired muscle-men serve strictly limited anatomical functions with respect to his text, and whenever he mentions his illustrations he does so in terms that restrict their role. However, as someone who was a prominent figure in the Medicean Rome of Pope Leo X, as the recipient from Raphael of a painting of Saint John the Baptist, and as a protégé of Aldus Manutius, Berengario was well placed to understand the value of stylish illustrations in making his book effective in its social, intellectual, and commercial environment. His poised *écorché* holding a noosed rope is the participant in an implied *historia* in the setting of Berengario's demonstration of the *theatrum* of the body (fig. 2.3). If Wächtlin's

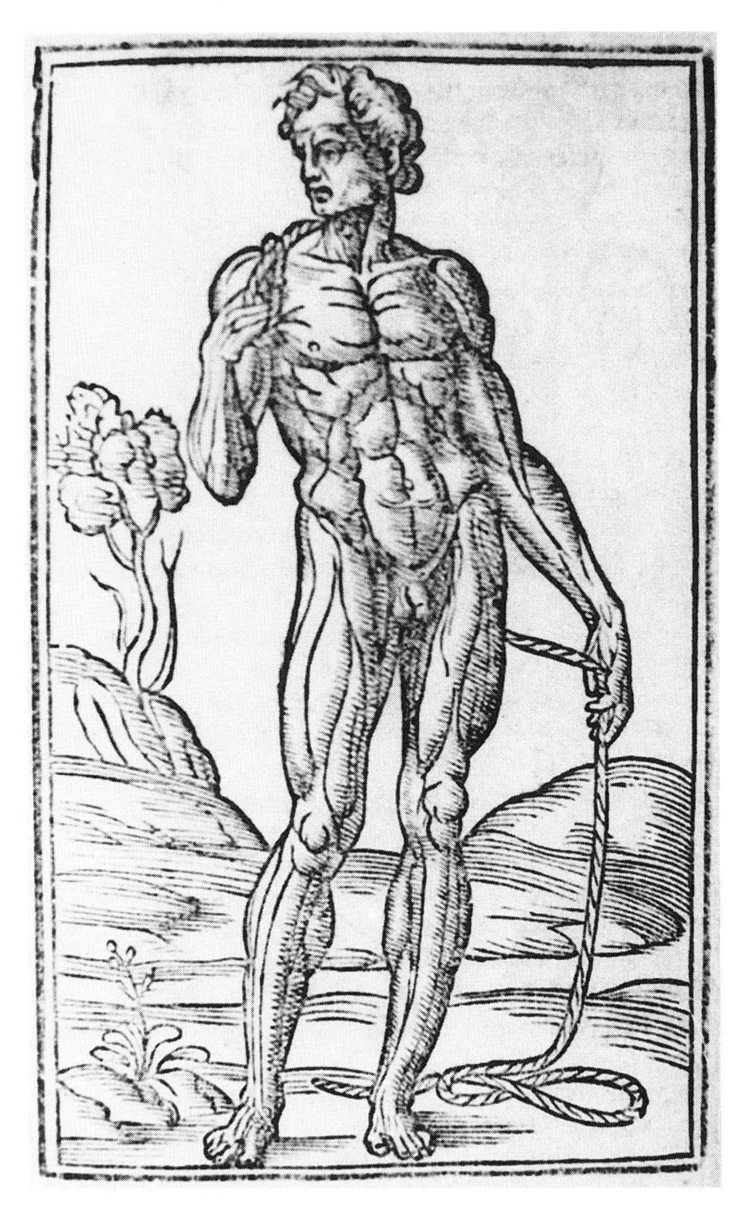

2.3 Muscle-man with rope (Berengario da Carpi 1521).

rhetoric of reality was of a rustic nature, Berengario's tends towards the nobly Roman.

Not surprisingly, in the humanist orbits of medical science in the Renaissance, it was the nobly tragic which became the dominant mode of illustration. The School of Fontainbleau stylishness of the illustrations in Charles Estienne's *De dissectione* (completed before Vesalius's *Fabrica* but only published in 1545 in Latin and translated into French a year later) has often been mocked for overwhelming their anatomical content, but the fancy presentation is far from gratuitous or merely decorative (Estienne 1545; for the artistic sources, see Kellett 1964 and Kornell 1989). The anatomized men and women (fig. 2.4), performing the assigned roles as dying warriors or violated Lucretias, testify to the drama of human beings who have been placed in the world by God to contemplate the heavens, to 'investigate the divine works of nature,' and to give due purpose to the creation through their deeds. If we read the introduction to the first book of *De dissectione*, 'containing the argument of the whole work,' in which he debates the purpose of man with Anaxagoras, with due references to Chrysippus and Zeno, we gain a sense of the Stoic foundation of his enterprise, in which man as observer and as the 'measure of all things' gives value to God's creation through perception of his divine plan.

4. VESALIUS

Not the least of Vesalius's achievements was to embody all the existing varieties of the rhetoric of reality into a wonderfully functioning and complex whole. The title page of the *Fabrica* obviously sets the anatomist in the context of a great *historia*, in which the 'house of the soul, as Plato has it,' is explored in a *all'antica* temple or *theatrum* of anatomy (Vesalius 1538, letter of dedication, trans. Saunders and O'Malley 1950, p. 234). But the underlying message of Vesalius descending from the professorial throne and abandoning the textbook to conduct the dissection with his own hands, also aligns him with certain aspects of the German directness of Wächtlin and von Brachenau. He is overtly declaring his reliance upon the book of the body itself, which is itself embodied directly in his own book.[6] His insistent emphasis upon firsthand dissection, a practice in which Vesalius must have possessed remarkable skills, is visually underlined by the cluttered still life of instruments (fig. 2.5), many of which were common or garden tools used by other trades. Hans Baldung's illustration for Walther Ryff

2.4 Dissection of the abdomen of a woman (Charles Estienne 1546).

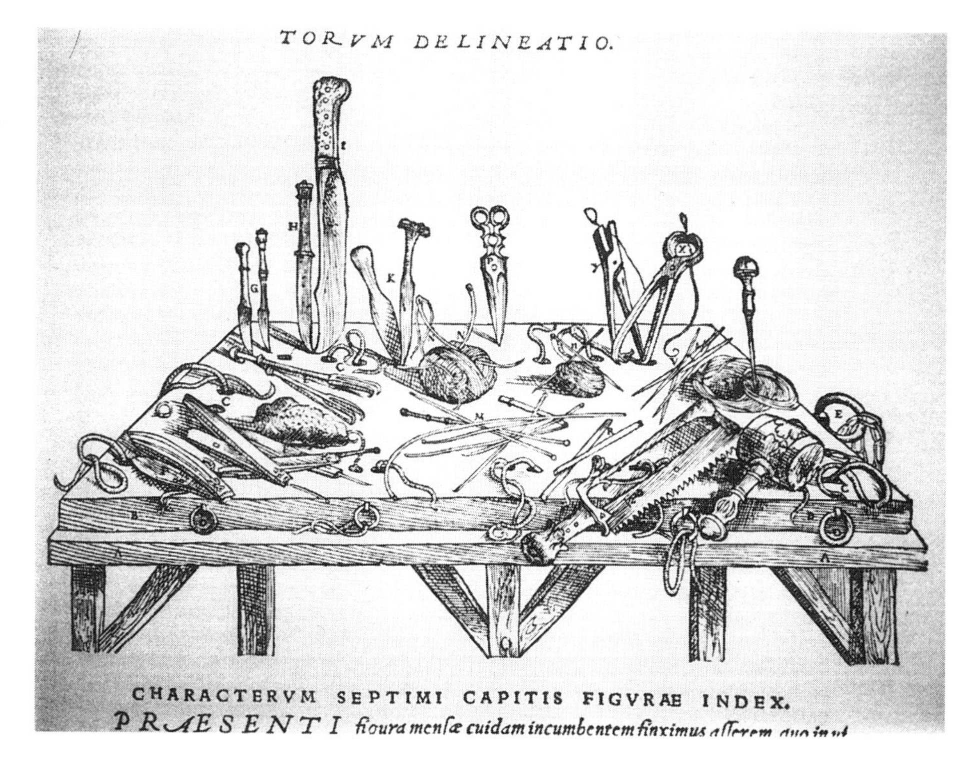

2.5 Tools for dissection (Andreas Vesalius 1543).

in 1541 had already included comparable tools in much the same spirit (see also Dryander 1537). The illustration of a tethered pig on a board, 'which we usually provide for the administration of vivisections,' appears at first sight to serve a similar purpose, but the text provides a rather different gloss, since it is concerned with Vesalius's conscious adoption of Galen's practice of vivisections of pigs for physiological investigations (Vesalius 1543, VII, xix, p. 661). This serves to remind us that the principles of anatomical investigation enunciated by Galen provided inspiration for Vesalius to study form in rigorous detail through first-hand dissections, rather than acting (as so often believed) as the dead hand of tradition.

The famous muscle-men sustain this air of actual dissection, as they perform their myological striptease, and the tone of the accompanying notes talks the spectator through the various procedures in much the same way as Vesalius must have done in the dissecting room (Kemp 1970 and 1993). Thus on the seventh plate he informs us that the rope from which the cadaver was suspended 'was diverted back to the occiput because of the muscles that are conspicuous in the neck.' However, the overall presentation is remote from the German manner, and clearly adopts and extends the more heroic mode of Berengario's Italian woodcuts. The frieze of gesturing figures in their continuous landscape act out a grand drama, gesturing like Old Testament prophets or collapsing in martyr-like death. Such a heroic presentation is fully justified as an appropriate (i.e., decorous) way to present 'the ingenuity and workmanship of the supreme artisan' (summi opificis sollertiam articiumque) (Vesalius 1538, letter of dedication, trans. Saunders and O'Malley 1950, p. 234). It is appropriate to this aspiration that he should have concentrated on displaying the normative male body, synthesized from his dissections and readings, rather than attempting to evoke the raw directness of a single specimen in the style of Wächtlin. With the highly contrived illustrations of the skeleton (fig. 2.6), which Vesalius acknowledges 'contribute more to display than to instruction,' the grand pseudo-history becomes literal, as we are informed that while 'genius lives on, all the rest will perish' – a motto taken from Virgil's *Elegiae in Maecenatem* (1.38), which makes particular sense in the context of Vesalius's self-conscious bid for enduring fame in his hugely ambitious project.

No book was ever planned more meticulously to effect an enduring reform of both the subject and its mode of presentation. The letter to Oporinus, published in the opening matter, is insistent both about the necessary visual quality – 'nowhere neglect the significance of the pictures' (nusquam picturae ratione ... neglecta) – and about following his intricate system of text, indices of figures, labelling, commentary, and cross-references. Whether the openly pictorial representations supplied by Jan Steven van Kalkar or his own more diagrammatic illustrations, the visual qualities of the blocks sent to Basel were to be meticulously respected.[7] The variety, insight, and maturity with which different kinds of visual material are exploited is astonishing. In addition to the large, pictorial representations of the main components in the fabric of the body, a series of small inset illustrations (fig. 2.7) graphically demonstrate structural principles. Some of the diagrams, such as the hinge, appear more than once, with a full annotation reserved for the first appearance. Sometimes the demonstration is

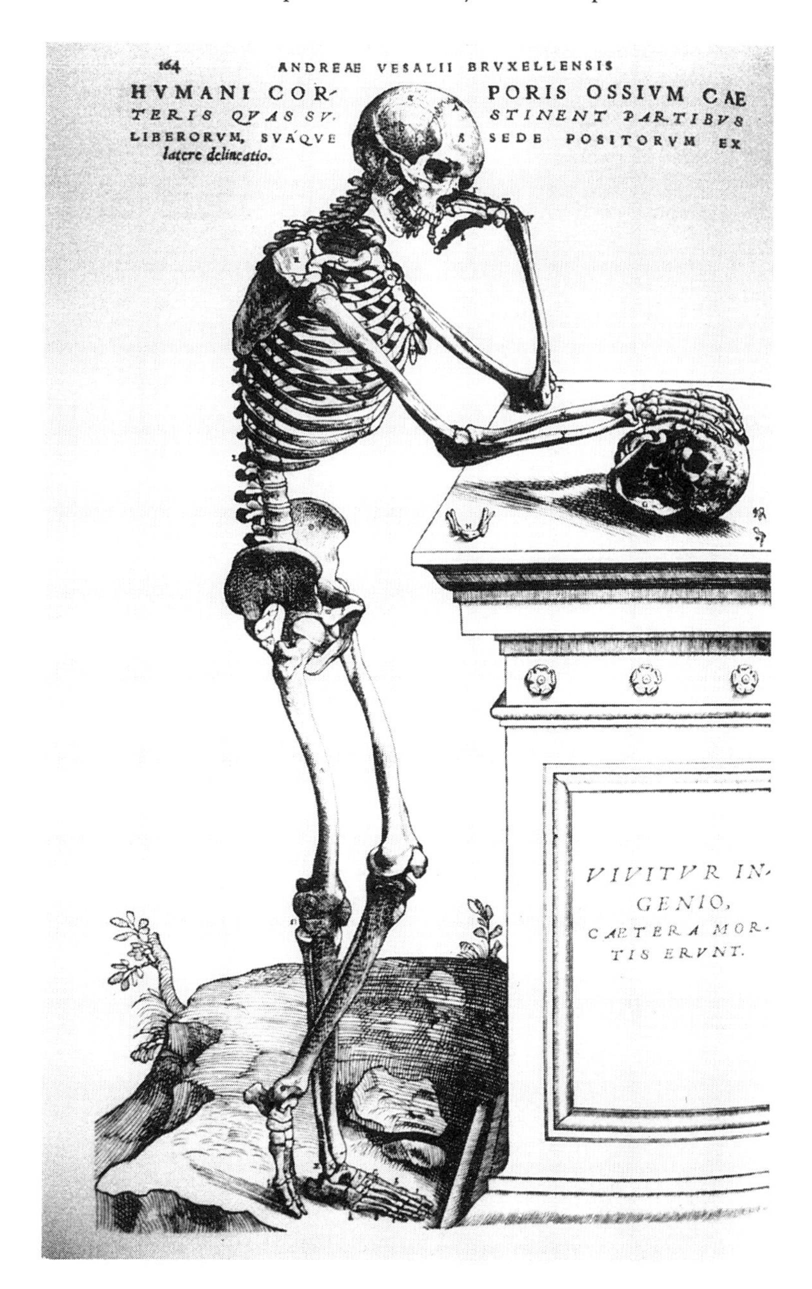

2.6 Skeleton from the side (Andreas Vesalius 1543).

γίγγλυμος.

ter flexionem & extenſione
duplicem uertebrarum ad ſ
actè ſimplicis motus gratia
cies cenſetur, manifeſtis ſin
ginglymos, quoties mutuo
caua ſubintrent: & rurſus u

Hic A ferrũ ſeu cardinem notat, parieti infixum: B autẽ, ferrũ quod oſtio ſeu feneſtræ nectitur.

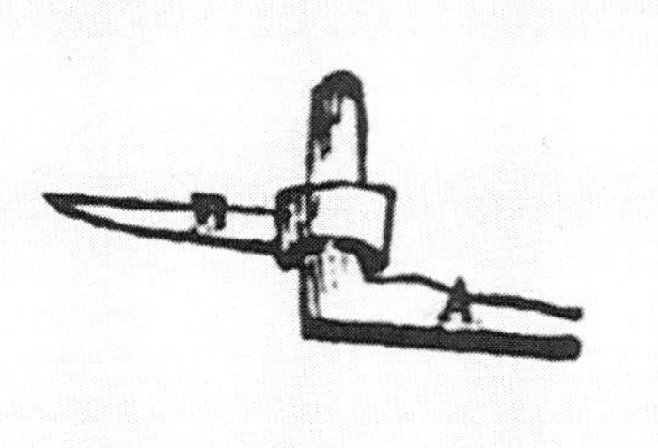

tuis ingr
confera:
cipiat, i
ne præſ
tibiæ en
minet: fe
dientia. dein in capitum me
dis & tertijs quatuor manu
internodio, & in ſecundo p
pita, in quorum medio ſinu
donatur ſinibus, tubere qu
gantiſſimè perficitur. Vlnæ
ri ſinus mirificè ſubeunt. A

2.7 Demonstration of a hinge (Andreas Vesalius 1543).

diagrammatic in the most schematic sense, using what he called delineations in a 'perfunctory,' 'rough,' or 'rudimentary' (rudis) manner (Vesalius 1543, I, xx, p. 93; and III, i, p. 358). In his *Venesection Letter* of 1539, he introduced his illustration of the veins with the words 'in this accurate though rather rudimentary figure' (in haec vera, quamvis rudiori figura), which shows clearly that he recognized the way in which the 'truth' of a particular illustration is dependent upon a correct reading of its conventions in relation to its designated function (Saunders and O'Malley 1950, pp. 230–1). A particularly nice example of his discussion of conventions is when he tells us that his section of the eye shows the forms 'in the manner in which we habitually depict the heavens and four elements on a flat surface' (atque hoc quoque modo caelos & quatuor elementa in plano depingere solemus) (Vesalius 1543, VII, xiv, p. 643). The following page shows how the three-dimensional components may be built up like a piece of precious jewellery by a 'supreme artisan.' He was also alert to the problem of exactly what woodcut lines represent in the more pictorial illustrations. Parallel lines could, after all, stand as shading or serve to indicate linear structures. Generally the lines serve to outline major contours and to shade within these contours, but on one occasion – specifically to demonstrate the muscle fibres (fig. 2.8) – Vesalius stresses that the lines signal the linear appearance of the forms rather than serving as elements in the artistic modelling of relief (Kemp 1993, pp. 100–1).

In this one great book, Vesalius essentially tested all the illustrative types that were to be available to anatomists during the sixteenth century. The only variations left were the choice of medium, such as the copperplate engravings used by Valverde, and the devising of different systems of reference and labelling. One of the most interesting of these variants was devised by Bartolomeo Eustachio for his anatomical tables, which were not finally published until 1714 (see Roberts and Tomlinson 1992, pp. 188–93). The style of the illustrations is consciously synthetic – that is to say, presenting the forms in a simple and clear manner which abstracts them from the flesh-and-blood reality of dissection – and the overall presentation is self-consciously poised (fig. 2.9). Eustachio has done away with even Vesalius's reticent letters, leaving the figure totally unmarked. Reference to individual parts is achieved through a Ptolemaic system of coordinates in the marginal scales, which necessitates the use of two straight edges, just as demonstrated in Apianus's *Cosmographicus* (1524). This ultra-cool and cerebral system of mapping the topography of the body did not prove popular,

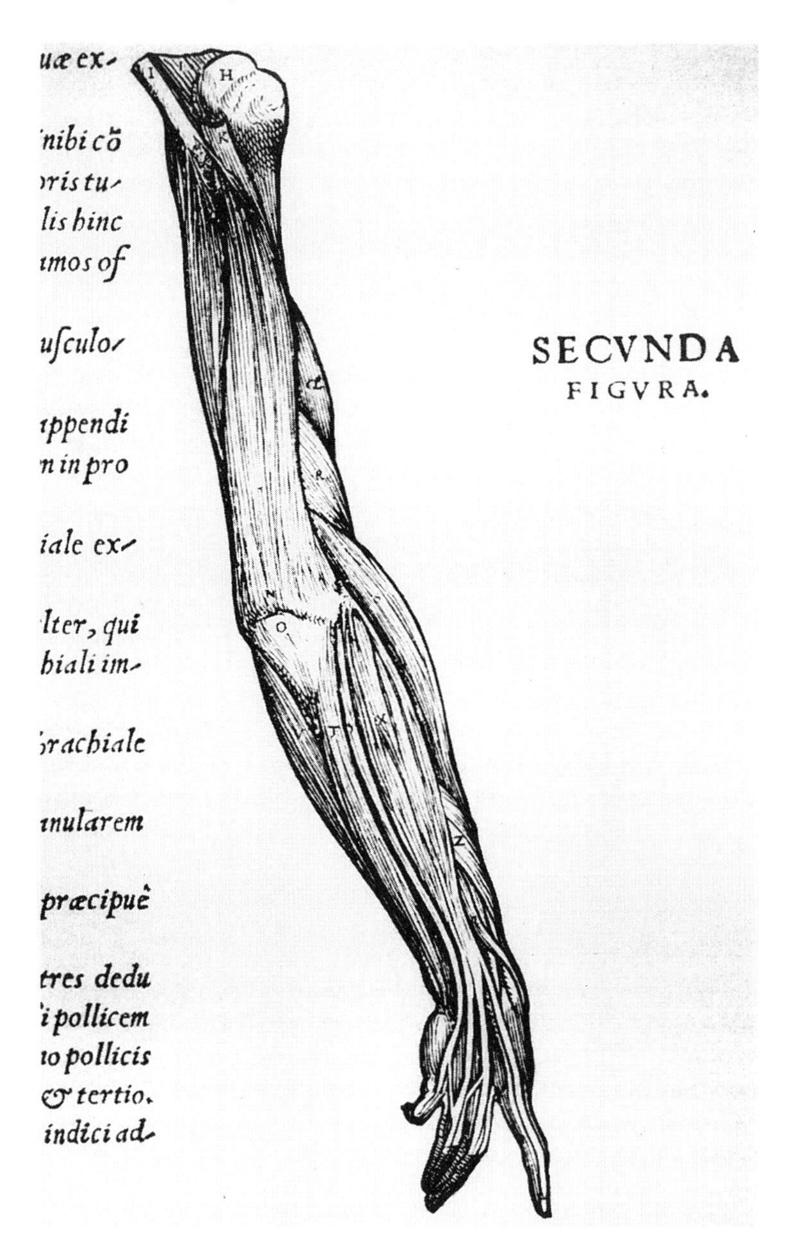

2.8 Muscles of the upper arm and forearm, and tendons of the wrist (Andreas Vesalius 1543).

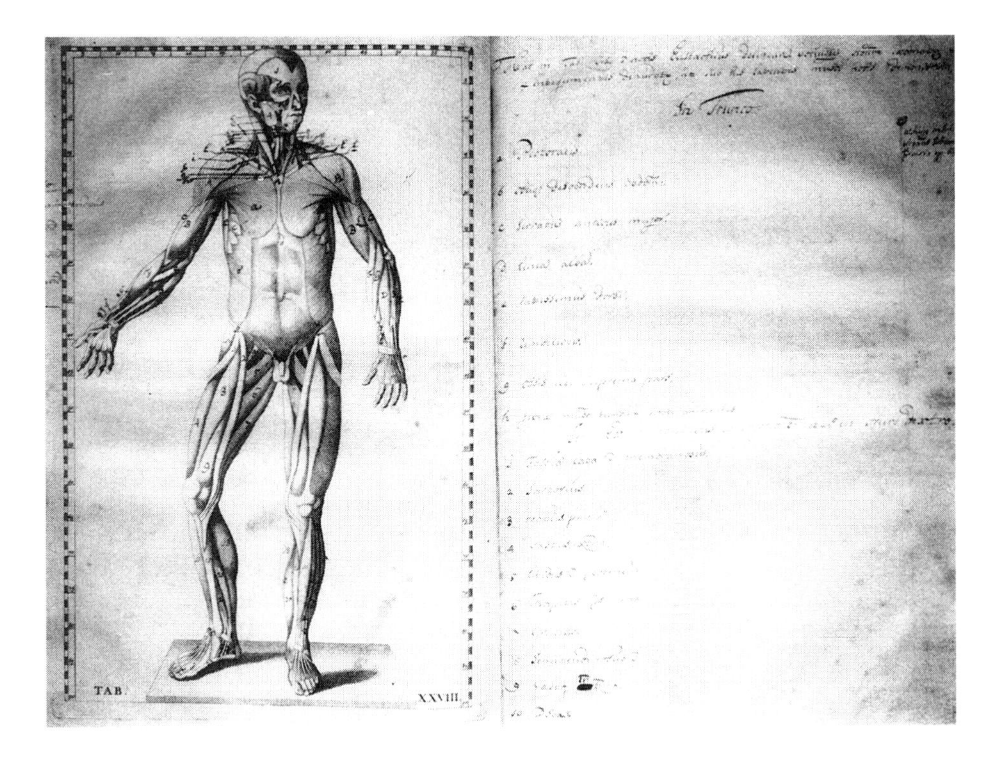

2.9 Superficial dissection of the muscles from the front (Bartolomeo Eustachio 1722).

and one of the owners of the copy I consulted obviously ran out of patience with it, adding labels in a conventional manner.

The actual 'look' (or visual quality) of anatomical illustrations in the sixteenth century is, as these necessarily few examples have shown, far from being simply determined by the need to portray the forms accurately. However, the core of the endeavour does reside in a belief in the value of the veridical portrayal of tangible objects in space and in due proportion, so that there could potentially be a direct process of visual matching between the actual forms and their depictions.

5. COPERNICUS

I emphasized at the outset that such a system of veridical portrayal could not stand at the heart of the astronomer's strategy of research and exposition. However, many of the figurative and metaphorical images evoked by Renaissance writers on astronomy rely upon exactly the same kinds of criteria of formal structure as characterize the treatises by anatomists. The most famous of these is the much cited bodily simile in Copernicus's dedication of *De revolutionibus* to Pope Paul III in 1543. He characterized those astronomers who had relied upon the proliferation of such devices as epicycles, eccentrics, and equants as failing to

> elicit or deduce from the eccentrics the principal consideration, that is the structure of the universe and the true symmetry of its parts. On the contrary, their experience was just like some one taking from various places hands, feet, a head, and other pieces. Very well depicted, they may be, but not for the representation of a single person. Since these fragments would not belong to one another at all, a monster rather than a man would be assembled. Hence in the process of demonstration or 'method' as it is called, those who have employed eccentrics are found either to have omitted something essential or to have added something extraneous and wholly irrelevant. (Copernicus 1543, preface, iv, trans. *Works*, 1972, II, p. 4)[8]

To some extent, the resort to a bodily analogy was a standard strategy in medieval and Renaissance thought, just as Apianus in 1524 followed Ptolemy in comparing geography to the portrayal of a complete head, while chorography was equivalent to the portrayal of an eye or ear in isolation (fig. 2.10). But, considered in the context of its humanist audience in the Rome of the Farnese Pope (who was a notable patron of Michelangelo), this passage is full of weighty allusions.[9] The idea of a body perfectly proportioned according to the principle of *symmetria* was by this time deeply embedded in Renaissance aesthetics, and was an integral part of the key doctrine of decorum, according to which every part should be appropriate to the form and significance of the whole. This doctrine could be gleaned directly from ancient poetics and rhetoric, or from such Renaissance authors as Leon Battista Alberti, whose *De re aedificatoria* had transmogrified the Roman ideas of Vitruvius into Renaissance form. As someone who had participated in humanist poetics as a translator of Theophylactus Simocatta from the Greek, including a letter which centres upon Parrhasius's portrait of Helen of Troy (Copernicus, *Works*, 1972, III, p. 31), and as someone who was reputed to have been sufficiently competent in painting to produce a self-portrait (a version of which was once owned by Tycho

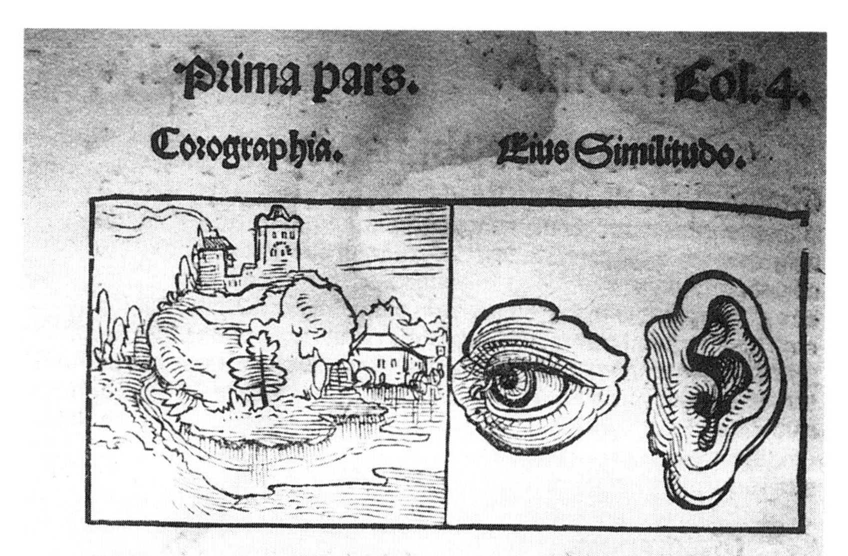

¶ Antequam quis ipsius Cosmographie studium aggrediatur/fundamentum inprimis seu Astronomie principia que sunt circulorū sphere noticia/quibus tota utitur Cosmographia disquirat necesse est. Quod in sequentibus q̄ brevissime manifestabitur.

De motu sphaerarum Coelorumq; divisione. Caput secundum.

Mundus bifariam partitur in Elementarem regiōem et Aetheream. Elementaris quidem assidue alterationi subiecta/quattuor elementa Terrā/Aquam/Aerem et Ignē continet. Aetherea autem regio (quā philo: quintā nuncupant essentiam) Elementarem sua cōcavitate ambit: invariabilisq; substantia semper manens/decem Spheras complectitur. Quarum maior semper proximam minorem sphericé (eo quo sequitur ordine) circūdat. Imprimis igitur circa spheram ignis deus mūdi opifex locavit sphērulam lune. Deinde Mercurialem. Postea Veneream. Solarem. deinde Martiam. Jouiam et Saturniam/quelibet autē istarum unicam tantum habet stellam/que quidem stelle Zodiacum metiētes semper primo mobili seu decime sphere motui obnituntur/alias sunt corpora diaphona: hoc est omnino perlucida. Mox sequitur firmamentum: et quod stellifera sphera est: queq; in duobus parvis circulis

2.10 Chorography compared to pictures of the eye and ear (Petrus Apianus 1540).

Brahe), Copernicus is unlikely to have been using visual analogies in an innocent manner.[10]

Indeed, in the next paragraph, he reinforces the meaning by emphasizing that the 'movements of the world machine' were 'created for our sake by the best and most orderly artisan of all' (ab optimo et regulariss. omnium opifice), and he later refers to *divina haec Opt. Max. fabrica* (Copernicus 1543, preface iii, and I, x, p. 10, trans. *Works*, 1972, II, pp. 4, 22). The terms *opifex* and *fabrica* are already familiar to us from the Vesalian lexicon of significant words. Rheticus, that faithful promoter of Copernican ideas, glosses the bodily analogy by direct reference to Galen's *De usu partium*, to the effect that 'Nature does nothing in vain' (Rheticus 1541, trans. Rosen 1971, p. 137).[11] He then asks rhetorically, 'should we not attribute to God, the creator of nature, that skill which we observe in the common makers of clocks? For they carefully avoid inserting in the mechanism any superfluous wheel.' The idea of perfection such that nothing can be added or taken away without detriment to the *symmetria* of the whole conforms to the standard Renaissance concepts of visual beauty and structural necessity, expressed in their canonical forms by Alberti.[12] It was from such a standpoint that Copernicus decried those who 'either ... omitted something essential or ... added something extraneous and wholly irrelevant' (Copernicus 1543, preface iii, trans. *Works*, 1972, II, p. 4).[13]

On the page of the manuscript of the first book of *De revolutionibus*, facing his key visual statement of his new system of orbits (fig. 2.11), he further extends the visual analogies into specifically architectural and social contexts:

> At rest ... in the middle of everything is the sun. For in this most beautiful temple, who would place this lamp in another or better position than that from which it can light up the whole thing at the same time? For the sun is not inappropriately called by some people the lantern of the universe, its mind by others, and its ruler still by others. The thrice Greatest [Hermes] labels it a visible god, and Sophocles's Electra, the all-seeing. Thus indeed, as though seated on a royal throne, the sun governs the family of the planets revolving around it. (Copernicus 1543, I, 10, pp. 9v–10r, trans. *Works*, 1972, II, p. 22)[14]

If we put together the principles that are emerging – *symmetria*, decorum, perfect economy and necessity of design, hieratic social order with respect to supreme authority, and the human observer as the agent through which the whole system becomes apparent – we are in precisely

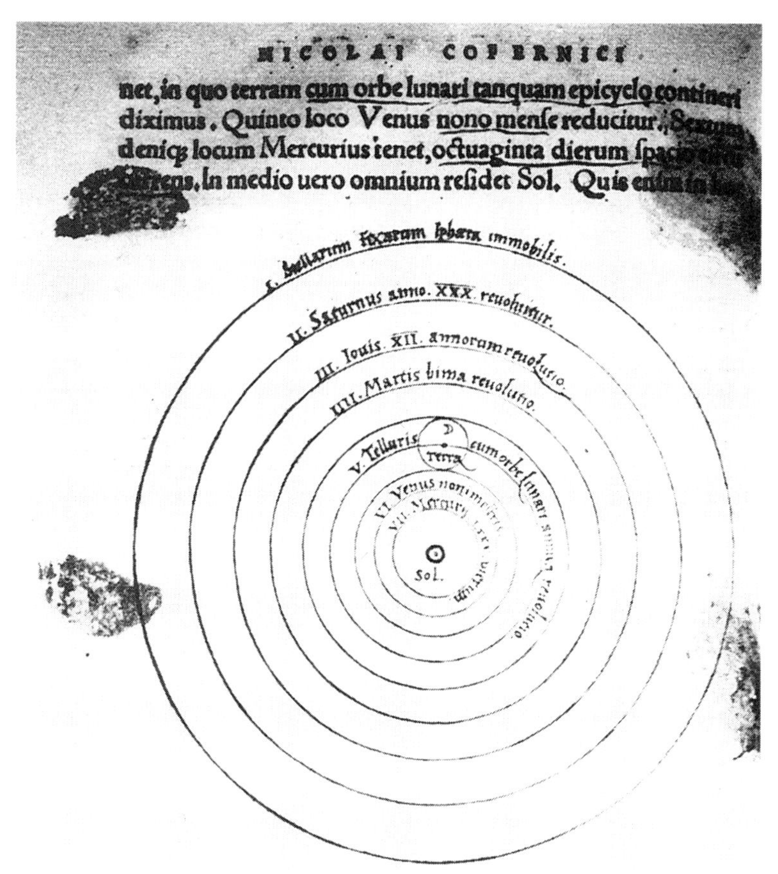

NICOLAI COPERNICI

net, in quo terram cum orbe lunari tanquam epicyclo contineri diximus. Quinto loco Venus nono mense reducitur. Sextum denique locum Mercurius tenet, octuaginta dierum spacio circu currens. In medio uero omnium residet Sol. Quis enim in hoc

pulcherimo templo lampadem hanc in alio uel meliori loco poneret, quàm unde totum simul possit illuminare? Siquidem non inepte quidam lucernam mundi, alij mentem, alij rectorem uocant. Trimegistus uisibilem Deum, Sophoclis Electra intuentē omnia. Ita profecto tanquam in solio re gali Sol residens circum agentem gubernat Astrorum familiam. Tellus quoque minime fraudatur lunari ministerio, sed ut Aristoteles de animalibus ait, maximā Luna cū terra cognationē habet. Concipit interea a Sole terra, & impregnatur annuo partu. Inuenimus igitur sub hac

2.11 Diagram of the orbits of the earth and planets (Nicolaus Copernicus 1543).

the kind of world enunciated in Alberti's writings, and signalled in a less sustained way in Estienne's preface.

The vision being formulated by Copernicus embodies a number of the central tenets of humanist philosophy as reflected in Renaissance aesthetics. At the centre is the kind of neo-Stoic outlook that saw the orderly and rational pursuit of human affairs as founded on a proper perception of the underlying order of nature. This notion of order in its turn was laced with Neoplatonic idealism and Pythagorean metaphysics. From Neoplatonism came a reverence for geometry in particular and mathematics in general as reflective to some degree of divine ideas, while from the kind of Pythagorean theories that had provided the standard base for musical theory came notions of the harmonic proportionality of universal design – notions that were to become particularly vital in Kepler's thought. I am not concerned here to debate whether Copernicus's system actually lived up to its ideals of perfect economy, decorum, *symmetria*, and harmony – which seems doubtful – but rather to characterize the nature of the vision that underlay his aspirations.[15]

Not the least important of these ideas was the conception of the central role of man as the observer, and, indeed, in the form commonly formulated in the Renaissance, as the reason why the whole set-up had been created by God. The paradox at the heart of the Copernican system was, of course, the fact that this central observer had been removed from the physical centre of the system. And inhabitants of the earth were now in a position where all planetary motions were relative to the motion of the body on which they were standing. To make this point, Copernicus quotes from Virgil's *Aeneid*: 'forth from this harbor we sail, and the land and the cities slip backwards' (Copernicus 1543, I, 8, p. 6r, trans. *Works*, 1972, II, p. 16; Virgil III, 72). This notion of the appearance of shapes and motions as irredeemably relative to the position of the observer was essential to the Renaissance revolution in the depiction of the visible world. A theorist like Leonardo could stress that the point at which parallel lines appear to converge (the 'vanishing point') moves with any motion of the observer's point of view, and that two horses running away from us along parallel tracks appear to be converging.[16]

However, this relativity was not taken to mean that visual experience must collapse in subjective confusion. Rather, the science of perspective leads to a rational understanding of the principles of systematic depiction such that the true shape, position, and motion of an object can be determined unambiguously from proper analysis of the image. Thus, the centrality of the observer is if anything strengthened by his or her role within a system of relative perceptions. There are clear signs in Copernicus himself, and in some of his more realistically inclined successors, that the position of the astronomer on a mobile body was seen as presenting an opportunity to record the motions of the whole system around the static sun in such a way that the astronomer could capture the physical reality rather than merely formulating mathematical hypotheses which were analogous to the appearances. It was much in this sense that Kepler asked and answered his question: 'in what manner were the earth's dimensions adapted to the size of the solar globe?' (Kepler 1618–22, in *Werke* 1938–88, VII, p. 277). He answered that it was 'in terms of vision. For the earth would be the home to the contemplating creature, and it was for him that the entire universe had been created.'

Although such concepts as the proper visual principles of the body or temple of the universe and the role of man as the 'mean and measure of all things' are deeply shared by the two sciences of 1543, the relationship of the overall vision and the illustration of the phenomena was necessarily quite different in each discipline. Any Copernican could not but be aware of the obvious problem that what we actually 'see' is the sun rising, moving across the skies, and setting. We may understand the point of relativity, but, in terms of how our perception actually works, our eyes and body do not bear obvious witness to the motion of the earth. This was the dilemma which Kepler endeavoured to overcome in his paper written as a student, which postulated the appearance of the system as seen from the perspective of an observer on the moon, and in his posthumously published *Dream*, in which it is the earth which appears to move from a station point on the planet Levania (i.e., the moon) (Kepler 1634, trans. Rosen 1967).[17] This notion of being able to envisage the appearance of a physical set-up from the perspective of an observer located anywhere in the system was precisely what painter's perspective could accomplish in the hands of its supreme operators. What Kepler has done is analogous to what would happen if we decided to move the observer standing centrally on the floor in Mantegna's *Camera degli Sposi* – looking up at the illusionistic oculus on the ceiling vault (fig. 2.12) – to the position of one of the women who peer downwards. A perspectivist of Mantegna's skill would have been well able to accomplish the relocation of the viewpoint, should it have been required.

2.12 Oculus in the ceiling of the *Camera degli Sposi* (Andrea Mantegna 1465–74).

However, to make the point visually in the illustration of an astronomical treatise required a different strategy from that of a perspectival picture. The obvious one was to represent the system diagrammatically as if characterized by an Olympian viewer who could stand outside the system. This was of course the stock method adopted for the geocentric system in earlier publications, and Copernicus's diagram contained no new mode of visual presentation.[18] In fact, his manuscript could hardly be more unadventurous in its visual presentation, containing inset or marginal diagrams of an entirely linear and traditional kind. In this format he does no more than to show the basic geometrical components of the motion of the system in a sequential and accumulative manner.

When he did attempt to characterize one of the more complex, compound motions – that of the pole around the mean position (I) –

the 'twisted line' is not easy to read in terms of the resulting motion, and in the printed edition is mistakenly transposed into two separate ovals (fig. 2.13) (Copernicus 1543, III, 3, p. 66v, trans. *Works*, 1972, II, p. 124).[19] Copernicus had admitted earlier in the same section 'that these matters are not easily explained adequately with words. Hence they will not be understood when heard, I am afraid, unless they are also seen by the eyes. Therefore let us draw on a sphere the ecliptic ABCD ...' (Copernicus 1543, III, 3, p. 66r, trans. *Works*, 1972, II, p. 124). But the diagrammatic resources available to him were not visually eloquent to anyone who had not already cultivated an ability to visualize in the mind in non-verbal form (as described by Einstein) the complex consequences of the relative motions of bodies moving in orbits and epicycles with eccentrics. I think it is fairly clear that an astronomer of Copernicus's and Kepler's levels of visionary insight must possess abilities of spatio-temporal visualization of an astounding order – at least astounding to me – if he or she is to envisage in a coherent manner what would happen if 'any part thereof were to be moved from its place' in such a way that it would not produce 'confusion in all the other parts and of the Universe as a whole' (Copernicus 1543, III, 3, p. 66r, trans. *Works*, 1972, II, p. 123). But this spatio-temporal visualization is not reflected in any new visual configurations in Copernicus's illustrations.

This is not to say that the key diagram of the orbits lacked a certain kind of potency. It is significant that at least two astronomers, including Kepler, tabulated the distances of the planets on this same page in their copies of Copernicus, since it is easier to envisage the numerical values in juxtaposition to a visual key.[20] And a copy of the first edition in St Andrews University shows clear signs of paste marks of a sheet once stuck over the heliocentric system (fig. 2.11), presumably to substitute the printed universe by a less offensive arrangement.[21]

When we consider more generally the kinds of visualization demanded by either Ptolemaic or Copernican astronomy, we may assume that professional practitioners were acquainted with a wide range of visual sciences and their associated instrumental devices. The nexus of required learning is neatly encapsulated in the one-page catalogue issued by Regiomontanus in Nuremberg in 1474, containing books already available and titles he was intending to publish.[22] The range extends from pure geometry, such as Euclid's *Elements*, Archimedes on the sphere and cylinder, Apollonius's *Conics* and a treatise on the five regular bodies, through the *scientiae mediae* of music, astronomy (Puerbach,

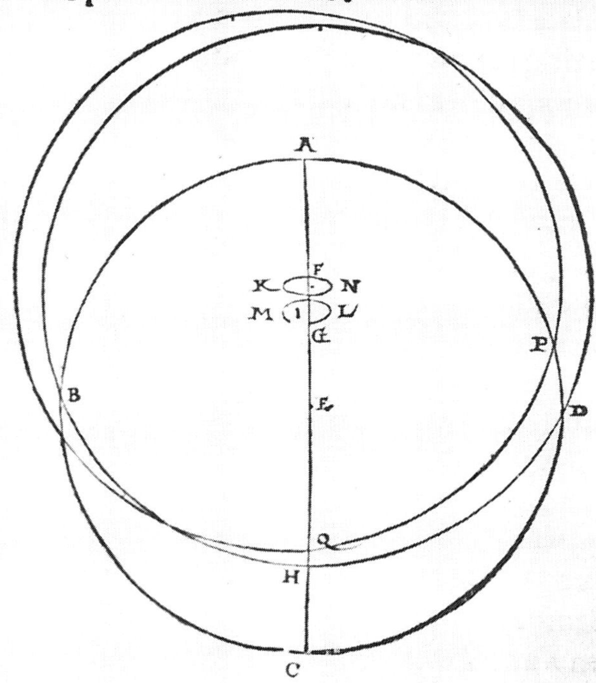

2.13 Diagram of the motions of the pole around a mean position (Nicolaus Copernicus 1543).

Ptolemy, Proclus et al.), and optics (Witelo and Ptolemy), to the practical sciences of engineering and instruments. The availability of actual instruments for the practice of astronomy is signalled in large print at the base of the prospectus.

6. ASTRONOMICAL INSTRUMENTS AND PHYSICAL MODELS

It was through the use of astronomical instruments that the essential mediation between the observed phenomena and their geometrical analysis could be accomplished, and it was through astronomical models that representation could best be achieved for the purposes of instruction. Instruments such as armillary spheres, orbaria, and torqueta, as illustrated by Apianus (fig. 2.14), could provide aids to spatial understanding in a way that was impossible with Copernicus's illustrations, although the schematic orbits could still only deal with the rudiments of the system rather than the full complexity of apparent motions.[23] Most instruments were not of course direct attempts to model the celestial machine in fully spatial terms, but a number may be seen as serving as kinds of analogue models. The most common was the astrolabe, which results from a conical projection from the centre of the north celestial pole in such a way that the *rete* acts as a star map, which is laid over coordinates, lines of equal latitude, and (often) hour lines. Astrolabes were designed for practical observation and mathematical calculation, and were not well suited to serve as aids to the visualization of the actual spatial configurations, any more than Copernicus's diagrams had been – however visually compelling and 'concrete' a finely made astrolabe may at first sight seem as a model of the physical set-up. Such physical models also in their turn stood in a symbiotic relationship to depictions, since they were themselves subject to explanatory illustration in a variety of diagrammatic and perspectival techniques, most spectacularly, as will be seen, in the publications of Tycho Brahe.

Copernicus was of course well familiar with the varieties of highly appealing models available in the Renaissance, and a magnificent set of instruments, comprising an astrolabe, torquetum, and celestial globe were presented to the Jagiellonian University in Cracow by Martin Bylica in 1494, the final year in which Copernicus was a student.[24] The astronomical globe is particularly nice as an aid to visualization since it can be adjusted so that the rotation of the globe models the observed rotation of the stars around the celestial pole at the particular location

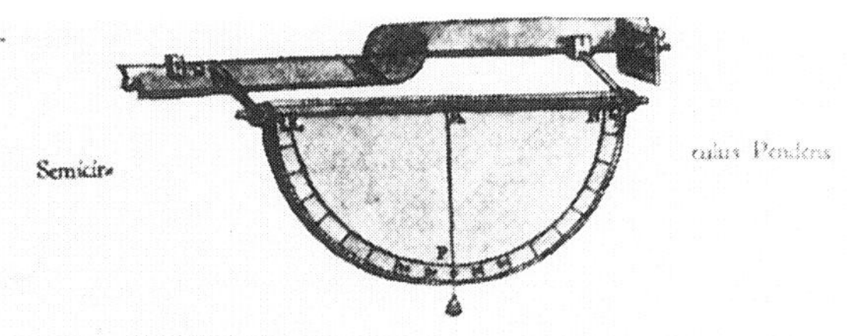

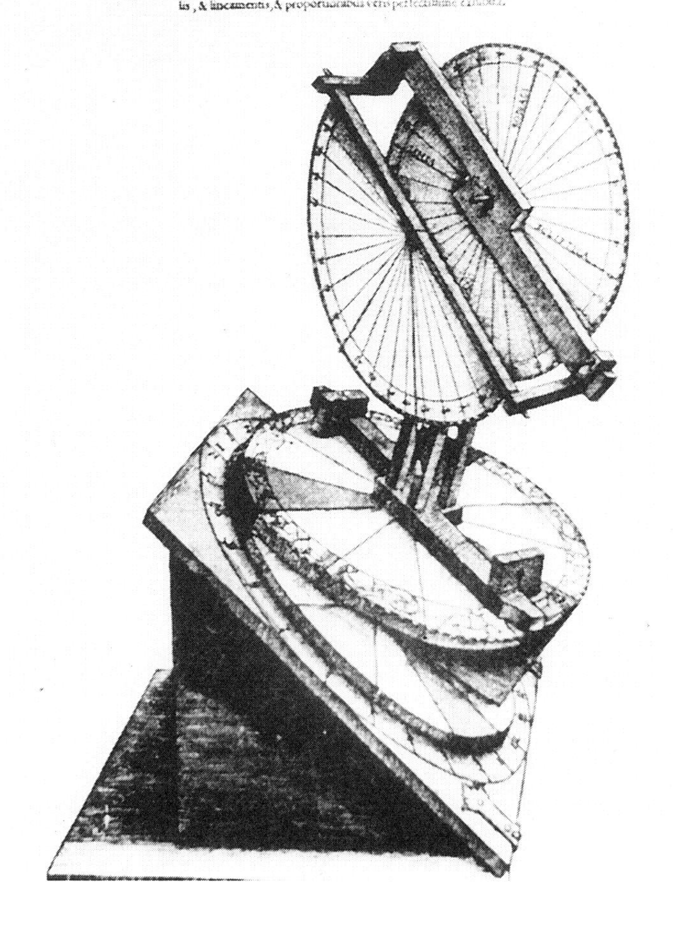

2.14 Torquetum (Petrus Apianus 1524).

in which it is used, while the astrolabe can be used to work out related problems of an astronomical or astrological nature. These are luxury instruments, which could only be made by a supreme *opifex* of the worldly variety – in this case probably Hans Dorn. The quality underlines the aesthetic of economy and perfection which Copernicus shared with humanist theorists of the visual arts, such as Alberti and Leonardo, for whom the 'fittingness' of form and function was a keystone to the understanding and representation of nature. The expensive perfection of such devices made them especially suited to flourish in the courtly culture which did so much to ensure the triumph of humanism across Europe. The technical success of instruments in modelling the motions of the heavens obviously played a major role in what I have called the 'rhetoric of irrefutable precision,' but they also gave astronomers a kind of opportunity of participating in the 'rhetoric of the real' that was not open to them through veridical depiction.

In fact, for one of the major contributors to the reform of astronomy, Tycho Brahe, instruments became the keystone in the construction of the real edifice of the heavens, and the chief means of personalizing astronomy in terms of the heroic observer. No one had ever placed such weight upon the explanation and illustration of his instruments. Apianus's *Instrumentum primi mobilis* in 1534 and *Astronomicum caesareum* in 1540 provide only very partial precedents for the way that instruments are described in Tycho's *Astronomiae instauratae mechanica* in 1598 and in his other publications.[25] By demonstrating the mechanisms by which his observations were achieved, Tycho was certifying his practice in terms of the concrete reality of his personal procedures – 'so that certainty of the form and use of the instruments might be apparent,' as he said in his treatise on the new star of 1572 (Brahe 1573, trans. 1623). He explained that 'the construction and use of the instrument is understood by careful study of the accompanying figure quicker than through more elaborate verbal explanation' (Brahe 1598, p. 67). The level of his personalization of instruments was equally strong. He cherished a parallax device once owned by Copernicus, and which was, 'it was said, made by him with his own hand.' Although it was wooden and not convenient to use, Tycho recorded that 'I was so delighted because it reminded me of the great master,' and he was moved to compose a heroic poem in its honour (Brahe 1598, pp. 44–5).[26]

The personalizing of his own equipment is vividly apparent. His great *Mural Quadrant* or *Quadrans Tychonicus* (fig. 2.15) served as an emblem of his endeavour. He explains that the pictorial adornments, including

2.15 *Mural Quadrant* or *Quadrans Tychonicus* (Tycho Brahe 1598).

his own portrait by Thobias Gemperlin and landscape by Johannes of Antwerp, were 'only added for the sake of ornament, and in order that the space in the middle should not be empty and useless,' but this should not lead us to underrate their significance to Tycho's agenda, any more than ornament would have been regarded as redundant in rhetoric (Brahe 1598, p. 30). The whole set-up casts Tycho in the role of a new Ptolemy, or perhaps even more ambitiously as a personification of astronomy itself. The emphasis upon the instruments for observation stresses the reality of his procedures, while the mobile brass globe in the niche signals the process of envisioning which lead to his own peculiar conflation of the Ptolemaic and Copernican systems. He explained that a large celestial globe he had constructed allowed him to 'determine mechanically, with very little trouble and without calculations, all the details concerning the doctrine of the sphere.' It was the invention of new instruments, upon which Tycho set such store, that permitted some Renaissance thinkers to claim most decisively that the ancients had been both emulated and surpassed.

Throughout his account of his instruments, Tycho explains their manufacture and use in highly individualized terms, although it should be stressed that his representations are not primarily aimed at providing instruction in the actual making of the devices – which would require different kinds of technical illustration. Not infrequently, he outlines the iconography of their ornamentation in some detail. Thus he explains that his *Quadrans minor* (fig. 2.16), which was mercury gilded 'so that it stays beautiful and clean,' was adorned so that it might 'offer some instruction' – in this case, through an allegory which contrasts a life of higher contemplation (his own life by implication) with the vanity of worldly things (Brahe 1598, p. 13). The first of the accompanying inscriptions – 'Vivimus ingenio coetera mortis erunt' – is virtually identical to the Virgilian tag in Vesalius's illustration of the skeleton, while the other says that 'in Christ we live, all the rest perishes.'[27] Another device, his *Armillae equatoriae*, was adorned with paired portraits of Ptolemy and Albategnius (al-Battani), and Copernicus and Tycho, as a way of underlining his place in the heroic succession (Brahe 1598, p. 57). However, as a corrective to the ideal picture conveyed by this contemplative and productive life of observation, he warns that the vagaries of patronage are such that the astronomer should always ensure that the instruments can be dismantled for transport to another site. As he says, 'the astronomer, as well as a student of other branches of knowledge, has to be a citizen of the world' (Brahe 1598, p. 27).

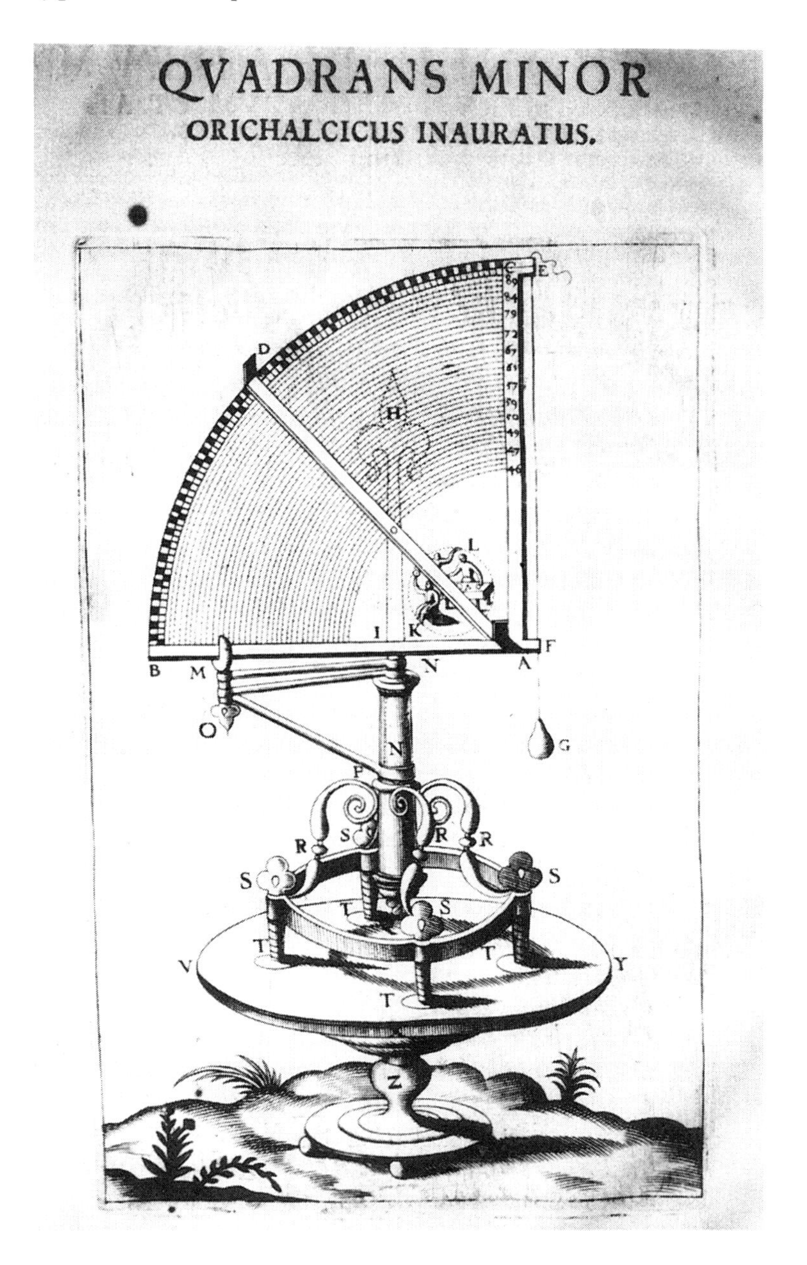

2.16 *Lesser Quadrant* (Tycho Brahe 1598).

The most complete expression of Tycho's world in visual terms was of course his remodelling of the island of Hven. His castle of Uraniborg (fig. 2.17), with its surrounding plantations and ponds, was contrived as a microcosm of the universal harmonies.[28] The central building, as he explained, was 'strictly symmetrically arranged, as required with architecture if the work is to be executed in a proper manner according to the rules of art' (Brahe 1598, p. 131). The key rule, here as in Copernicus's vision of the body of the universe, was *symmetria* – a rule which Tycho saw as embodied in Dürer's books on human proportion – though the architectural vocabulary in which the *symmetria* was expressed by 'my architect, Johannes Stenwickel [or Steenwinckel] of Emden,' is actually remote from the strict requirements of Renaissance theorists and practitioners. A more architecturally literate realization of an *all'antica* temple for the Danish astronomer's muse was provided for Kepler's publication of Tycho's *Rudolphine Tables* (fig. 2.18). Kepler himself seems to have been notably literate in the visual arts, an expert in stereometric estimation – the highly useful merchant skill in the visual judging of volumes – and a decently accomplished draftsman in his own right, as the sketch for the frontispiece to the *Tables* suggests.[29] In the printed version, incorporating changes apparently demanded by Tycho's heirs, the slow perfecting of astronomical science is represented by an architectural progress from rustic supports at the rear, through crude piles of stone blocks and archaic Doric columns, to the more polished Tuscan pillar of Copernicus, and climaxing in the beautiful Corinthian column, beside which Tycho points to a diagram of his own version of the heavenly system engraved on the ceiling of the temple. The fact that Tycho did not adopt the full-scale heliocentric theory of Copernicus confirms that Renaissance aesthetics did not in themselves decisively predispose their adherents to the Copernican system rather than any other which promised the required elegance and economy.

Kepler's own manner of astronomical visualization represents the climax and consolidation of the various strands of visual modelling we have seen progressively developing in the writings of Copernicus and illustrations of Tycho. His most famous visual shaping of the planetary system was, of course, the characterization in his *Mysterium cosmographicum* of the ratios of the orbs in the Copernican system as corresponding to the arrangement of a set of Platonic solids nesting one inside the other (fig. 2.19).[30] We know from his own account in *De stella nova* that the idea came to him visually in 1495, when he was drawing 'quasi-triangles, in the same circle, in such a way that the end of one was the

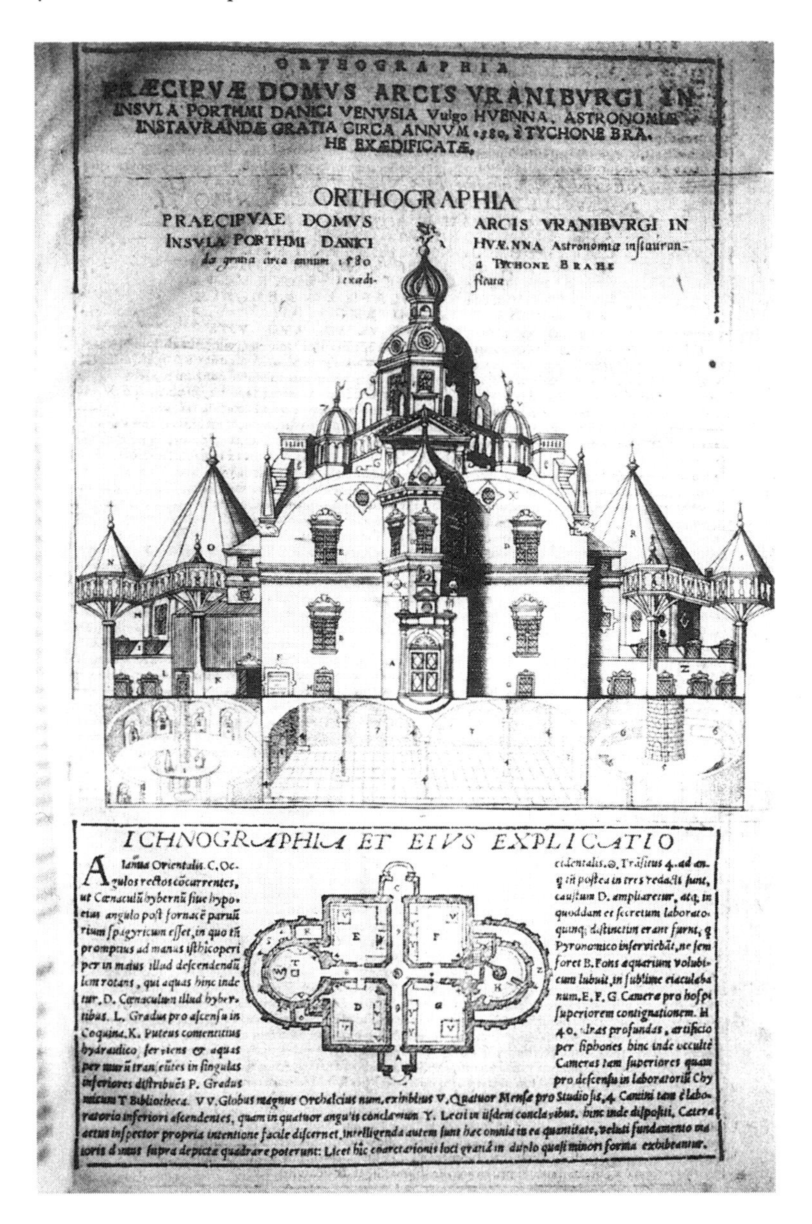

2.17 Elevation and plan of the palace of Uraniborg on the island of Hven (Tycho Brahe 1598).

2.18 Temple of the Astronomers (Johannes Kepler 1627).

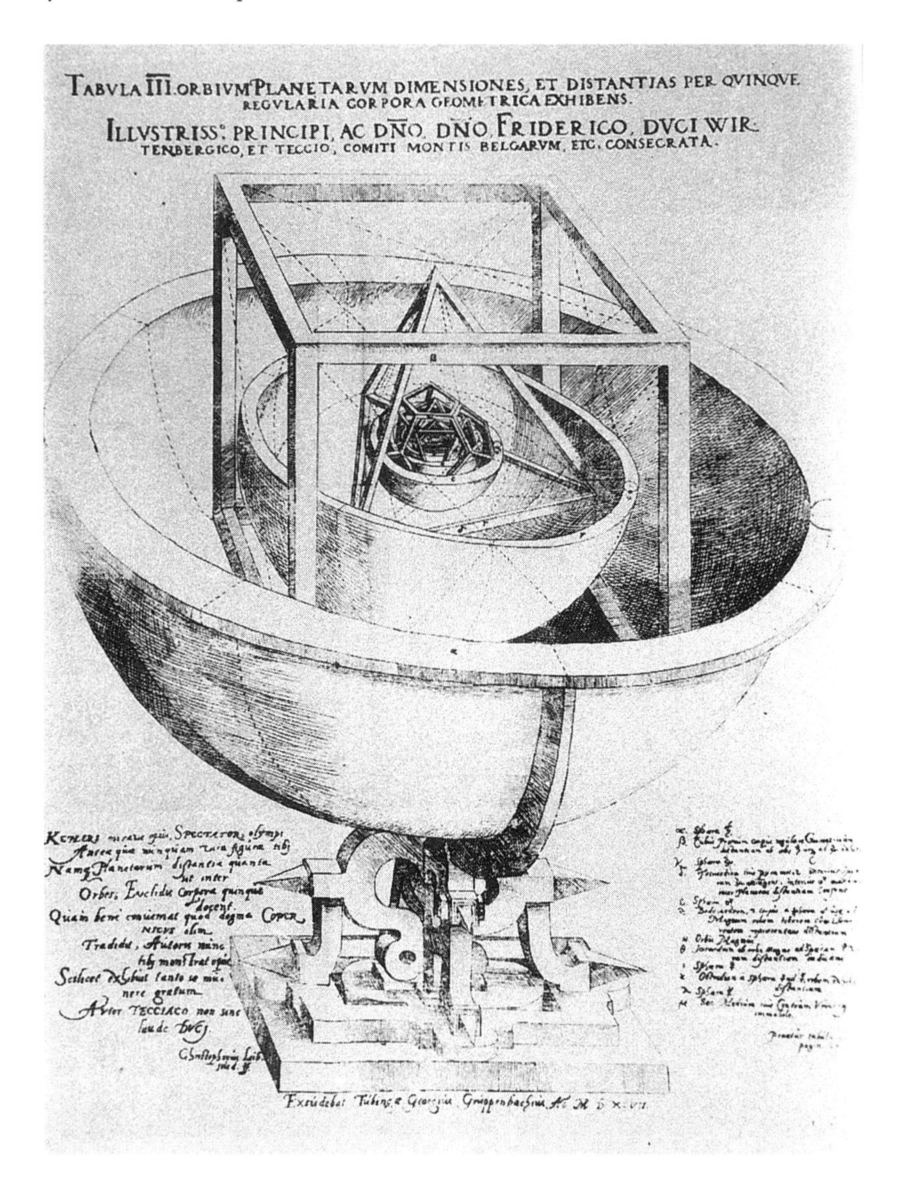

2.19 Demonstration of the orbits of the planets (Johannes Kepler 1596).

beginning of the next' for the instruction of his students (Field 1988a, p. 47 and pp. 45–51). The full-scale visualization is presented in a folding plate as a perspectival picture of considerable sophistication, in which the system is characterized in terms of an elaborate piece of mannerist metalwork, of just the kind that his noble patrons enjoyed.[31] In fact, the dedication of this plate to Duke Frederick is closely related to his unavailing attempts to fabricate the system in three dimensions. He promoted his *inventum* to his patron by explaining that 'the whole work and the demonstration thereof can be fittingly and gracefully represented in a drinking cup of an ell in diameter which would be a true and genuine likeness of the world and model of the creation in so far as human research may fathom' (Kepler 1938–88, XIII, p. 51; see Prager 1973).

The hollow armatures of the Platonic solids were each to be filled with appropriate beverages, which could be drawn off through taps at the rim. This bizarre scheme was dropped in favour of a plan for a model operated by clockwork, and he hoped to find a master *opifex* who could construct one with such precision that it would have an error of only one degree in a hundred years (Kepler 1938–88, XIII, p. 218ff). Even if this ambitious object was never to be realized, his dedication of the plate to his noble protector did have one fortunate consequence. Mästlin reported that theologians were deterred from voicing open criticism of Kepler's Copernicanism by the identification of the scheme with Duke Frederick (Kepler 1938–88, XIII, p. 151).

For Kepler, the conceit of remaking the universe in a working, physical model was no mere intellectual and technical game. At the heart of his enterprise – and of his discovery of the elliptical orbits – lay a desire to harmonize the Platonizing geometry which he valued above all other forms of mathematical truth with an understanding of the physical mechanics of the motions of the planets. Metaphysics alone would not suffice: 'the celestial machine is not so much a divine organism but rather a clockwork' (Kepler, letter to Hewart von Hohenburg, 10 Feb. 1605; 1938–88, XV, p. 146). It was in this spirit that he transformed one of the stock metaphors of astronomy into a functioning analogy in mechanics. This metaphor, used by Ficino (1493, cap. XIII, p. 255) amongst others, envisaged the heavenly bodies steered as by a pilot. The title page to Sebastian Münster's *Organum Uranicum* of 1536 picks up this metaphor in visual form (Münster 1536). Kepler, in one of the diagrams of a planet orbiting the sun in his *Astronomia nova,* depicts a pilot in the 'magnetic' stream emblematically, and in another (fig. 2.20) adds schematic oars with rippling waves to the diagram which explains the physical geometry of the orbit. Here the process of visualization is joined to a sense of physical action which is very like the muscular empathy described by Einstein – even if the consequence of the physical analogy in this case hardly shows Kepler at his most efficacious.

7. TELESCOPIC POSTSCRIPT

In all this discussion of visualization in astronomy, however strong the visual model adopted for particular purposes, we have seen nothing to compare with the direct practice of veridical representation in anatomy. Such representation only became effective in astronomical science when the celestial bodies could be observed as bodies, that is to say, as objects with discernible, individual features. It is this condition that explains why the invention of the telescope occasioned a new branch of visual astronomy, namely that concerned with the actual anatomy of the individual planets and the sun. The two key episodes – the dispute over the apparent irregularities in the surface of the moon, and the nature of the spots observed on (or not on) the sun – have been discussed elsewhere, but it might help towards the conclusion of this paper to remind ourselves of the rather different nature of seeing and knowing which the new sights down the telescope occasioned (Edgerton 1985; and Kemp 1992, pp. 93–4). The first controversy involved how to interpret the pronounced lights and darks on the moon, particularly at the interface between the shaded and illuminated portions. Galileo, well versed in the science of perspective and the artist's systematic understanding of cast shadows, was able to argue that the most rational way to interpret the changing patterns of light and dark was in terms of shadows cast by huge topographical features, including mountains (Galilei 1610). The other incident concerns the patches which were seen to progress across the image of the supposedly immaculate sun. Galileo argued from the perspectival foreshortening of the spots as they neared the edge of the sun that they were integral parts of the surface and not shadows of intervening bodies. His method of argument, as he explained, was 'in virtù di perspettiva' (Kemp 1992, pp. 94–5). Galileo's advanced understanding of the principles of artistic representation, which informed his method of analysis and exposition in these two cases, is undoubtedly important more generally to his theory and practice of observation, but the accompanying techniques of veridical

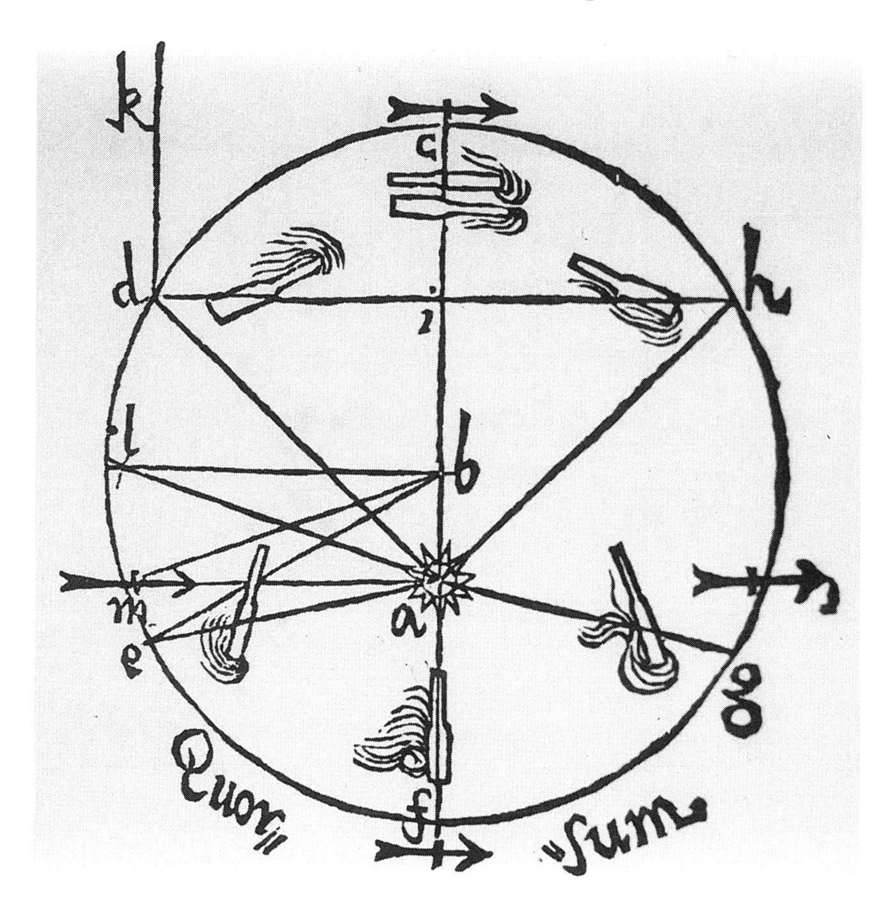

2.20 Demonstration of an orbit by analogy to a boat in a stream (Johannes Kepler 1609).

representation could still only be brought to bear upon a very narrow range of problems in astronomy as a whole. Galileo's innovations in other of his sciences, such as dynamics and statics, were conducted with quite different forms of visualization, experimentation, and proof, and he did not sustain the pictorial mode in his own later work in astronomy (Winckler and Van Helden 1992).

8. PROVISIONAL CONCLUSIONS

Looking back over this necessarily selective survey, what conclusions might be drawn at this stage of our understanding about the role of illustrations and its relationship to the process of visualization?

For astronomers in the Renaissance, the fundamental processes of representation do not seem to have been essentially different from those of Ptolemy or his Islamic successors. The visual qualities of the illustrations bore only schematic relationships to the visualization demanded of the astronomer. Scientific instruments come closer to the hypothetical mental models, but only with respect to the gross characteristics of the arrangement of the basic armature of the celestial machine. Where more specifically Renaissance modelling can be discerned is in the humanist metaphors and analogies used to characterize form and function, relying upon beauty, economy, and decorum (intellectual, visual, and social). The form and function were now related to a sense of the real arrangement of the system as it can be perceived, in contrast to the kind of modular system of mathematical modelling which exempted the Ptolemaic hypotheses from considerations of strict realism. The new realism involved a re-characterization of the heroic observer, in which the objects were defined relative to the observing subject – a move which was crucial if the Copernican and Keplerian systems were to become acceptable. But Copernicus's own diagrammatic expositions retained the standard forms of the Ptolemaic treatises.

One field in which the new practice of perspectival representation did eventually become crucial was the depiction of instruments. The publications of Apianus and Tycho Brahe gave astronomers a chance to participate in the kind of broadcasting of secrets and marvels that had become typical of the prestigious books of mechanical devices. The other major aspect of astronomy that was radically affected by new pictorial means was the depiction of celestial bodies as viewed in the telescope. The new features, such as the topography of the moon, involved the new vocabulary of perspective and light and shade, but they remained somewhat peripheral to the major changes in astronomical science.

For anatomists, the visual power of naturalistic representation was a powerful and central tool in the rhetoric of the real, and could be used as an expression of the impulse to reconstruct the fabric of the body on the basis of direct, hands-on experience. The representations served as a powerful form of visual pointing, both to their own features and, potentially, to those of the actual object. However, we should remain alert to the fact that this visual pointing could draw apparently convincing attention to what was not there, and that the process of matching expectation to experience was (if anything) rendered more complex and challenging rather than less so. We should also remain continually aware of the way in which the representation of the human body, in the eyes of its major investigators, was designed to demonstrate the wonderful artifice of the maker of the bodily 'temple' for the soul.

In sum, I do not see any obvious prospect of a grand, unifying theory based on new forms of representation as corresponding directly to (or precipitating) some great overarching reform of the means of visualization. The relationship between illustration and visualization seems quite different in the various sciences, though we can frequently observe intricate conjunctions in the structure of metaphor, analogy, and 'aesthetics' that is used to locate a specific field of study within its broader intellectual, theological, and social nexus. I have to say, as far as I am concerned, the lack of conformity to a grand theory makes matters more interesting to me as a historian of visual representation rather than less so.

NOTES

1 For reworkings of the Panofskian standpoint, see especially Edgerton 1980, 1985b, and 1991.
2 For recent contributions that make some inroads into these matters, see particularly Westman 1984, Mahoney 1985, Ashworth 1989, Tufte 1983 and 1990, Lynch and Woolgar 1990, Winkler and Van Helden 1992, and Mazzolini 1993. See more generally Cohen (1980) and Ford 1992.
3 Compare *De revolutionibus*, II, introd., p. 27 (Copernicus 1972, II, p. 51). See Gingerich 1973.
4 More generally for artists and anatomy in the Renaissance, see Schultz 1985 and Kornell 1992.

5 See Koch 1974, Boudreau 1978, and *Hans Baldung Grien: Prints and Drawings* (Morrow 1981).

6 I owe this formulation to Karen Bassi, who spoke at a symposium on 'Pre and Early Modern Anatomies' at the University of California, Santa Cruz, on 24 April 1993.

7 Confirmation that the assigning of the authorship of the illustrations to Titian cannot be sustained is provided by Kornell 1993, pp. 72–5.

8 For Copernicus generally, see Beer and Strand 1975; and Verdet 1991.

9 For an instructive interpretation of Copernicus in the tradition of Renaissance rhetoric, see Westman 1990. See also Rose 1975, pp. 153–8; Prera and Shea 1992; and Moss 1993.

10 For further discussions of Copernicus's humanism, see Hutchinson 1991 and 1993. For the standard likeness of Copernicus, see Beer and Strand 1975, figures 5–7. The evidence regarding the possible self-portrait(s) is assessed by Westman 1990, pp. 184–6.

11 Quoting Galen, *De usu partium,* X, 14.

12 See Alberti 1486, especially the prologue and the introductions to books I and VI.

13 Compare *De revolutionibus,* I, 10, p. 9r (Copernicus 1972, II, p. 22), where it is asserted that nature 'avoids producing anything superfluous or useless.'

14 See Nebelsick 1985, pp. 200–73, Hatfield 1990, and, in a postmodern vein, Hallyn 1990. For a more circumscribed interpretation of this passage, see Drake 1975.

15 For a detailed assessment of Copernicus's astronomy, see Swerdlow and Neugebauer 1984.

16 Leonardo da Vinci, MS A 36r and MS K^3 120v, in Kemp 1989b, p. 55.

17 In his *Astronomia nova* (1609), Kepler envisages an observer on Mars; Kepler 1938–88, III, p. 22.

18 For a survey of cosmological representations, see Heninger 1977.

19 The MS illustration is on fol. 74. See Neugebauer 1968, p. 96.

20 Westman 1975, pp. 318–19, for annotations by Kepler and the Scottish philosopher Duncan Lidell.

21 The annotated Copernicus in St Andrews was, as an inscription indicates, the property of the 'German Nation' in 1626. Owen Gingerich tells me that it was then in the University of Padua.

22 Illustrated by Gingerich (1975, fig. 70).

23 For reviews of such instruments, see Zimmer 1987, Bennett 1987, and Turner 1991.

24 Illustrated and discussed by Maddison 1991. For a suggestive discussion

of the nature and use of such instruments, see Field 1988b. For the courtly context, see Kaufmann 1993, especially pp. 188–93. See also Kemp 1991, pp. 135–52.

25 See Gingerich 1971; and Brahe 1598. The illustrations of instruments are also found in Brahe's *Progymnasmata* (1602). For Tycho's career and achievements, see Thoren 1990. See also the suggestive discussion by Eisenstein 1983, pp. 207–25.

26 The poem is in Brahe 1913–29, VI, p. 266.

27 Appendix Vergiliana, *Elegiae in Maecenatem*, I, 38: 'vivitur ingenio, cetera mortis erunt' (ref. courtesy of Professor H. Hine).

28 See Thoren 1990, pp. 106–13, for the architecture, though overestimating the classicism and Palladianism of the enterprise.

29 The sketch in the Archiv der Kepler-Kommission, Munich, is illustrated in Kepler 1938–88, X, p. 279; and Beer and Beer 1975, figure 3.8. For Kepler's stereometry, see Kepler 1615. For Kepler generally, see Caspar 1959, Field 1988a, Beer and Beer 1975, and Shea 1991b.

30 Kepler's treatise is also unusual in that he represents the actual paths of the planets in addition to their orbs.

31 For apsects of the 'aesthetics' of the Platonic solids in perspectival depiction, see Kemp 1989a.

3. Descartes's Scientific Illustrations and 'la grande mécanique de la nature'

BRIAN S. BAIGRIE

1. INTRODUCTION

Descartes's expressed reservations about visualization, at least on the face of it, are intimately connected to his overall project to demonstrate the capacity of the unaided reason to deduce the composition of nature from first principles that are plain to the attentive mind.[1] Visualization involves the fabrication of mental images, which are then exhibited by means of pictorial devices. Since the contemplation of these mental images seems to involve perception, visualization (and its associated false beliefs and prejudices) is targeted by Descartes as a potential source of error in science.[2]

What, then, are we to make of the many illustrations incorporated by Descartes into his scientific treatises? Why does Descartes seem to place such epistemic weight on picturing? In raising this question, we are entering new territory since, to my knowledge, Descartes's scientific illustrations have been treated by scholars as textual ornaments. Those few scholars who recognize that art is in fact used by Descartes as science have held that he is simply inconsistent with respect to visualization. Since Descartes describes his physics as proceeding *more geometrico* from first principles, the presence of these illustrations is standardly taken as evidence that, if anything, he needs to introduce these and other perceptual elements to give any empirical substance to his explanations of natural phenomena. For these scholars, the pictures in Descartes's science sustain the charge that he was a rationalist in name but not in practice (see Dijksterhuis 1986, p. 407).

I will argue here that such allegations are invalidated by the supposition that, if these illustrations have any value, it is to help us see what the world is like; that is, they must be perceptual resources in the context of Descartes's substantive theories. My own view is that the pictures in Descartes's science are not meant to depict a world but are designed to help us to conceive how it might work (in mechanical terms); that is, they are viewed by Descartes as resources that can enhance human cognition – the artifice of drawing enables natural philosophers to explore the plausibility of postulated mechanical arrangement of insensible particles, and thereby to develop the intuitions that are needed to grasp the working of things that exceed our perceptual grasp.[3]

2. THE DISAPPEARANCE OF DESCARTES'S ILLUSTRATIONS

In the first Latin edition of the *Principia philosophiae* of 1644, illustrations are incorporated into the main body of the text in the manner of a contemporary treatise in mechanics or physics. These illustrations are appended to the end of the text in the French edition of 1647, prepared by Abbé Picot. Charles Adam and Paul Tannery – the editors of the canonical edition of Descartes's works – followed the practice of Picot, which in turn means that they were compelled to add a number of awkward footnotes that link the pictures to the appropriate bits of text. Though the editors of the recent English edition – Miller and Miller (1983) – maintain that their edition is based on the original Latin text, this edition adopts Picot's practice of placing the pictures at the end of the text. Numerous annotations of phrases and passages are added in footnotes that are often helpful with the text, but not one reference to the illustrations is offered that is not to be found in Descartes's original text. It is as though these illustrations do not exist.

This same pattern is to be found in secondary sources. Though commentators have expended enormous energy getting straight on the meaning of the text, virtually no one has studied the pictures in Descartes's scientific treatises; no effort has been made to examine their role in Cartesian science or to tackle the many philosophical and historical questions raised by their very presence.[4] This omission is remarkable on three distinct counts:

(1) Evidence relating to the preparation of the illustration for Descartes's *La Dioptrique, Les Météores* (1637), and the *Principia philosophiae*

is sketchy. They were prepared by Frans van Schooten the Younger (1615–61), editor of the Latin edition of Descartes's *Géométrie* and responsible for the well-known portraits of its author prepared for the first (1649) and prefixed to the second (1659) edition of this work. Schooten's illustrations are the tangible result of an intellectual collaboration between Descartes and an important Dutch scholar who served, through Descartes's influence, as professor of mathematics at Leyden from 1646. Not only is this collaboration an important area of study in its own right, but it also may shed light on the positive reception given to Descartes's ideas in the Dutch universities.

(2) During his stay in Amsterdam, Descartes lived a short distance from Rembrandt (1606–69) during the most prosperous days of the Dutch state. With Frederick Henry (1625–47) at the helm during this period, its navigators dominated world commerce and the Dutch school of painting reached its apex. Rembrandt's *Nightwatch* was completed less than a year after the publication of Descartes's most important work, the *Meditationes de prima philosophia* (1641). Rembrandt was apprenticed for a time with Joris van Schooten, father of Frans, so that it seems probable that Rembrandt would have been known to Descartes. In 1632 Rembrandt was commissioned to paint his first great work, the *Anatomy Lesson of Dr Tulp,* depicting the celebrated Dutch anatomist, who orchestrated a series of public anatomies that attracted a large audience that numbered Descartes on occasion. There is no hard evidence that Rembrandt and Descartes crossed paths at one of these events, but Tulp's public anatomies, as immortalized in Rembrandt's painting, seem to indicate that they moved in the same circles.[5]

With the exception of *Meditation I,* which introduces a famous supposition – that God might deceive – by reference to painters, Descartes's writing are silent about art.[6] This is to be expected. During the seventeenth century, little was written about Dutch art. As Joshua Reynolds would later remark (1809), one can only look at Dutch painting, not tell a story about it (cited by Alpers 1983, p. 1). The Dutch school of art excelled at observations – still lifes, landscapes, and detailed studies of domestic life – authoring a new style of descriptive art that was divorced from the narrative ends that dominated the painting of the Italian school (see Alpers 1983). Though Descartes offers no systematic theory of art, there are hints that he was interested in painters and in art.[7] Besides his association with van Schooten, during his stay in Utrecht in 1640, Descartes visited the painter and language scholar Anna-Maria Van Schuurman (see *AT* III 230; *CSMK* 156). Moreover, he tried his hand at some intriguing illustrations of muscle antagonism – one of these (fig. 3.1) can be found in *l'Homme de René Descartes* (Clerselier's edition, 1664). Descartes's own illustrations are an untapped source of evidence about his collaborations with Dutch artists. What's more, the possible impact of Dutch art on Descartes's representational practices has not been seriously studied, despite the fact that the problem of representation vexed the entire Cartesian tradition and still vexes Descartes scholars.

(3) We know that Descartes conducted numerous 'experiments' during his 'retirement' in Holland. He observed meteorological phenomena, conducted experiments on the weight of the air and on vibrating strings, and he designed an apparatus to study the motions of vapours (fig. 3.2). Descartes replicated Christopher Scheiner's experiments, which involved removing the opaque layers at the back of extirpated eyes from freshly dead animals and humans (see Crombie 1967). With his knowledge of the sine law of refraction, he was able to confirm Scheiner's suspicion that the lens accommodates to distance by changes of shape, rather than the position of the lens, as Kepler had originally supposed (fig. 3.3). The solution that Descartes fashioned to the problem of the size of the rainbow in *Les Météores* is almost unknown, yet it displays the patient experiment and laborious calculations that we now readily identify with a healthy experimental outlook. His anatomical studies on a wide variety of animals are well known and rightly praised for their attention to detail, as evidenced by the detailed illustration of the human heart (fig. 3.4) that appeared in *l'Homme* (1664). Descartes's correspondence reports countless dissections of dogs, cats, rabbits, codfish, and mackerel, and of eyes, livers, and hearts obtained from animals slaughtered in an abattoir. Many of his writings reflect these activities and show that, at least in physiology and anatomy, his experimental knowledge was firsthand and sound.

Descartes designed and constructed new experimental devices and scientific instruments, such as the microscope and a compass for the construction of geometrical curves. Vrooman (1970, p. 20) reports that Descartes even tinkered with the use of wheelchairs for the handicapped. His activities are what we have come to expect from science in action and not science in the imagination. Descartes's passion for the laboratory life may seem inimical to the emphasis that his method placed on reasoning from first principles. However, his writings testify

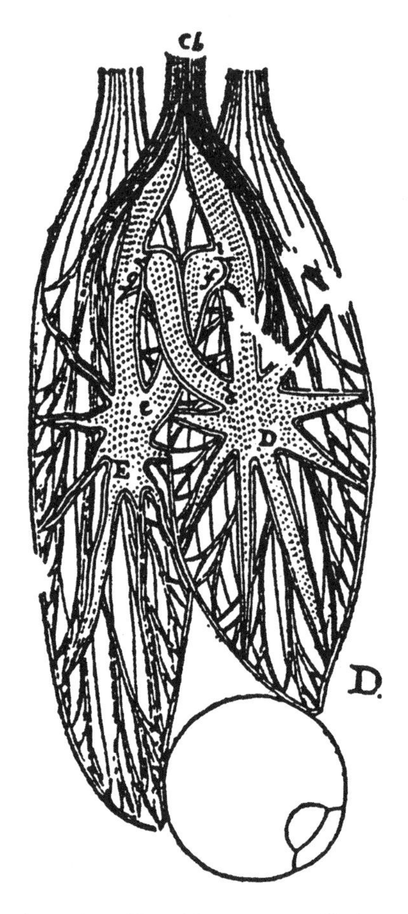

3.1 Reciprocal muscle action (René Descartes 1664).

to a certain contempt for bookish culture, and throughout his life he prided himself on 'seek[ing] no knowledge other than that which can be found in myself or in the great book of the world' (*AT* VI 9; *CSM* I 115). In this preference, he seems to have followed Leonardo da Vinci and Bernard Palissy, the famous French potter, that the book of nature is richer and more complex than any other book, especially the books written in Latin by philosophers. Indeed, it is characteristic of experimental philosophy of this period to polemicize against an excessively bookish culture imparted by the schools, in favour of an experimental philosophy that values knowledge only if it is useful to practice (Houghton 1957, pp. 379–80).

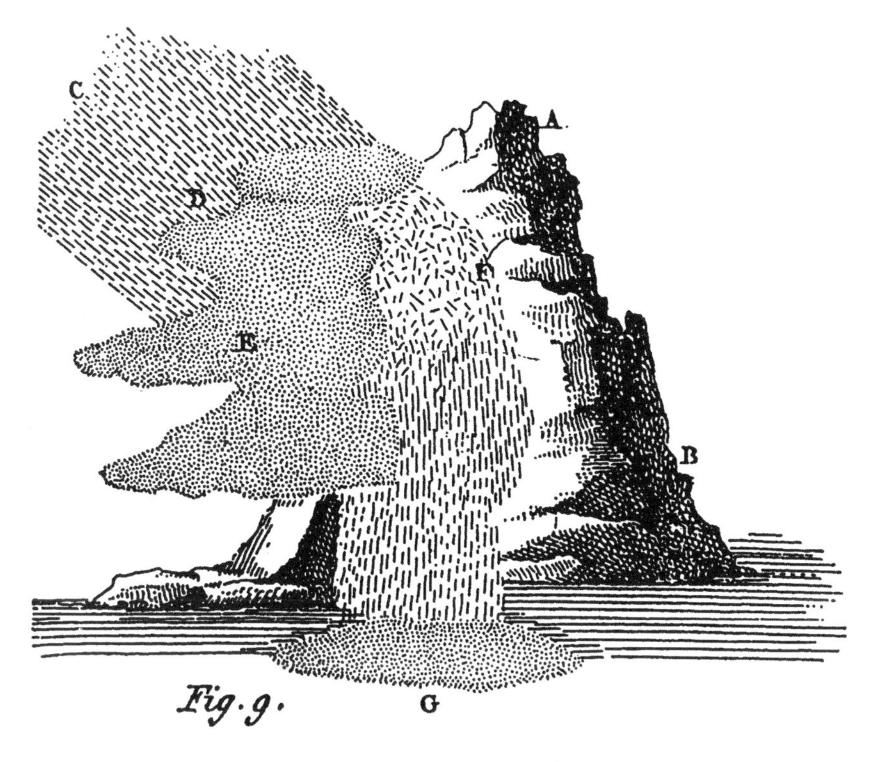

3.2 The formation of vapours (René Descartes 1637).

Descartes's accounts of his experimental practices are often accompanied by detailed illustrations (fig. 3.5). My concern here is tangibly with the role that these illustrations play in Descartes's new science. Are they included merely to help the reader come to grips with the text or, more substantially, are they involved in some way in the creation of knowledge? One thing that is certain is that these illustrations are not recipes for anyone who wants to reconstruct Descartes's scientific practices. The ninth discourse of *La Dioptrique*, for instance, contains designs for two kinds of microscopes – a simple unit with a bi-convex lens, and a second construction with a plano-convex lens and a metal reflector. Descartes also describes another immense instrument (fig. 3.6), which is the earliest known drawing of a compound microscope. The optical part consists of a bi-convex lens and a plano-convex objective worked to hyperbolic curves. The object is illuminated by a hyperbolic mirror surrounding the objective. There is also a condensing lens in the axis of the instrument for illuminating transparent objects.

3.3 Perception of distance through binocular vision (René Descartes 1664).

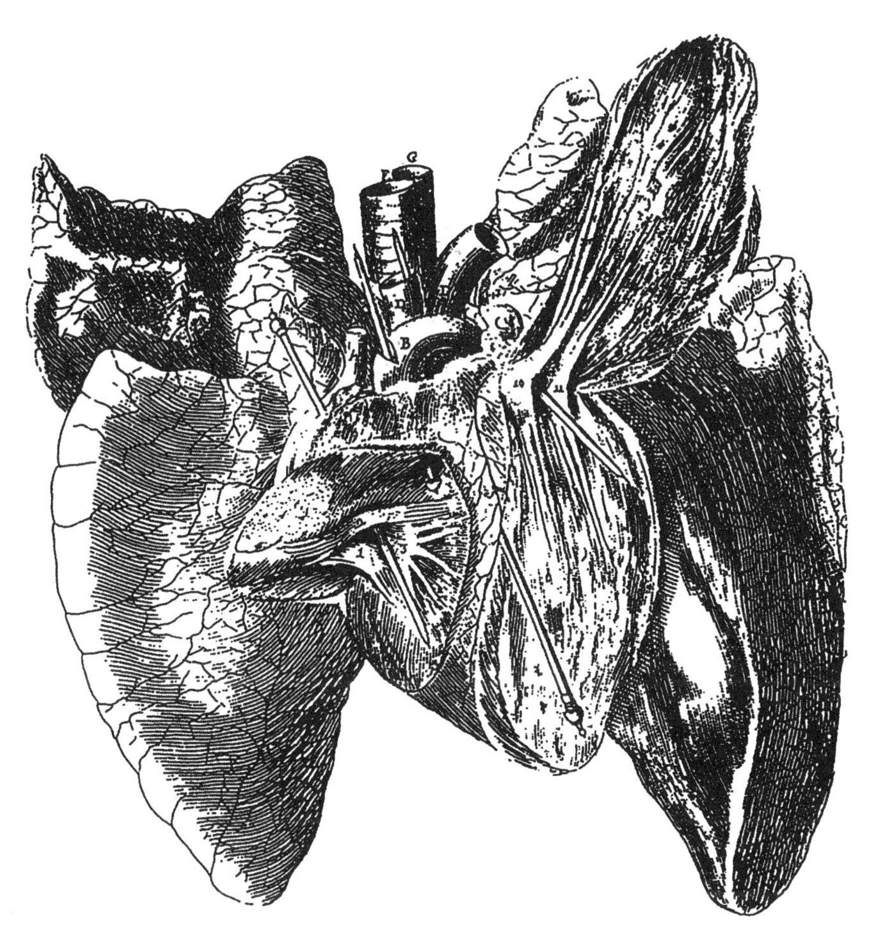

3.4 The human heart (René Descartes 1664).

As an indication of its size and impracticality, the entire instrument is mounted on a stand the height of an observer, in the manner of a telescope, such that any object placed at its focus would be immediately destroyed by sunlight. This explains the angle of the instrument, which seems positioned in the illustration for astronomical rather than microscopic work. If the stand is as high as an observer, then clearly the eye in this illustration is out of proportion – perhaps pointing to some confusion on the illustrator's part about the purpose of such a large instrument. In many eighteenth-century treatises, this eye is replaced by a full-sized human figure, equal to the height of the stand, as indicated

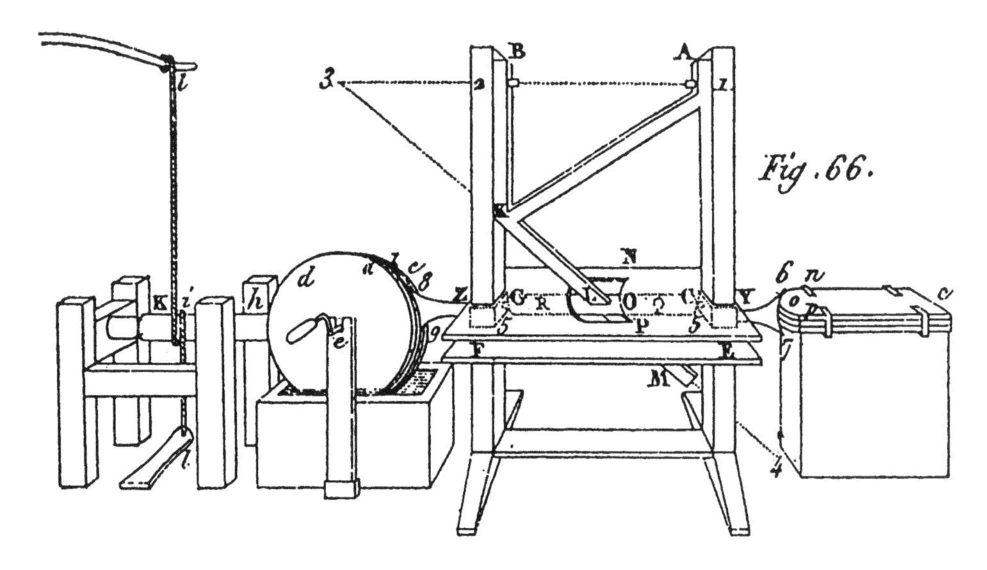

3.5 A device for preparing glass (René Descartes 1637).

by Descartes – but virtually anyone who glances at these corrected illustrations will suppose that they are looking at a telescope. This impression is nurtured by the title of the ninth discourse – 'The Description of Telescopes.'

Granted that Descartes's illustrations are not always reliable, I do not want to claim that they can serve as shop manuals for anyone wanting to reconstruct his experimental practices. The relationship between Descartes's use of art as science and seventeenth-century experimental technology is subtler and a great deal more interesting than this. In Holland, Descartes's adopted homeland, one finds a deep connection between art and experimental technology – countless picture of clocks, water supply systems, chimneys with improved drafts, etc., testify to the intimate connections between the artifice of drawing and the artifice of craft (see Alpers 1983, p. 5). Not only do these pictures point to an assumed identity between drawing and seeing – that is, between the artifice of drawing and nature – but they testify that drawing itself was an integral part of the creation and the ongoing refinement of new experimental technology. For the Dutch, there was an important sense in which all the mechanical arts (including drawing) are experimental. The illustrations in Descartes's scientific treatises, it seems to me, are best seen in the manner of Dutch art as experimental studies that (in

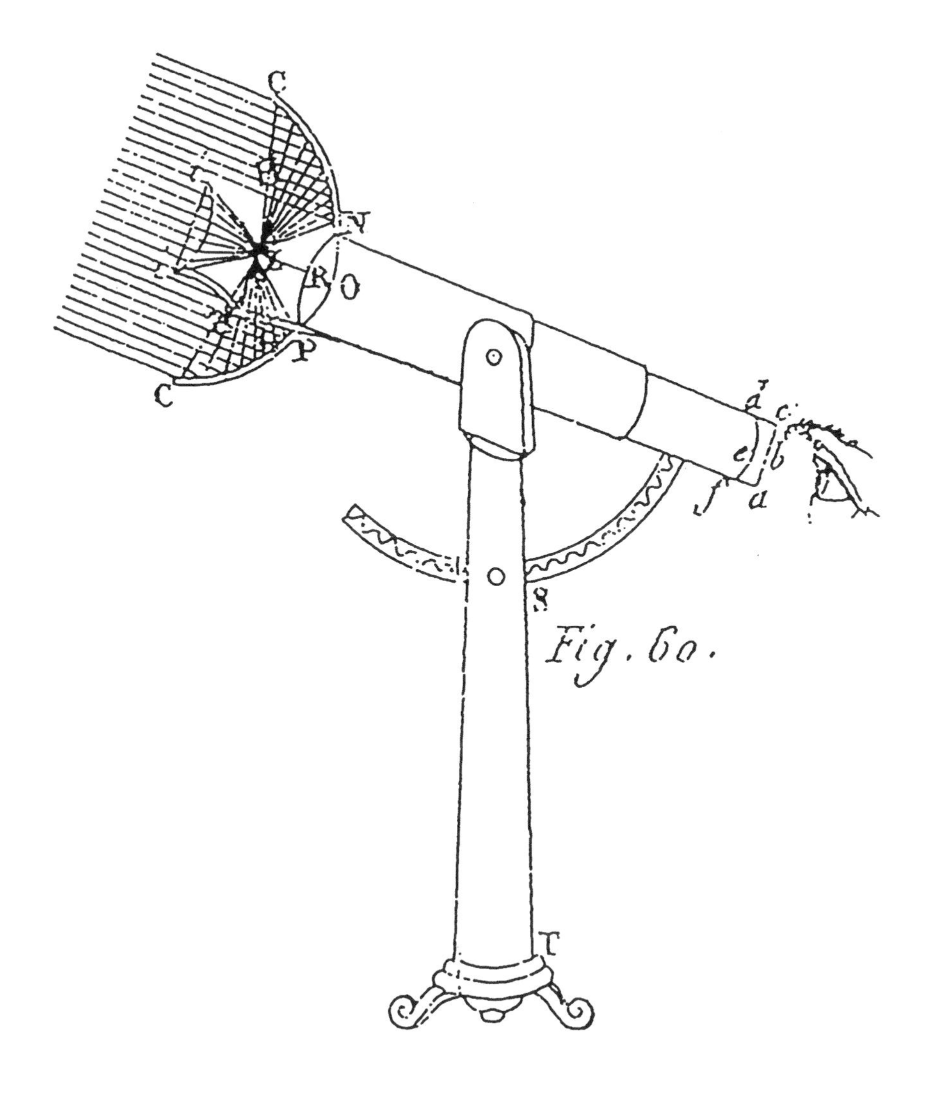

3.6 An unusual microscope (René Descartes 1637).

collaboration with various theoretical and material resources) are meant to play an important role in the creation of new knowledge.

What is the explanation for the disappearance of Descartes's illustrations in the history and philosophy of science? Though I cannot furnish a comprehensive answer to this difficult question here, I can distil one factor from the foregoing considerations that deserves critical scrutiny: it is the narrow correspondence theory of representation that most scholars inadvertently invoke in their attempt to make sense of the pictorial elements in scientific treatises. Just as propositions are thought by many scholars to have meaning for us in consequence of their representing facts, illustrations often are treated as having meaning for us on account of their representing experimental devices and activities, and even substantive entities and natural processes. The consequence for those who subscribe to this theory is that the pictures in Descartes's scientific treatises are regarded as aids to help the reader to see what is being asserted – to clarify the meaning of the text – but as nothing more. The result is that the illustrations in science are placed on a par with a view of experiment that is now decidedly out of fashion – namely, as something that is carried out in the service of theory. For Descartes – who is often caricatured as holding that science is something that is best done in one's study with pure reason alone – this is a disaster, since these illustrations are then seen to have no real value at all.

Though I will hazard a few global remarks about the generalized concept of representation that sustains Descartes's work, my concern here is more concretely with what his use of art as science can tell us about his *grande mécanique de la nature* and with some philosophical issues concerning the consistency between Descartes's views concerning the authority of reason and the presence of these pictorial elements. Before I turn to these issues in more detail, however, a great deal needs first to be said about the intellectual and social milieu in which these illustrations are to be situated, namely, one in which the scholarly community was being increasingly interpenetrated by a comparatively new technology that was the inspiration for a new mechanical philosophy of nature rooted in the production of machines. It is his take on this artifice of craft and his associated convictions concerning its epistemological significance that hold the key to Descartes's pictorial practices.

3. THE MATHEMATICAL SPIRIT OF HUMAN INQUIRY

Adrien Baillet (1693, p. 267), in his biography of Descartes, submitted that 'never did man under the scope of Heaven, manifest that which we call, a Geometrical Spirit; and an exactness of Wit and Solidarity of judgment in a higher degree.' Descartes's deliberations embodied this geometrical spirit in two interwoven senses. It is reflected, firstly, in Descartes's conviction that natural philosophy can be transformed so as to produce certain knowledge if it emulates mathematics – because mathematics is so perfectly transparent, it is something about which we can have unconditional knowledge. This geometrical spirit is reflected, secondly, in Descartes's insistence that the ultimate nature of all that is material is to be determined by the purely geometrical characteristic of extension; thus, in the *Principles* (Part II, §64), he states that 'I openly acknowledge that I can point to no other kind of thing than that which can be divided, shaped, and moved in all kinds of way, and that Geometers call quantity and take as the object of their demonstrations' (*AT* IX-2 102). Real science is fabricated in mechanical discourse; that is, it employs no other concepts than those found in mechanics – geometrical concepts such as shape, size, and quantity, which are employed by mechanics as a department of mathematics, and motion, which forms its specific subject (see Dijksterhuis 1986, pp. 414–15).

In the previous section, I submitted that the rise of a new technology of mechanical devices holds the key to Descartes's use of art as science. The issue that concerns me here is whether the above gloss on Descartes's 'geometrical spirit' – though admittedly crude – is robust enough to help us get clear on the conception of mechanism that sustains and informs Descartes's philosophy of nature. For many scholars, Descartes's mechanism is synonymous with the so-called geometrization of nature – with the denial of activity to bodies. Descartes's assertion to Fromondus (3 October 1637), for example, that 'my philosophy may seem too "crass" ... because, like mechanics, it considers shapes and sizes and motions' (*AT* I 420; *CSMK* 64) has been widely interpreted by Descartes scholars (see Merchant 1980, p. 195) as tantamount to the claim that bodies are inert and utterly bereft of activity, powers, forces, and the like.

Alistair Crombie (1967) has drawn on Descartes's study of Kepler's optics both to sustain this view and his contention that the key to Descartes's mechanism is Kepler's optical researches. Kepler had treated the living eye as far as the surface of the retina as a 'dead' (inanimate) optical instrument, but, says Crombie, Descartes took the additional step, in *l'Homme* and *La Dioptrique* of treating the entire living animal body as a dead machine and by focusing exclusively on one question: what physical motions follow from each preceding motion?[8] Whereas Kepler had reasoned mechanistically about one particular organ (the eye) in his *Ad vitellionem paralipomena* (1604), leaving its functioning in relation to the entire system of the body untouched, Crombie submits that Descartes advanced a general physiology based on physics reduced to laws of matter in motion (see Crombie 1967, p. 67).

Crombie gives us a detailed and powerful account of what it meant for Descartes to reason mechanistically, at least insofar as his deliberations impacted on perception. While it is indisputable that some of Descartes's followers (e.g., Malebranche) denied activity to bodies, there are indications that Descartes did not. A letter to Henry More (August 1649), for example, states that

> the transfer that I call motion is no less something existent than is shape; it is a mode in a body. The power causing motion may be the power of God himself preserving the same amount of transfer in matter as he put in it in the first moment of creation; or it may be the power of a created substance, like our mind, or any other such thing to which he gave the power to move a body. (*AT* V 403; *CSMK* 381)

Finite minds can move bodies. Descartes's correspondence with More (*AT* V 347; *CSMK* 375) suggests that angels can move bodies as well, but I agree with Hoenen (1967, p. 359) that the 'any other such thing' of the above passage is a reference to bodies – an interpretation that squares with the many passages in Descartes's writings that point to an *activity* on the part of bodies.[9] Descartes even contends that a body can act on a soul – a possibility that is crucial to the argument of the sixth meditation, where he argues from the passivity of the senses that there must be an active faculty at work on the senses and, since this faculty is not in the mind, nor in God, nor in an angel, it must be found in bodies. It is the existence of this active faculty that Descartes parlays into a proof of the existence of bodies (*AT* VII 79; *CSM* II 55; cf. *Principles*, Part II, §1).[10]

Defining matter in terms of the geometrical property of extension and seeking the certainty of mathematical reasoning are important ingredients in Descartes's mechanism, but there is another, overlapping ingredient that may shed welcome light on many of the gray areas on Descartes scholarship (particularly, his use of analogical models and pictorial devices, and his reliance on experience at critical points in his attempt to deduce explanations of phenomena from first principles that are plain to the attentive mind). Descartes identified matter with extension and filled space with variously sized and shaped particles. To these particles, he added a fixed and finite quantity of motion that is distributed and redistributed among the particles that jostle against one another in the universal plenum. The cosmos contains innumerable particles that are constantly polished and refined in consequence of countless collisions. Descartes's universal law of nature, as given in Part II, §37 and §39, of the *Principles*, stipulates, however, that God's immutability is best expressed by motion in a straight line and in a perpetually undiversified manner. The world, therefore, contains many kinds and shifting modifications of curvilinear motions. What it does not contain is the perpetually undiversified motion in a straight line mandated by Descartes's supreme law of motion. The sole exception is light which, for Descartes, points to a pressure in a medium and not a genuine movement of particles at all.

This discrepancy between the fundamental laws of physics and the behaviour of discrete bodies highlights a number of issues that are pivotal to our getting clear on Descartes's mechanical philosophy of nature. Is there a way to look at the cosmos that will compose clear and distinct mathematical intuitions with the behaviour of particles? Part II, §39, of the *Principles* submits that 'each part of matter, considered individually, never tends to continue its movement along curved lines, but only along straight lines' (*AT* IX-2 85). In a plenum, such movement can never be realized. Newton will later declare that the planetary orbits are rectilinear to a near approximation, but this option was not open to Descartes, who was disinterested in the many imprecise and gray areas of experience (see Baigrie 1992). There is also an associated worry about the kinds of phenomenological models that are best suited to making intelligible the processes whereby particles mutually interact to produce a world such as ours. Finally, it is God who maintains in the sum total of matter a fixed and determinate quantity of motion. Descartes maintains that besides God only humans have the capacity for self-initiated movement, despite the fact that many other objects seem to be self-activating. No less a thinker than Gilbert had attributed a soul to the magnet in consequence of its ability to move and to be moved (see Jaynes 1970, p. 249). Is there a way to compose Descartes's insistence that movement is the consequence of innumerable collisions with what appears to be the self-instigated movements of bodies of ordinary experience?

What I will suggest in the following pages is that mechanism furnished Descartes both with a blueprint for working his way through these general issues and a cluster of fertile but ultimately limited models for reconciling clear and distinct mathematical intuitions with the behaviour of bodies. Where I break with standard attempts to characterize Descartes's mechanism, however, is that I take Descartes quite literally when he states, in conversation with Burman, that 'we are not sufficiently accustomed to thinking of machines, and this has been the chief source of error in philosophy [nos autem machinas non satis assueti sumus considerare, et hinc omnis fere error in philosphia exorsus est]' (*AT* V 174; *CB* 73).[11]

4. MACHINES AND COGNITION

Well, then, what did Descartes think of machines? References to machines in his published corpus are surprisingly rare. The word 'automate' occurs in Part 5 of the *Discourse* (*AT* VI 55–6) but clearly as a term of art. There is also §203, Part IV, of the *Principles* (*AT* IX-2 321), where Descartes states that 'I know of no distinction between these things [made by human skill] and natural bodies, except that the operations of things made by skill are, for the most part, performed by apparatus large enough to be easily perceived by the senses.' References in the unpublished works are more abundant – in the *Regulae* (XIII, *AT* X 435–6) and in *l'Homme*, Descartes enthuses about the Royal Gardens at Saint-Germain with its articulated clocks and hydraulically powered garden figures that moved, danced, and sang 'selon le caprice des Ingénieurs qui les ont faites' (*AT* XI 130–1; see Jaynes 1970).[12]

Though references in the unpublished writings are more abundant, the fact remains that they give us little concrete information about just what Descartes thought about machines and the craft of *ingénieurs* generally – what he might have learned from his two years (1614–16) at Saint-Germain preoccupied with the elaborate grottoes and mechanical statues built by the celebrated Francini brothers; or from his stint in the late 1610s as an engineer in the army of Maximilian I of Bavaria; or from collaborations with artisans such as Guillaume Ferrier. Still less do we know what Descartes may have thought about treatises on machines

– Ramelli's bilingual (French and Italian) work *Le Diverse et Artificiose Machine* (1588), Heron's *Pneumatics* (translated from the Greek into Latin in 1589 by Aleotti d'Argentina), Giambattista della Porta's *Pneumaticorum libri tres* (1601), and Salomon de Caus's *Les Raisons des forces mouvantes* (1615), just to mention only a few of the most eminent works that must have been known to Descartes. Much work remains if we are to reconceptualize Descartes's mechanism as a very particular take on machines and the mechanics of machinery.

Between 1400 and 1600, numerous elaborately illustrated books were published in France, Italy, and Germany that depicted hundreds of machines that were extrapolated from existing technology (see Basalla 1988, pp. 67–8). These books were a celebration of technological possibility and were given the title *theatrum machinarum* (theatre of machines). Some of these machines had not been built – some could not be built – but they were depicted with such authenticity that the reader was persuaded that they might possibly be constructed in the future. Along with the *theatrum machinarum*, there were technological visions – fantastic schemes that date from the fifteenth century, such as Conrad Keyser's *Bellifortis* (1405), noted for its many fantastic war machines, and Leonardo's notebooks, which depicted flying machines, parachutes, armoured tanks, multi-barreled guns, and a steam engine. Whether fantastic or real, such devices were spectacles – tokens of human ingenuity. What counted was the fact that they could be envisioned and perhaps built, and not that they could be used in a Baconian spirit to relieve human labour by increasing our control over nature.

The marriage of natural philosophy and the machine-shop created an off-spring in the form of an associated epistemological thesis that made a profound impact on the construction of modern science – namely, that the artifice of craft has a direct bearing on the creation of knowledge (see Rossi 1970, p. 31; cf. Crombie 1958, p. 318). No doubt, the spirit of invention that led to the appearance of spectacles, mechanical clocks, and the scale had tugged at the medieval imagination. It is nonetheless true, however, that with the early modern period this work, which had been performed by illiterate artists, was now taken up by a new generation of philosophers that included Leonardo da Vinci, Bernard Palissy, Pierre Gassendi, Marin Mersenne, and Descartes who discerned in the artifice of craft, and especially in machinery, a powerful model of cognition – a new way of reasoning about natural processes.

This epistemological thesis rose in concert with a new theistic conception of God as artificer, one that usurped the Neoplatonic account of God as a geometer in many quarters. This theistic conception prepared the fertile intellectual soil for a remarkable analogy between human contrivances and God's own machines. As early as the fourteenth century, it had been suggested that the stars are a piece of clockwork. However, the Renaissance analogy between God's own machines and human contrivances went much deeper than this: the methods employed by artisans to manipulate nature can be exploited to acquire a real knowledge of the cosmos, since the artifacts produced by hammer and forge, though less grand, were no different than the artifacts produced by God.

Descartes was profoundly affected by this interpenetration of natural philosophy and the artifice of craft. In his mind, there was no significant difference between mechanisms of all kinds, whether the giant celestial apparatus readily arranged in a pure state of nature, the *corps humain* described by Descartes in the *Discourse* as 'a machine which, having been made by the hands of God, is incomparably better ordered than any machine that can be devised by man ...' (*AT* VI 56; *CSM* I 139), or the insensible mechanisms responsible for many mundane phenomena that he explicitly asserts can only be divined indirectly through an acquaintance with devices that we can build and manipulate with our own hands. God's own machines are incomparably better ordered, but there simply is no difference between the kinds of machines that exist in a pure state of nature and the devices created by hammer and tongs. As Descartes explicitly asserts in Part IV, §203, of the French edition of the *Principles*, 'all the things that are artificial are furthermore natural [en sorte que toutes les choses qui sont artificielles, sont avec cela naturelles]' (*AT* IX-2 321). Nature is *not an ideal* that artisans seek to imitate. Although artificial combinations of bodies enable natural philosophers to comprehend bits and pieces of the world that exceed our grasp because of their size and remoteness, machines are not designed to explore alien parts of the world.[13] Nature is synonymous with the artifice of craft, and so Descartes reckons that 'it is certain that all the rules of Mechanics apply to Physics ...' (IV, §203; *AT* IX-2 321).

5. DESCARTES AND THE MECHANICS OF MACHINERY

Descartes writes with boundless enthusiasm about machines, but his writings contain nothing of a systematic character on the mechanics of machinery. Indeed, it is a characteristic of mechanical treatises on the Continent – though not in England – that little is advanced by way of systematic theory. Since Renaissance engineering treatises proceed largely in the absence of theory as well (see Panofsky 1955, p. 243; cf. Rossi 1970, pp. 32–3), perhaps this indicates that the coupling of the mechanical arts and natural philosophy was more thoroughgoing on the Continent. Setting this issue aside, what's clear is that, by our contemporary standards, even the most rudimentary distinctions are absent. Contemporary treatises on machinery, for example, draw a distinction between a machine and a mechanism – the former being a device that is designed to modify energy and to carry out work; whereas the latter modifies motion. A draw-bridge is a mechanism but not a machine. Descartes was interested in mechanisms, not in machines. As it happens, some of the mechanisms that interested Descartes involved sources of energy, such as the muscular effort of animals, but there is no recognition in his work that the primary function of some machines is doing work. Though the term 'mechanism' is closer to the sense in which he portrays the world as a machine, I will follow Descartes's practice of blurring the line between machine and mechanism.

There are other distinctions that we cannot overlook, however, if we want to reconstruct Descartes's mechanism as a very particular take of machines. Contemporary treatises on machinery typically open in the manner of Newton's *Principia* with a series of definitions – body, force or power, weight or resistance, motion, linear velocity, acceleration, uniform and variable velocity, etc. These definitions are followed by general laws – statements of Newton's laws of motion, the parallelogram of forces, resolution and composition of motions and velocities, and, finally, the parallelogram of velocities. Only after this preamble is the reader introduced to the all-important distinction between *free and constrained bodies*, that is, between a body that has no material connection with other bodies, and another body that has a material connection with other bodies, permitting motion relative to this body along certain restricted paths. A planet is a free body – its orbit is the resultant of all forces acting upon it, with every disturbing action or force altering its path.[14] The crank-pin of an engine is a constrained body – if motion occurs under the action of any force, it must be in a fixed path and no force, whatever its direction, short of one that will injure the machine, can cause motion in any other path.[15] The critical feature of a constrained mechanical system is that all points in the body have definite paths in which they move, if motion takes place under the action of any force whatsoever. The stresses occasioned in the material connections between parts supply the components of force required to combine with the primary force to give a resultant in the direction of the prescribed path. So long as these connections are robust enough to resist the maximum stress, no further adjustments are necessary.

It is this distinction between constrained and free bodies that demarcates the mechanics of machinery or the science of mechanism from the science of mechanics (sometimes called pure or abstract mechanics). The mechanics of machinery is sometimes characterized as 'applied mechanics' (i.e., the application of the principles of pure mechanics to deformable bodies of ordinary experience), but this label is inappropriate. Though the mechanics of machinery involves the application of the principles of pure mechanics in the design, construction, and operation of machinery, its subject is a cluster of theoretical and practical problems that are peculiar to machinery. The science of mechanism is concerned almost exclusively with constrained bodies and, with the possible exceptions of anatomy and physiology, its central principles have almost no relevance to bodies of ordinary experience. The mechanics of machinery consists of two principal components, each with a number of subdivisions: the first component includes pure mechanism or the theory of machines considered simply as modifying motion; applied dynamics or the theory of machines considered as modifying both motion and force; and applied energetics or the study of sources of power. The second component deals with the structural features of machines – the composition of machinery, the materials of machinery, conditions of stability and strength, etc.

It is the case of a constrained mechanical system that concerns me here: to reiterate, if motion takes place under the action of any force, it must be in a definite path and no force, whatever its direction, can cause motion in any other path.[16] The only exception to this definition is when the pressure exerted by the force is so strong that an injury occurs to the machine. My central thesis on Descartes's mechanism is this: without exception, *his explanations of natural phenomena are mechanistic precisely in the sense that they are treated as instances of constrained mechanical systems.*[17] The following excerpt from a letter to More (5 February 1649) seems to bear out this thesis: 'since art copies nature, and people can make various automatons which move without thought, it seems reasonable that nature should even produce its own automatons, which are much more splendid than artificial ones ...' (*AT* V 277; *CSM* II 366). It is true that Descartes is here referring to animals but, in the *Discourse* and other writings, he makes it plain that the organic and the inorganic are continuous; God is immutable and acts always in the same way – namely, as a supreme artificer (see Rodis-Lewis 1978, p. 159). The continuity between the organic and the inorganic – both are treated as solutions to engineering problems – is reflected in his description of the formation of the foetus (see Rodis-Lewis 1978), which Descartes started in 1648 and abandoned shortly thereafter. The formation of the various parts of the foetus – the formation of the various organs by the assembling of the thicker parts, including the blood vessels and nerves that carry the animal spirits – is explained by vortices of matter and the principle of inertia.

My thesis constitutes a break with standard accounts of Descartes's mechanism which hold, in contrast, that machines are involved in Descartes's science in two restricted ways: substantively, as models of corporeal phenomena (living things literally are machines); and, formally, as a general set of constraints on explanations of natural phenomena (nature may be machine-like in its fundamental features but the physics of natural phenomena is not synonymous with the mechanics of machinery). The suggestion that, so far as mechanism is concerned, Descartes really elaborates two sciences – a science of living systems based on mechanism and a science of nature that is mechanical only in the formal sense that its explanations appeal to those features of mechanical systems (i.e., magnitude, figure, and motion) that are consistent with the stipulation that nature is not synonymous with artifice – has made Descartes the target for some unfair criticism. He has been widely portrayed, for example, as an ultra-mechanist who failed to appreciate the importance of force for the emerging science of mechanics. This allegation is fuelled by an equivocation between the mechanics of machinery and the science of mechanics: on the assumption that his explanations of phenomena are held by Descartes to be grounded in abstract mechanical principles, historians of physics have taken pains to point out that his appreciation of the fundamental principles of mechanics was deficient.

Consider the account of circular motion elaborated in §39, Part II, of the *Principles.* Descartes invokes the sensation that we have when we whirl a stone in a sling in order to illustrate the centrifugal tendency of a body moving in a circle (fig. 3.7). The stone in the sling tends to recede along the tangential line *ACG* at each moment it circulates about the centre of rotation, as specified by Descartes's second law of motion. The sling resists this tendency, however, which gives rise to a second effect, namely, an endeavour to recede radially along the line *AD* from the centre of rotation *E.* If we consider only the part of the

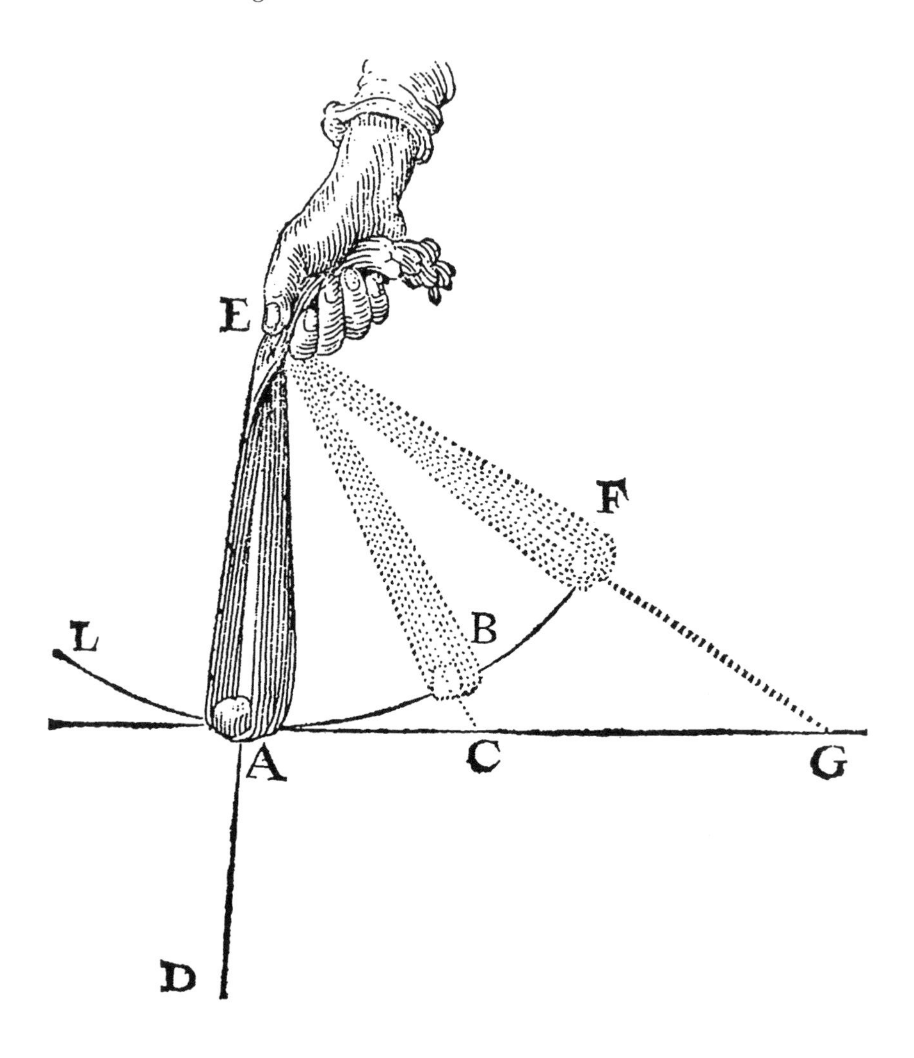

3.7 Centrifugal force (René Descartes 1644).

stone's motion that is impeded by the sling, we can say that the stone endeavours to recede radially from its centre outward along a straight line, even though it is in fact impeded by the sling and describes a circle. Scholars have been unanimous in interpreting this model as Descartes's attempt to frame the (pure) mechanical principles of circular motion. Interpreted in this way, Descartes's account is deeply flawed since it fails to discern that the tendency to recede along the line *AD* from the centre of rotation *E* is an imaginary one and that the stone in the sling, in fact, is accelerated by the force acting on it. In his defence, Descartes scholars have argued that, since he did not regard force as essential to matter, it is hardly surprising that no rigorous development of this notion is to be found in his writings. Moreover, even if Descartes were to have granted the notion of force a more central place in his cosmology, his interest was restricted to force as that property of a body by virtue of which it acts on another body. While his analysis of circular motion pointed to an external constraint that diverts a body from its inertial path, Descartes's conception of force did not point in the direction of Newton's dynamical theory.[18]

This defence has some merit, but it does not get to the heart of the matter. The fact is that the study of the forces acting is generally not involved in the mechanics of machinery. Since Descartes was *exclusively* concerned with constrained mechanical systems, and not with the principles that underwrite the motions of free bodies, it was reasonable for Descartes to have set considerations of force aside. There are exceptions to this rule – for example, escapement mechanisms that are used in clocks and watches must be explicated in terms of the force exerted by gravity – but it is generally true that machines can be explicated without considering the magnitude of the forces transmitted. In the case of a simple lever balanced on a fulcrum, for instance, the ratio of the motions is determined by the length of the lever arms, independently of the actual forces involved. The ratio of these motions is determined *by purely geometrical considerations and can be understood without taking anything else into account.* This approach, which is grounded in the mechanics of machinery, gives us a deeper account of the sense in which Descartes attempted to reconceptualize nature in geometrical terms – one that is a great deal more promising than the standard gloss on Descartes's mechanism as grounded in the so-called geometrical features of matter.

In their haste to take Descartes to task for failing to anticipate the development of modern mechanics, scholars have failed to appreciate the downside of the confidence that he placed in the machine as a model for scientific explanations. A machine has a number of characteristics: it involves (a) a collaboration of bodies; (b) the constrained motion of the bodies; (c) modification of force and motion and the performance of work; and (d) resistant bodies (see Barr 1911, pp. 30–4). Descartes's underlying conception of machines is consistent with (a) and (b); an isolated particle tends to move along a straight line but, in the universal plenum, other particles collaborate so as to constrain its movements. We can also set aside (c) granted that the concept of work had not yet been clearly distinguished from the concept of force. But consider (d) or the stipulation that machines consist of resistant parts. Any machinist knows that the parts of a mechanism must be resistant if it is to transmit and modify motion. These parts need not be rigid – confined fluids can transmit motion under compression – but they must be able to bear a load under stress. The notion of resistance is given scant attention in Descartes's writings. The sole exception is §132, Part IV, of the *Principles* (*AT* IX-2 270), where the 'rigidity' of glass and matter in general is contrasted with its elasticity but, by any stretch of the imagination, this brief discussion does not amount to a theory of mechanical resistance. Descartes's account of the creation of our present cosmos is sensitive to the fact that all materials are deformed under stress – that is, the particles of the cosmos are constantly being reshaped and polished in consequence of countless collisions – but there is no recognition in his writings that the nature of the constraint depends on a number of factors, such as the form of the contact surfaces, the dimensions and material of the constraining particles, etc.

The reason why Descartes failed to develop a systematic account of mechanical resistance and other concepts that are pivotal to the science of mechanism is difficult to discern. I've already remarked on the tendency on the part of philosophers on the Continent to avoid systematic mechanical theory. I have no general explanation for this disposition but, in Descartes's case, it would appear that he relied on extant lore about machines which was insensitive to the technical problems involved in the construction of machinery. To be charitable to Descartes, there simply wasn't a great deal of systematic knowledge about machines – works by Ramelli, de Caus, and Vittorio Zonca were really shop manuals for those who wanted to try their hand at building machines. They were celebrations of the untapped potential of the new machine technology. One can readily see why seventeenth-century natural philosophers came to see in machine technology a powerful new resource for understanding natural phenomena. Still, there is nothing of a systematic character in this literature that would lend plausibility to Descartes's contention that natural structures can be explicated in the same way as mechanisms.

6. THE DECEPTION OF SELF-INSTIGATED MOVEMENT

To this point, I've argued that Descartes's mechanical philosophy of nature is not grounded in a systematic theory of machines and that his inspiration for his mechanical philosophy of nature is to be found in lore about machines. We can trace two different ways that machines were conceptualized during the seventeenth century. The first is closest to the Greek μηχανη, something that has the power of self-instigated movement; the second is closest to the Latin *machina*, signifying a device that *simulates* the actions of animate beings and therefore only appears to be capable of self-instigated movement (see Jaynes 1970). The distinction is that the former conceives machines as things that move automatically, the latter as things that only appear to do so. The distinction is critical for our understanding of Descartes's mechanism: in the latter case, machines are tricks or devices that *deceive* us, especially our senses, which tell us that machines are self-moving when they are not. My contention is that Descartes writes almost exclusively with the Latin sense of the word in mind.[19] It is this Latin sense that informs the following passage in a letter to Reneri for Pollot (April or May 1638):

> Suppose that a man had been brought up all his life in some place where he had never seen any animals except men; and suppose that he was very devoted to the study of mechanics, and had made, or helped to make, various automatons shaped like a man, a horse, a dog, a bird, and so on, which walked and ate, and breathed, and so far as possible imitated all the other actions of the animals they resembled, including the signs we use to express our passions, like crying when struck and running away when subjected to a loud noise. Suppose that sometimes he found it impossible to tell the difference between the real men and those which had only the shape of men and had learned by experience that there are only the two ways of telling them apart ... first, that such automatons never answer in word or sign, except by chance, to questions put to them; and secondly, that though their movements are often more regular and certain than those of the wisest man, yet in many things which they would have to do to imitate us, they fail more disastrously than the greatest fools. (*AT* II 39–40; *CSMK* 99)

This passage offers support for my contention that Descartes identified mechanics with the mechanics of machinery; what he refers to here as 'the study of mechanics' is nothing other than the science of mechanism. Machines are made by artisans (*artifex*), and God is repeatedly described in just these terms by Descartes (see *Meditation III*, *AT* VII 51; IV 55, 56). Further to this, the machines that he writes about do not move automatically – they preserve the deception that bodies are self-activating, while furnishing us with models that allows us to construct and deconstruct this deception at our leisure.[20] Louis de la Forge's *Remarques*, which accompanied his edition of Descartes's *l'Homme*, is instructive. A machine is described by la Forge as 'any body composed of several ... parts which being united conspire to produce certain movements of which they would be incapable if separate' (Descartes 1664, p. 132; cf. Rodis-Lewis 1978, p. 156).[21] La Forge's teleological language calls for caution, of course, but it sheds welcome light on Descartes's mechanism. When we isolate the case of a single particle in the laboratory of our minds, our reason informs us that once in motion, the only direction in which it can move is in a straight line. In a densely packed cosmos, however, the task is to discern how particles collaborate so as to produce movements that would otherwise be impossible.

The suggestion that the cosmos involves deception is rooted in developments in the discrete mathematical sciences, especially in optical technology. This claim is borne out by Constantijn Huygens's remarkable *The Day's Work*, a lengthy poem of over two thousand pages that, among other things, grapples with a host of new problems raised by the new optical technology – especially how we perceive distance and estimate relative size in view of the new theatres of nature opened up by the telescope and the microscope (see Alpers 1983, p. 10). Which is the true view of the little animals that are seen through Leeuwenhoek's microscope – the animal that now fills our visual horizon or that animal that cannot ordinarily be seen at all? Since our vision evidently deceives us by presenting physical bodies as though their proportions are absolute and their distances easily computable, can we trust our eyes? If we are so easily deceived in vision, must our eyes not be machines of a kind and sight itself an artifice? And what of the veracity of pictures, not only the images on our retinas, but the lifelike representations produced by the great Dutch artists? What do these pictures convey to us, if not the truth?

In terms of the issues raised by visual deception, Huygens and Descartes went their separate ways – Huygens worrying about the truth of representations produced by the camera obscura and by the painter, and Descartes about the truth of our intellectual representations of nature given that nature itself (and our means of observing it) is an artifice. For Descartes, it is the artifice of craft that will help us to make intelligible the process whereby innumerable particles are organized into a world that seems to be self-activating, as though each of its parts were animated with some spirit that directs its movements. It is in this spirit that Eudoxus – the character in *La Recherche de la vérité* who voices Descartes's own views – asserts that

> ... I shall lay before your eyes the works of men involving corporeal things. After causing you to wonder at the most powerful machines, the most unusual automatons, the most impressive illusions and the most subtle tricks that human ingenuity can devise [les plus subtiles impostures, que l'artifice puisse inventer], I shall reveal to you the secrets behind them, which are so simple and straightforward that you will no longer have reason to wonder at anything made by the hands of men. I shall then pass over to the works of nature ... (*AT* X 505; *CSM* II 405).

It is mechanism or the theory of this craft that enables us to synthesize the totality of these moving particles into a unified description of nature, one that manifests itself in the image of nature – both organic and inanimate – as a cosmic machine (Collins 1971, p. 29). Machines furnish the *deception* of self-initiated movement when, to the trained eye of an engineer, or to a philosopher conversant with the workings of machines, their movement is seen to be engendered by an extraneous cause, whether the wheels and mainsprings of a watch or, in the case of the cosmic machine, God's *concursus ordinarius.* Without a knowledge of the mechanical principles that govern these movements, we might be deceived by them, in the way that we are often deceived by God's own machines and suppose that they are self-activating.

7. PICTURING KNOWLEDGE

I've already made reference to §203, Part IV, of the *Principles.* I now want to quote this passage more fully in order to highlight the role that illustrations play in Descartes's own scientific practices:

> Some readers may perhaps ask how I therefore know what they [imperceptible] bodies are like ... to this end, things made by human skill helped me not a little: for I know of no distinction between these things [les machines que ont

les artisans] and natural bodies, except that the operations of things made by skill are, for the most part, performed by apparatus large enough to be easily perceived by the senses [les effets des machines ne dependent que de l'agencement de certains tuyaux, ou ressorts, ou autres instrumens]: for this is necessary so that they can be made by men. On the other hand, however, natural effects always depend on some devices so minute that they escape all senses. (*AT* IX-2 321)

The view of the cosmos in this passage is that it is simple and self-consistent – that which occurs insensibly at the level of *petites parties* occurs sensibly in the workshop, the observatory, and on the battlefield. In principle, God's own machines are no less knowable than the contrivances of artisans. The relevant difference is that we cannot manipulate God's own machines in the same way that we can manipulate human contrivance, either because (in the case of the solar system) they are too large or (in the case of *petites parties*) too small. Descartes is not recommending in the above passage that philosophers consult mechanical devices in some theoretical capacity as a way of intellectually grappling with the finer insensible structure of things. When he informs Burman that 'things made by human skill helped me not a little,' he is referring to artifacts that he has built with his own hands – to the dancing man, a flying pigeon, and a spaniel chasing a pheasant that Poisson (1671) claimed Descartes had planned to build as early as 1619 (see *AT* X 232), as well as a 'little machine representing a rope-dancer [une petite machine qui representoit un homme dansant sur la corde].'[22] There is also the important passage in conversation with Burman:

[it is] scarcely possible to understand this figure [illustrating movement in a vortex] without the help of eight or so little balls to demonstrate the movement. The author, despite the fact that he has accustomed his mind to imagining, was scarcely able to conceive of it without the balls. So others will find it much more difficult. For these things depend on mathematics and mechanics, and can be demonstrated better in a visual demonstration than they can in a verbal demonstration. (*AT* V 172; *CB* 67)

With the possible exception of a few mathematical intricacies in the *Dioptrique*, Descartes explicitly declares that 'we do not ... need Mathematics in order to understand the author's philosophical writings ...' (*AT* V 177; *CB* 79). What we do require is facility with mechanical devices, whether these are two-dimensional pictorial models or three-dimensional working models fabricated in the workshop. What Descartes is advocating, in effect, is that natural philosophers don the hats of the *horologier*, the instrument-maker, the military engineer, and the illustrator and become conversant with the workings of machines of all sorts. For Descartes, then, it is the artifice of craft (and, by association, the artifice of drawing) that enables us to grasp nature; working with mechanical models, whether constructing or drawing them, is a form of cognition that engenders favourable dispositions for grasping the insensible workings of nature.

In order to fully appreciate the sense in which the artifice of craft is supposed to help us to frame explanations for natural phenomena, we need to get clear on the kinds of equivalences that are expressed by Descartes's mechanical models. Scholars standardly portray Descartes as suggesting that the structure of things that lie outside the range of our sensory apparatus can be grasped by *analogy* with the mechanical properties, so-called, of macro phenomena; substantive explanations of such diverse phenomena as magnetism, heat, and the rainbow require empirical substance that Descartes furnishes through analogies with mundane experience. This contention only makes sense if it can be shown that for Descartes some of the properties of bodies of our experience are genuinely representational; that is, that we obtain genuine knowledge of bodies of mundane experience via sensation. It does not require that all of the properties identified by sensation are representational, but only that some are representational, namely, those geometrical properties that Descartes seems to identify as essential to matters.[23]

There are passages that seem to sustain this interpretation of Descartes as advocating an epistemology that is congenial to a variety of sensationalism. The most favourable passage is to be found in Part IV, §198, of the *Principles*, where Descartes submits that 'we perceive by our senses nothing in external objects except their movement, figure or situation, and the size of their parts' (*AT* IX-2 317; cf. *AT* VII 80). This passage has been widely interpreted as tantamount to the claim that the geometrical properties of natural bodies are preserved in the act of sensation. If this interpretation is allowed to stand, then it is damning for Descartes's entire project for delivering explanations of natural phenomena *more geometrico* from first principles, since it would be properties identified at the macro level that inform his explanations for particular phenomena.

In order to ascertain whether this interpretation is warranted, we need to turn to the *Dioptrique*, where Descartes elaborates an intriguing account of vision – an account that needs to be bracketed with the thesis, elaborated by Kepler in his *Ad vitellionem paralipomena*, that the human eye is a kind of machine – a picture-taking machine – and that the act of seeing should be regarded as a means of picturing: 'Vision is brought forth by a picture [*pictura*] of the thing seen being formed on the concave surface of the retina' (Kepler 1604, 2:153; cited by Alpers 1983, p. 32). This startling thesis was fashioned by Kepler in answer to a riddle generated in 1600 by Brahe's use of the pinhole camera in lunar observations – namely, the lunar diameter as formed by the rays in the pinhole camera appeared smaller during a solar eclipse than at other times. Could the moon have changed size or moved further away from the earth during the solar eclipse? Kepler rejected the possibility that the puzzle was an astronomical one and submitted, instead, that the solution to the riddle was to be found in optics – that the issue involved the optics of the images (which he called pictures) formed behind small apertures in the pinhole camera. The changing diameter of the moon was caused by the intersection of the optical mechanism with the rays of light; in Kepler's own words, 'the deception of vision [*visus deceptio*] arises partly from the artifice of observing ... and partly just from vision itself' (Kepler 1604, 2:143; cited by Alpers 1983, p. 34). As it turns out, deception is built into the pinhole camera and, by implication, into the human eye, which, Kepler argued, was an optical mechanism furnished with a lens that has focusing properties. If our optical mechanism mediates the world, it follows that the world would be seen differently through the eye of an insect. One can readily grasp why, as Kepler's views gathered momentum in the seventeenth century, scientists such as Leeuwenhoek become obsessed with studying the eyes of other animals and in reconstructing the world as pictured by their optical mechanisms.

Crombie is right, in my view, that Descartes fastens on Kepler's treatment of the human eye as an optical instrument (not a 'dead' optical instrument but, as Alpers suggests, a picture-making device). Kepler's account of vision laid the foundation for a wonderful analogy between nature and artifice, one that Descartes elevated into the cornerstone of his natural philosophy. As to particulars, Kepler had presumed that the visual faculty receives an image that *resembles* its object in much the same way that a picture resembles what it depicts.[24] Descartes sternly rejected this thesis in consequence of its failure to explain how an image or physical motion produced by external stimuli in the sensory mechanism is transformed into a sensation in the brain.

Against Kepler, he submitted that objects are not known to us by virtue of the resemblance between these mental images and their objects – that is, we are not equipped with an inner set of eyes that enables us to see images in our heads. The objects of perception, rather, act directly on the soul and are ordained to give it sensations. The visual image itself is therefore not the object of sensing but, as Crombie (1967, p. 74) claims, *just the means of sensing*. What Descartes submitted, in effect, is that questions as to how physical motions cause sensing need to be firmly demarcated from questions concerning the physical and psychological clues that determine different images in the mind (sensations). We know that in vision the image need not be a strict representation of the object; the image is more like a two-dimensional engraving that suggests many different qualities, including some that are not visual at all, with just a few strokes of the pencil. In hearing and in the other senses, the image is even less representative. What this suggested to Descartes was that the pictorial resemblance of the retinal image to its object was incidental in affecting sensations.

The question, then, is how does the mind come to know its objects? The claim that Descartes relies on sensation for information abut the geometrical properties of bodies of mundane experience is tantamount to the thesis that the sensible image is the object of sensing, and not just the means of sensing (see Crombie 1967). The view that squares with Descartes's treatment of sensation in the *Dioptrique*, however, is that the mind only has a symbolic knowledge of objects in virtue of the motions they set up in the nervous system. So far as sensing objects is concerned, our position is no different than the blind who make their way around in the dark by means of a stick – the blind 'see with their hands,' but the knowledge that they possess of external objects is of a symbolic kind that helps them to frame a conception of these objects. It is the interaction between our nervous systems and external bodies that gives rise to different kinds of sensations; clues for sensations are given by the rapid and lively movement transmitted to our eyes and other organs of sensation (see Crombie 1967, p. 75).

To those who insist that Descartes advocates an account of sensation that is congenial to an empiricist epistemology, we can concede that he does consult experience, with the rider that he does so in a very deliberate and circumspect way. Though he often compares celestial vortices with eddies of water, for example, he never uses this analogy as justification for the claim that celestial vortices are eddies on a grander scale. Descartes's pictorial models are not homoeomorphs, in the sense that a toy train is fashioned after a particular train (see Harré 1970, ch. 2). Further to this, Descartes's pictorial models omit many of the properties of material particles and idealize some that they retain. Perhaps the most striking example of Descartes's use of idealization is that the particles in his illustrations are perfectly rigid; their plausibility as mechanisms is parasitic on the supposition that they are sufficiently rigid to bear a load. In some cases, his use of idealization borders on the extreme: his use of a bouncing tennis ball to illustrate some properties of light in the *Discourse*, for example, presumes that the ball travels at a finite speed, even though this is impossible on Descartes's own principles.

The view that I favour is that Descartes's illustrations are best viewed as isotropic with reality – that is, *the artifacts of craft are equivalent to the artifacts made by God only with respect to their general manner of acting* (i.e., they work mechanically). Mechanisms behave like God's own machines, or at least close enough to help us get clear on the workings of natural phenomena. Extending this line of argument to the picturing in Descartes's scientific treatises, my central claim is that Descartes's illustrations are to be seen as two-dimensional models of how things might work, and not what they look like. They are marshalled to develop and sharpen our mechanical intuitions – that is, to make us better natural philosophers – and so there is no danger that we will take their various qualities to be real existents. Celestial whirlpools of matter (fig. 3.8) and the human body are both conceived as machines, the former a hydraulic device and the latter a clockwork mechanism with a pump for a mainspring, but as Rodis-Lewis (1978, p. 158) notes, the force of the mechanical analogy varies to such an extent from one scientific explanation to the next that the comparison drawn in any given case is neither here nor there. What Descartes seeks, rather, is so thorough an understanding of vortices and the human body *as mechanical contrivances* that the explanation will be technically correct. To tie this in with my characterization of Descartes's conception of mechanism, we will plainly see that the illusion of self-movement perpetuated by mechanical devices is caused by the disposition and the collaboration of their assorted parts.

What we are weighing is two distinct ways that we can view mechanical models. Both take mechanical models to be cognitive aids, but one sees them as helping us to discern what the world is like, whereas the other sees them as devices to help us conceive how it might work. The former regards mechanical models as representations of the world, whereas the latter conceives these models as devices that are designed to stimulate our minds into conceiving how the world (regarded now as a machine) might work. For the former, it is critical that Descartes's

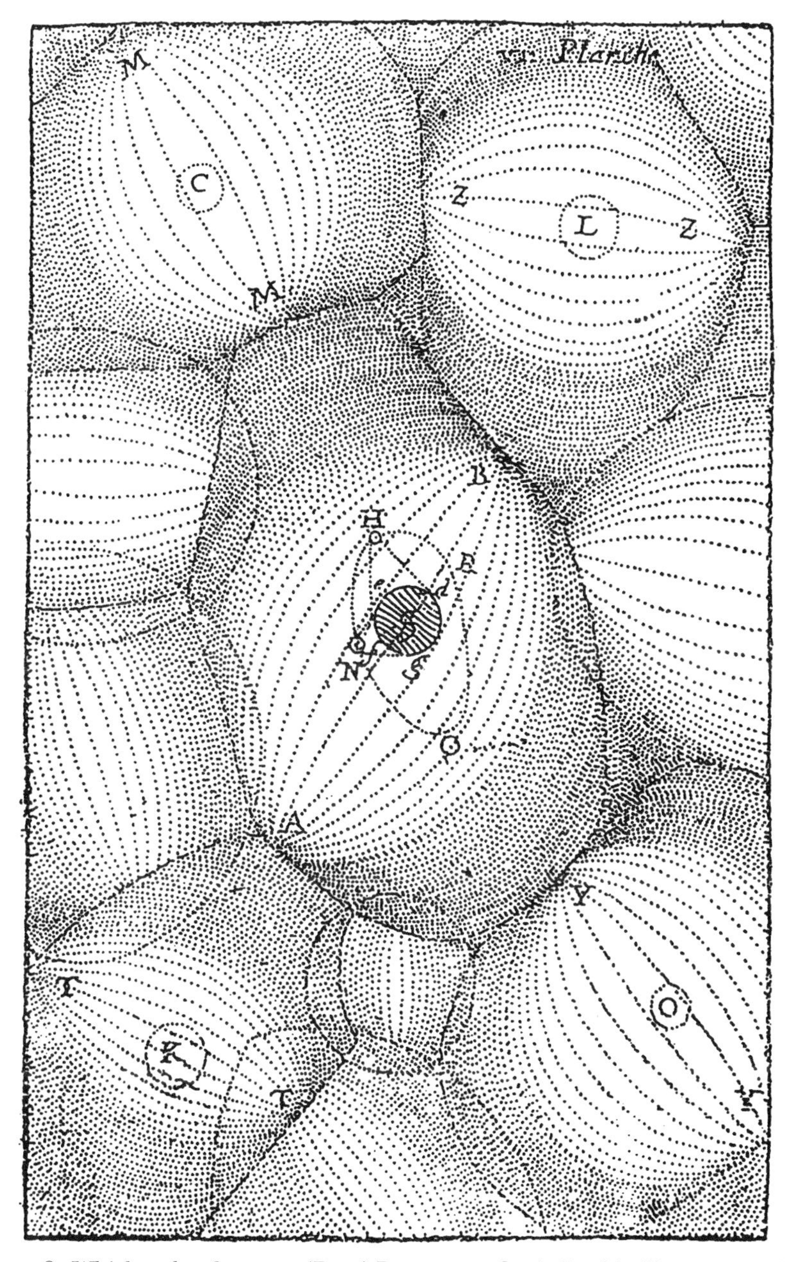

3.8 Whirlpools of matter (René Descartes 1644). In this illustration, the sun *S* is situated in the midst of the vortex *AYBM*, pressed on all sides by continuous vortices, *C, K, O.*

models preserve some of the world's properties; that is, they help us to discern what the world is like by representing its basic geometrical structure. For the latter view, it is beside the point whether these models represent reality in the sense that they preserve its so-called mechanical structure. It is my contention that Descartes regards models almost exclusively as devices to help us see how things work. Indeed, it is his conviction that, in seeing how things might work, we thereby understand their reasons.

Evidence for this thesis is given by Descartes's treatment of light in his letter to Reneri for Pollot (April or May 1638). In the first of three analogies, he draws on the familiar example of a blind person's use of a walking stick to clarify the propagation of light by a pressure that eventually impinges on our eyes. In the second, he uses the example of wine leaking from a vat to clarify the linear transmission of light:

> *I did not say that light was extended like a stick, but like the actions or movements transmitted by a stick.* And although the movement does not take place instantaneously, each of its parts can be felt at one end of the stick at the very moment (that is to say, at exactly the same time) that it is produced at the other end. Moreover, I did not say that light was like grape juice in a vat, but like the action whereby the parts of the juice at the top tend to move towards the bottom: these parts tend to move towards the bottom in a completely straight line, though they cannot move exactly in a completely straight line ... (*AT* II 42; *CSMK* 100; my emphasis)

The message in Descartes's adoption of the Stoic analogy of the stick to account for vision is that light is not like a stick at all. The analogy between the stick and light, rather, is drawn narrowly between their 'actions or movements' – light is transmitted in a medium from an object to the human eye *in much the same way* that movement across pavement is transmitted from concrete to the human hand. The blind man perceives differences in objects as well as anyone but 'in all these bodies the differences are nothing but the various ways of moving the cane or resisting its movements' (*AT* VI 85; *CSM* I 153). By the same token, the only way that light can be compared to grape juice in a vat is with respect to the action whereby 'the parts of juice tend to move towards the bottom' (see fig. 3.9). These models, then, merely serve to make plain the way that light works, but they do not tell us anything about the qualities of light per se. Instead, they help us to conceive movements by revealing the mechanics of mundane situations that sharpen our mechanical intuitions.

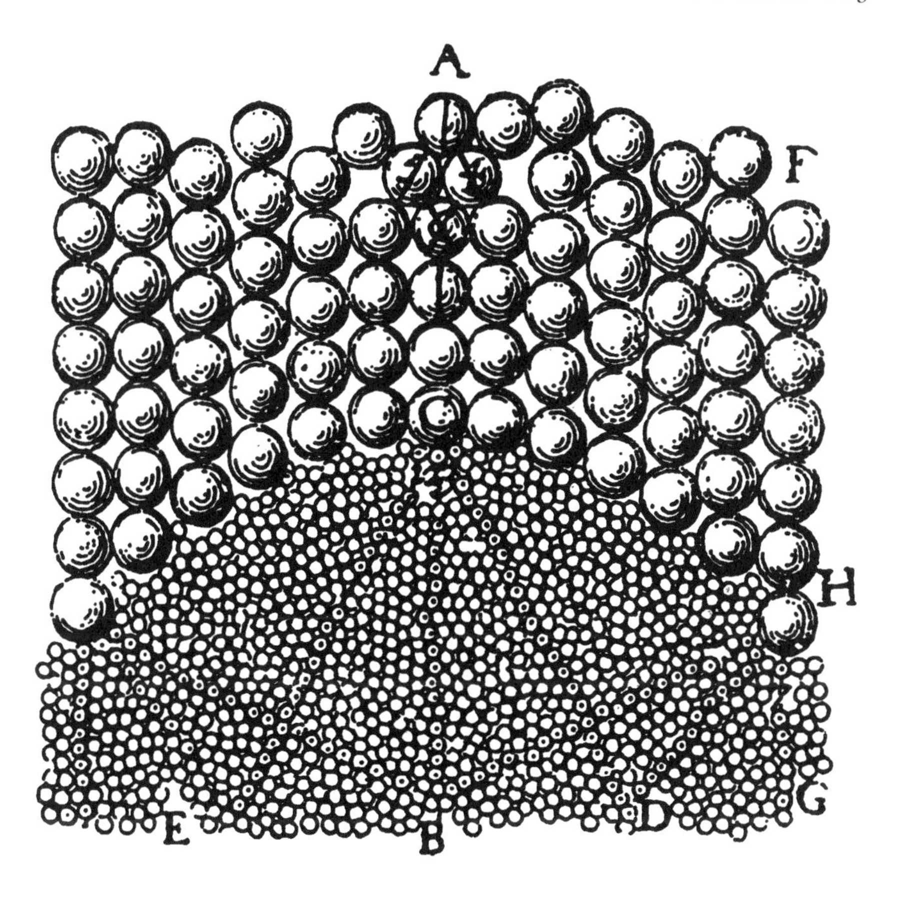

3.9 Grapes in a vat illustrating the properties of light (René Descartes 1637).

8. PICTURING

Descartes uses the expression 'imagination' literally to signify the act of forming a mental image of some sort – what I've referred to as picturing, as opposed to reasoning. In the *Discourse*, for example, Descartes states that imagining 'is a way of thinking specially suited to material things' (*AT* VI 37; *CSM* I 129). This thesis is reiterated in the second mediation: 'imagining is simply contemplating the shape or image of a corporeal thing' (*AT* IX 22; *CSM* I 19). And the fuller statement:

... the difference between this mode of thinking [imagination] and pure understanding may simply be this: when the mind understands, it in some way turns towards itself and inspects one of the ideas which are within it; but when it imagines, it turns toward the body and looks at something in the body which conforms to an idea understood by the mind or perceived by the senses. (*AT* IX 73; *CSM* II 51)

These passages seem to be opposed by the following statement reported by Burman:

The fact that there are some people who are clever at Mathematics but less successful in subjects like Physics, is not due to any defect in their powers or reasoning, but is the result of their having done Mathematics not by reasoning but by means of imagining – everything they have accomplished has been by means of imagination. Now, in physics there is no place for imagination, and this explains their [i.e., mathematicians who work through the imagination] signal lack of success in the subject. (*AT* V 177; *CB* 79)[25]

Cottingham (*CB* 117) is right to remark that it is odd that Descartes here should portray physics as a study where the imagination is of no value whereas, in a letter to Elizabeth, he asserts that 'the study of mathematics ... exercises mainly the imagination in the consideration of shapes and motions ...' (*AT* III 692; *CSMK* 227). One would suppose the opposite to be the case: mathematics would appear to be a study that involves pure understanding, as opposed to physics, which, one would suppose, would require images, models, pictures, and other sorts of cognitive aids.[26] What is the explanation? Cottingham offers an interesting answer to this question. He reminds us that for Descartes the imagination is closely allied to perception: imagining and perceiving both involve the depiction of images in the brain. The only relevant difference is that the imagination performs this function 'with the windows shut' (*AT* V 162; *CB* 42). Moreover, sensible images of perception are bound up with the false beliefs and prejudices that the metaphysician needs to eradicate in the quest for truth. The physicist is in a similar position with respect to external objects and so must avoid attributing to matter the qualities identified by the senses. And so Cottingham suggests that

the formation of images which we may be tempted to foist on the world of reality can be, for Descartes, a source of disastrous error in metaphysics and physics. In mathematics, on the other hand, there is no question of the objects of study 'really existing'; mathematical entities are considered merely as possibilities. Thus the image is a help to the mathematical student in marshaling his thoughts, rather than a tempting picture of reality. (*CB* 117)

Though Cottingham's claim is plausible, the fact is that it renders Descartes's own scientific practices inconsistent. Descartes plainly concedes that physical models can be indispensable for grasping the structure of things, and one would suppose that this concession would apply with equal weight to the many pictures of these physical devices scattered throughout Descartes's scientific treatises. In order to rescue Descartes from the charge of inconsistency, Cottingham draws a distinction between 'the heuristic and expository roles of the physicist,' suggesting that 'perhaps Descartes means that it is in discovering new truths that the imagination may be treacherous; in explaining a theory once arrived at, visualization and the use of models is a valuable bonus' (*CB* 118). The difficulty, however, is that there is no reason why picturing should be a 'valuable bonus' at the expository stage and not at the discovery stage as well. If imaging is a source of error for each of us in our quest for truth, then this principle should apply as well to others who are attempting to duplicate our discoveries for themselves. Imaging, one would suppose, would engender the very dangers that we sought to avoid by not imaging in the first place.

Nevertheless, I accept Cottingham's general line – that for Descartes picturing constitutes a real and present danger in physics because, in considering things that are supposed to exist, there is the danger that we will take the attributes possessed by the picture to have some basis in fact. 'We must,' Descartes advises, 'conceive the nature of these images in an entirely different manner from that of the philosophers' (*AT* VI 112; *CSM* I 165). The nature of objects consists in supposing, on the grounds that a picture can easily stimulate our mind to conceive the objects depicted by it, that our mind is stimulated 'by little pictures formed in our head' (*AT* VI 112; *CSM* I 165) to conceive the objects that affect our senses. Descartes objects that our mind is likewise stimulated by the signs and words 'which in no way resemble the things they signify' (*AT* VI 112; *CSM* I 165). Moreover, even if we persist in the view that objects really do send an image of themselves to the brain, we must observe that 'in no case does an image have to resemble the object it represents in all respects, for otherwise there would be no distinction between an object and its image (*AT* VI 113; *CSM* I 165).

It is enough, Descartes contends, 'that the image resemble its object in a few respects' (*AT* VI 113; *CSM* I 165). Indeed, for most practical purposes, it is imperative that the image not resemble its object 'as much as it might.' It is instructive that Descartes advances the case of ink engravings in support of this thesis:

> ... a little ink placed here and there on a piece of paper, they represent to us forests, towns, people, and even battles and storms; and although they make us think of countless different qualities in these objects, it is only in respect of shape that there is any real resemblance. And even this resemblance is very imperfect, since engravings represent to us bodies of varying relief and depth on a surface which is entirely flat. Moreover, in accordance with the rules of perspective they often represent circles by ovals better than by other circles, squares by rhombuses better than by other squares, and similarly for other shapes. Thus it often happens that in order to be more perfect as an image and to represent an object better, an engraving ought not to resemble it. (*AT* VI 113; *CSM* I 165–6)

We can extract two lessons from this passage. The first is explicitly identified by Descartes – viz. the images formed in our brain stand in the same position with respect to their object as do the images forged by the engraver. There is no difficulty explaining how mental images correspond to objects – they don't.[27] The problem, rather, is to explain how our mental images enable our soul to have sensory perceptions of all the various qualities of the objects to which they correspond. The second lesson – one that is not explicitly identified by Descartes – is that three-dimensional representations – namely, physical models – stand in the same position with respect to their objects as do ink engravings. If the engineer builds an armillary sphere or a clockwork mechanism to represent the cosmic sphere, there is no reason to suppose that it corresponds to its object; the perfection of it as a model often depends on it not representing its object.

Mental imaging plays a very particular role in Cartesian science. There is no evidence for Cottingham's suggestion that this role is to be limited to an expository one. What fuels this suggestion is the long-standing conviction that pictures are to be seen in the ordinary manner as representations of physical objects that are marshalled to help us get clear on the meaning of associated text. It is rather the case that imaging for Descartes is an artifice for helping us to grasp the workings of God's own machines. The great advantage of imaging is that it can enhance our mechanical intuitions even when our mental images do not resemble things; indeed, visualization often works best when it does not resemble things at all.

9. LIMITATIONS OF DESCARTES'S ILLUSTRATIONS

Western scientific culture is infused with the idea that the fortunes of a science are directly tied to its capacity to deal with facts. Whether we endorse the received view that facts are detached from our interests and activities or the constructivism of a new generation of scholars who see science in the manner of the anthropologist as a discursive cultural practice, the bottom line for science is its fact-managing capacity. I won't dispute this claim in the closing section of my paper, but I do believe that it is greatly overblown, at least when applied to the natural philosophy of the seventeenth century, which places a great premium on the persuasive power of illustrations.

There are many avenues for assessing the virtues of a science. In closing, I want to highlight the persuasive power of scientific illustrations that collaborate with theory, experiment, instruments, and other resources in making the facts known to us. In recent times, scientists and laypeople alike have been held in thrall by the power of artistic images, such as Darwin's tree of life, Bohr's atom, and Watson and Crick's double helix. Descartes's illustrations were a startling success insofar as they furnished a style of picturing that enabled a new generation of natural philosophers to frame intelligible conceptions of phenomena – for example, gravitation, the attractive power of the lodestone, and the healing action of curatives – that were castigated by their predecessors as 'occult' and so as beyond the pale of knowledge. Using art as science, Descartes was able to persuade many of his peers that even these 'insensible' and allegedly unscientific qualities could be conceived by the attentive mind. The lesson that Descartes and his fellow mechanists extracted from the science of machinery was that insensible causes of natural phenomena can be 'seen' – in a manner of speaking – and therefore rendered intelligible by reconceptualizing these phenomena as systems of rigid parts that collaborate in the production of mechanical effects. Much of the persuasive power of this reconceptualization of experience in mechanical terms can be credited to Descartes's use of art as science, which gave the would-be mechanical philosopher reason to believe that the world had been fabricated by God as a solution to a fantastic engineering problem. Even if these illustrations did not engender the conviction that God's own machines were anything like the machines of the engineer, they sustained the belief that at least they afforded the natural philosopher an intelligible conception of the workings of nature.

Descartes's illustration of magnetism (fig. 3.10) signifies the first attempt at picturing magnetic lines of force. It is one of his more intriguing attempts to reconceptualize a scholastic occult quality in mechanical terms. Descartes reckoned that magnets are pervaded with continuous pores, through which a stream of corpuscles continually circulates. These corpuscles are screw-shaped, some with left-handed and others with right-handed threads. Attraction occurs when the corpuscles are able to enter into the correctly threaded pores of another magnet; repulsion occurs when oppositely threaded particles meet. Though this explanation struck many of Descartes's contemporaries as somewhat contrived, the real power of the illustration as science is that it reinforced Descartes's contention that all phenomena are to be explicated as mechanical effects. A second example that is rarely discussed is his mechanical model of the formation of the earth (fig. 3.11), one that influenced Anastasios Kircher in the composition of what is widely regarded as the first modern work of geology – the *Mundus subterraneus in XII libros digestus* of 1665. In this illustration, the earth has a molten core *I*, surrounded by a sphere that is made of the same matter as sunspots *M*, a solid crust *C*, and a lighter crust that floats on an internal sea *D*. As pieces of the crust tilt, they become immersed in the sea and protrude as mountains, a thesis that enabled Descartes to explain the presence of tilted strata that do not follow the curved shape of the earth. This illustration testifies to the rigour with which Descartes developed his central insight that heaven and earth are to be accounted for by the same set of principles; that is, the earth and the other planets were once stars, differing only in size, and the changes in the earth's crust are explained by the gradual and continual cooling of its central mass.

Illustrations, I submitted, are one avenue for assessing the adequacy of a science. Much of their persuasive power stems from their collaboration with other resources that scientists draw on to make the facts known. In Descartes's case, the persuasive power of his pictorial devices was tied to a number of other resources – especially the systematic theory (principles, laws of motion, etc.) that he carefully elaborates in the *Principles* – and the material resources (experimental devices and apparatus) that he employs in his study of natural phenomena. Scholars

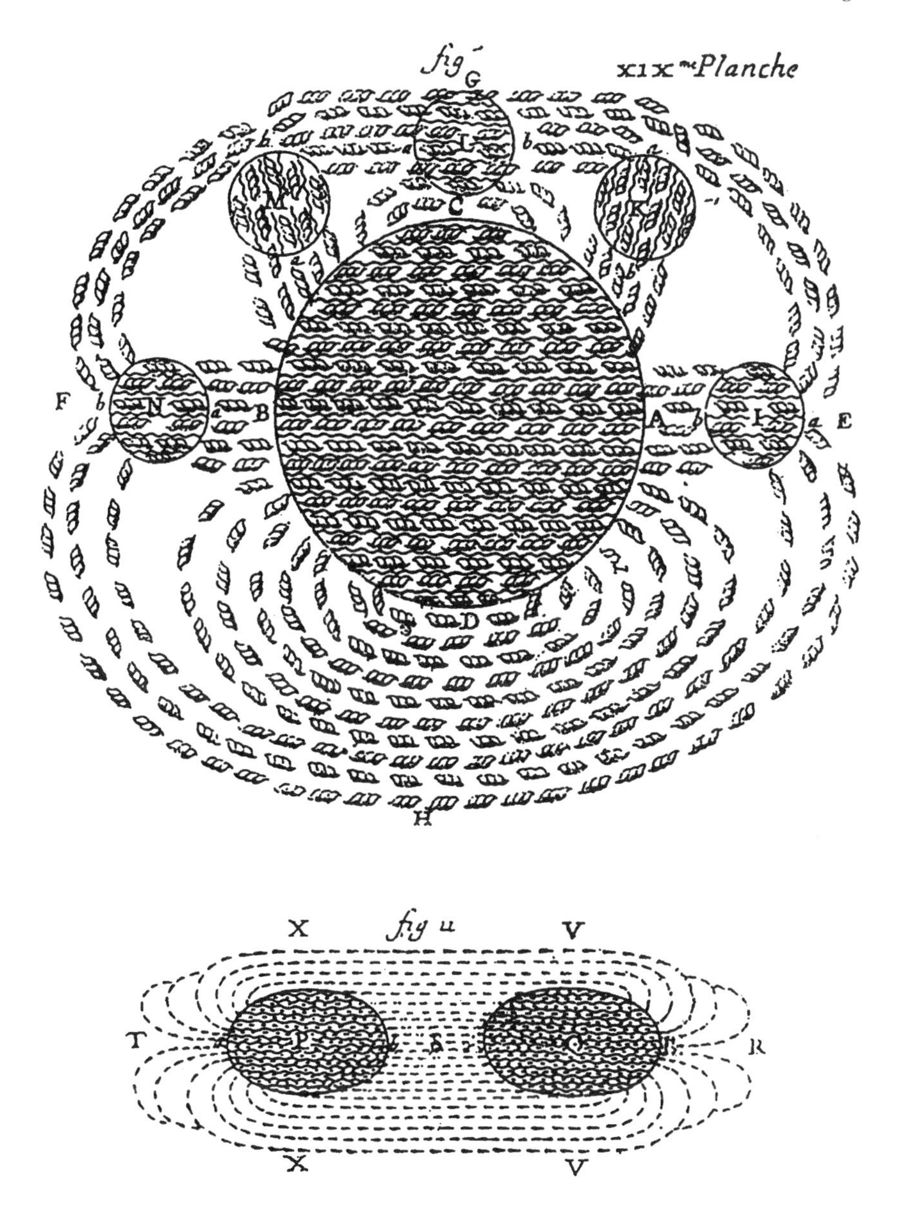

3.10 Magnetic lines of force (René Descartes 1644).

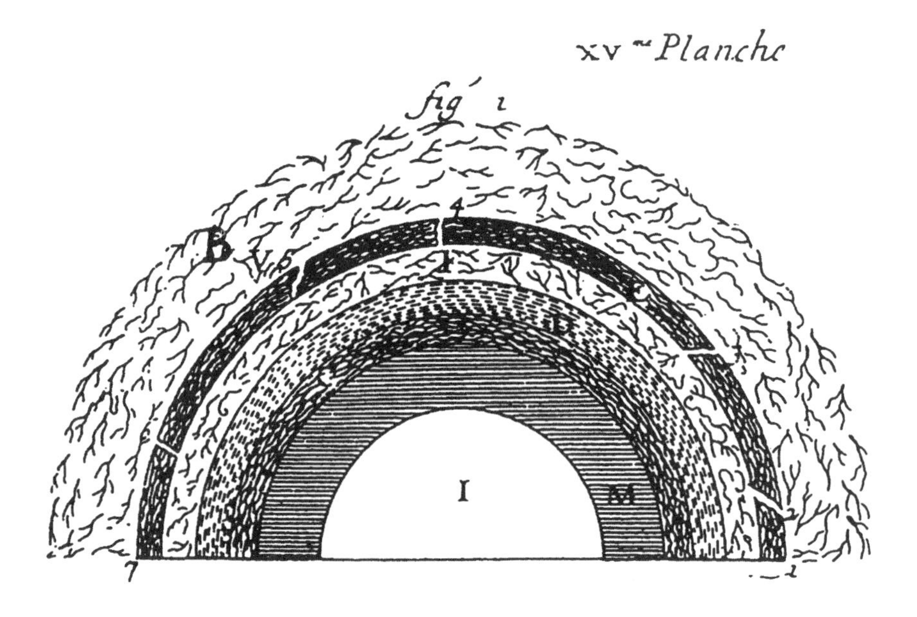

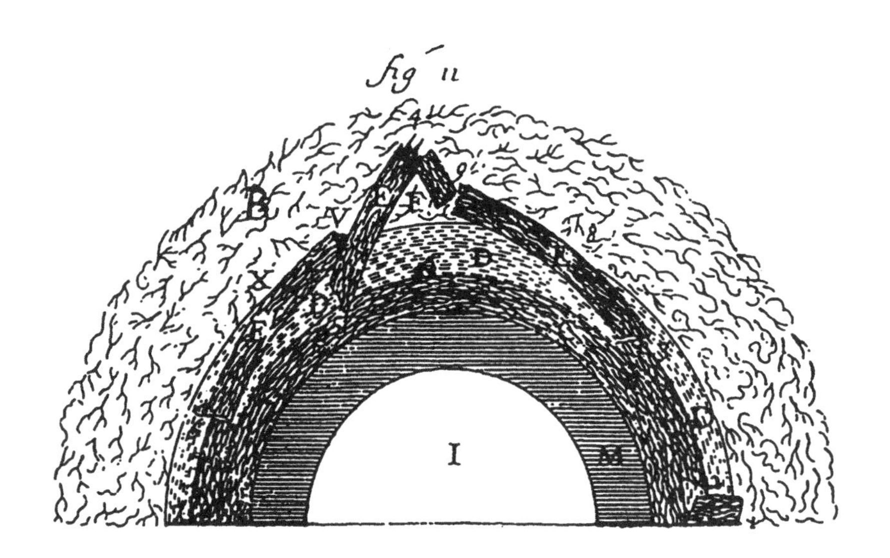

3.11 The formation of the earth (René Descartes 1644).

have documented the ways that systematic theory fails Descartes – for example, his third law governing exchanges of motion was shown to be deeply flawed by Leibniz, Huygens, Wren, and others. Scholars have also started to subject Descartes's experimental practices to careful scrutiny. What has emerged to this point is that these practices reveal a subtle bias against detailed experimental work in favour of simple experiments and observations about which many observers can be certain and in the interpretation of which there is less scope for differences of opinion (see *AT* II 542; *CSM* I 143; cf. Baigrie 1992, p. 175). In his preference for experiences 'presenting themselves spontaneously to our senses,' Descartes was guided by the engineer's vision of experience as the clear and distinct projection of the tools and the levers that were the stock and trade of artisans; his interest in phenomena was restricted to that aspect that appeared, at first sight, to be clear and distinctive to the attentive mind. The rest of the world is passed over in silence. This methodological preference for simple observations and straightforward experiments imputed an exactness to the phenomena that they do not possess. Even Nicholas Malebranche, who was an earnest supporter of Descartes's conception of science, protested that 'real levers and wheels are not the lines and circles of mathematics' (quoted by Lenoble 1964, p. 193; cf. Baigrie 1992, p. 175). Natural phenomena are a great deal less like constrained mechanical systems than Descartes was prepared to admit.

A third critical resource that collaborates with Descartes's illustrations in his attempt to make the facts known to us is lore about machinery, which played a central role in shaping Descartes's deepest cosmological convictions.[28] La Forge was essentially right to portray Descartes's conception of a machine as a system of constrained parts that produces a mechanical effect – viz., as a movement in some other particle that would otherwise be impossible. Even if we are prepared to accept this suggestion as a framework for Descartes's take on mechanism – that is, even if the seemingly self-instigated motion of God's own machines can be rationalized by an appeal to the mechanical arts – this mechanistic hypothesis was not robust enough to sustain any lasting analogy between God's own mechanisms and devices of human fabrication.

First, God's own mechanisms behave as though they are unconstrained – the bodies that captured the imagination of the majority of seventeenth-century natural philosophers – the planets – move in an indeterminate way. Even now, there is no closed expression for the movement of a planet in an elliptical orbit. Second, Newton demonstrated, in *Principia*, Book I, Propositions 70 and 71, that if a body is spherical in form and equally dense throughout, it attracts all other particles as though its entire mass were concentrated at its centre. If variable in density, he showed that a body can be regarded as made up of a series of spherical shells, having a common centre, each of uniform density throughout, different shells being of different densities. Planets can therefore be assumed to be homogeneous, with their masses concentrated at a central point. The consequence is that the factors that are central to the science of machinery are simply irrelevant.[29] The mechanical properties of celestial bodies can be studied without any consideration of the shapes of their contact surfaces and the many other factors that are critical in the fabrication of machine technology. In the case of mechanical effects produced by the action of gravity, the force upon any body does not depend upon the shape of the body, but acts in proportion to its quantity of matter. This thesis was first advanced by Newton nearly a half century after the publication of Descartes's *Principles.* It dramatized the gulf that exists between constrained and unconstrained systems, while Newton's success in treating planetary phenomena as instances of unconstrained mechanical behaviour discredited Descartes's conviction that the science of machinery was an epistemological reservoir for explanations in natural science.

Throughout the seventeenth century, we see the creation of specialized niches within the science of mechanics – the recognition that the science of mechanism is to be demarcated from mechanics, the division of mechanics into pure and practical parts, and the creation of a new niche in the early eighteenth century, experimental mechanics, which examines the mechanical properties of deformable bodies that we can manipulate in laboratory settings. The lore about machines on which Descartes capitalized to authorize his new science was replaced by a lofty body of mechanical theory of a very detailed kind that severed the epistemic connection between nature and the artifice of craft. It is true that this connection persisted for some time in some intellectual niches – in anatomy and in physiology, reverberations of Descartes's mechanism can still be detected – but the confidence that Descartes placed in the machine as the model for natural phenomena withered rapidly in physics in the wake of Newton's demonstration that Descartes's celestial mechanism, that great whirlpool of matter, could not be reconciled with the degree of freedom that is proper to planetary motion.

This break settled the fate of Descartes's illustrations once and for all. In the shadow of Newton's *Principia*, Descartes's pictorial devices seemed a lot less like science and more like works of art – symbolic renditions of natural things that bore little connection with reality. We are now back where we started with our question about the disappearance of Descartes's illustrations: they vanished because historians and philosophers who have been trained to regard science in the form that Newton gave to it could no longer see their point as science. As we have lost touch with Descartes's particular conception of mechanism and the take on imaging that goes hand in glove with his conviction that nature is to be grasped through the artifice of craft, we can no longer see the relevance of Descartes's pictures to his natural philosophy.

NOTES

1 The following abbreviations will be used throughout this paper:

AT *Oeuvres de Descartes.* Edited by C. Adam and P. Tannery. 12 vols. Paris: Cerf, 1897–1913. Reprinted with new appendices, Paris: J. Vrin, 1964–74.

CB *Descartes' Conversation with Burman.* Translated with introduction and commentary by J. Cottingham. Oxford: Clarendon Press, 1976.

CSM *The Philosophical Writings of Descartes.* Translated by J. Cottingham, R. Stoothoff, and D. Murdoch. Cambridge: Cambridge University Press, 1984.

CSMK *The Philosophical Writings of Descartes, Volume III: The Correspondence.* Translated by J. Cottingham, R. Stoothoff, D. Murdoch, and A. Kenny. Cambridge: Cambridge University Press, 1991.

References will be to the standard edition of Descartes's works by Adam and Tannery (abbreviated *AT*, followed by appropriate volume and page number). I have consulted a number of English-language editions, and doubtless have benefited from them, but translations are my own unless a second reference is given to one of the above editions.

2 The position that squares with the vast majority of Descartes's statements is that our perceptions have a purely mental status – that is, they *do not resemble* any quality that exists in the world. *Le Monde*, for instance, opens with claims about the unreliability of sensation as a source of knowledge (see *AT* XI 3–4). A sensation of tickling caused by a feather, Descartes asserts, does not resemble anything in the feather. By analogy, there is

no grounds for supposing 'that what is in the objects from which the sensation of light comes to us is any more like that sensation than the actions of a feather ... are like tickling ...' (*AT* XI 6). This position – first elaborated in the unpublished *Le Monde* – constituted a radical break with the earlier *Regulae*, which had suggested that physical objects impress their form on the imagination 'in exactly the same way as the shape of the surface of the wax is altered by a seal' (*AT* X 412). Descartes had even extended this argument to sensations of hot and cold, the examples exploited by Galileo in the *Assayer* to demonstrate that atoms occasion our sensations. Though perception only has a mental status for Descartes, it still plays a vital role in cognition – namely, it can be deployed to stimulate mechanical intuitions.

3 Descartes never wavered in his conviction that our mind gives us distinct ideas of matter in general, as well as particular determinations of shapes and sizes. See Descartes's letter to More of 5 February 1649 (*AT* V 275; *CSMK* 364).

4 Two recent studies of Descartes's science – Clarke 1982 and Shea 1991b – suffice to make this point. Clarke's is organized in terms of a cluster of issues that have dominated philosophy of science until very recently – the comparative weight of reason and experiment, the relationship between physics and metaphysics, the character of scientific explanation, etc. Descartes's illustrations are not identified as a resource for science and an object of intellectual scrutiny. Shea's contains more than one hundred illustrations, many drawn from Descartes's treatises, but there is no recognition that the pictures themselves are used by Descartes as science.

5 Though it is possible that the audience at Tulp's public anatomies included Descartes on occasion, there is no evidence whatsoever for this thesis.

6 I owe this reference to my colleague André Gombay.

7 It is instructive that Descartes's *La Recherche de la vérité par la lumière naturelle* – an unfinished work that was not published until 1701 – compares the role played by reason in correcting 'our imperfect senses' with a master painter who is 'called upon to put the finishing touches on a bad picture sketched by a young apprentice' (*AT* X 507–8; *CSM* II 406).

8 Merchant (1980, p. 195) has argued that mechanists regarded the cosmos as a system of 'dead' corpuscles, but Descartes nowhere says that the world is dead. Following Kepler, what he does say, in the fifth discourse of the *Dioptrique* is that dead systems can be used as the basis for understanding living systems. It seems to me that the distinction is critical to understanding Descartes's theory of the organism.

9 Garber (1992, p. 321) resists Hoenen's interpretation of Descartes's remark to More on the grounds that 'Descartes whole strategy for deriving the laws of motion from the immutability of God presupposes that God is the real cause of motion and change of motion in the inanimate world of bodies knocking up against one another.' There is a real tension in Descartes's writings between his 'whole strategy' for deriving the laws of motion from God's immutability and the many places where Descartes seems to attribute activity to bodies.

10 Garber (1992, p. 322) cautions that Descartes weakens this argument in the French and Latin versions of the *Principles* but, even if we set aside Descartes's claim that a body can act on a soul, there are good reasons to resist the contention that Descartes denied activity to bodies.

11 Historians of technology (see Price 1964) have long debated whether it was the rise of a new mechanistic philosophy that paved the way for the explosion of interest in machines in the seventeenth century, or vice versa. I offer no opinion on this general issue but am contending that of the two influences – machines and the mechanical philosophy – it was the former, in conjunction with lore about machines, that proved most important to Descartes.

12 Salomon de Caus's *Les Raisons des forces mouvantes avec diverses machines* (1615) features descriptions of water-clocks, turning machines, a water organ, a pump, and a machine that imitates the sounds of small birds with water and air. The second part, in particular, with its illustrations of grottoes and gardens gives the reader a good sense as to why Descartes may have been enamoured of the Royal Gardens at Saint-Germain. Cf. Jaynes 1970. See Rodis-Lewis 1978, p. 167, fn 17.

13 Descartes's work on vision is especially instructive: 'experiment,' he writes,'shows that ... the crystalline humor causes almost the same refraction as glass or crystal ...' (*AT* VI 106) – in short, that there is a continuity between the making of the instrument and our knowledge of the organ of sight. See Rodis-Lewis 1978.

14 This is an oversimplification. Unconstrained bodies admit of degrees of freedom. A falling body has one degree of freedom (the action of gravity), whereas a projectile has two degrees of freedom (a uniform velocity in a definite direction and the action of gravity).

15 Some exceptions are in order. The pendulum of a clockwork is driven by the intermittent action of gravity. The armature of a motor rotates in consequence of electromagnetic forces acting across space. In both cases, motion is transmitted and controlled but not through actual contact between parts of the machine. Here the motion of the parts of the machine is not fully constrained, and its motion as a whole cannot be

treated by purely geometrical methods, independently of the forces involved in its actual construction. Though Descartes often compares the human body with a clock (see *Meditation VI*; *AT* IX 67), this analogy is not consistent with his underlying theory of mechanism, which is targeted at constrained mechanical systems; indeed, Descartes's repeated use of this analogy highlights his insensitivity to the distinction between mechanism and machinery, and is an example of how 'the degree of perfection attained in clock-making at the time made the clock a privileged model for other automatic machines' (Mumford 1967, p. 177; cited by Rodis-Lewis 1978, p. 167n17).

16 The General Scholium of Newton's *Principia* (1934, p. 543) exploits the distinction between constrained and free bodies in its devastating critique of Descartes's theory of vortices: 'for comets are carried with very eccentric motions through all parts of the heavens *indifferently*, with a *freedom* that is incompatible with the notion of a vortex' (my emphasis).

17 The *foundations* of Descartes's new science encompass, among other things, a conception of eternal truths (see *AT* I 149–50; *CSMK* 24–5), including the existence of God as the mooring of the laws of motion, and the existence and nature of body. These foundations are not inscribed within the domain of the mechanics of machinery. Of course, many of these elements are not part of physics, in the ordinary sense of the word. I am not attempting to fashion a thesis about the essential nature of Descartes's physics, which, it goes without saying, relies on a number of resources besides mechanical models, but simply to convince my reader that an appreciation of the mechanics of machinery clarifies the distinctive quality of Descartes's scientific explanations.

18 These considerations do not discredit the importance of Descartes's analysis of the mechanics of circular motion to subsequent developments in mechanics. Christian Huygens and the young Isaac Newton came to regard centrifugal force, in the manner of Descartes, as the manifestation in circular motion of the inertial tendency of bodies. By 1679 Newton came to see orbital motion in terms of a force directed towards some centre of motion that diverts bodies from their rectilinear paths, an approach which suggested that centrifugal force was a reaction to the centripetal force, according to Newton's third law of motion. Though Newton eventually found his own way, his early dynamical papers testify to the influence of Descartes's account of circular motion.

19 Cf. Cottingham's (1976, p. 53) claim that in the seventeenth century '"automaton" carried no more than its strict meaning of "self-moving thing."' Despite the fact that Descartes sometimes portrays machines in

the Greek sense as devices that move themselves, I believe that Cottingham's thesis is false.

20 I need to thank my colleagues André Gombay and Calvin Normore for alerting me to this link between the sceptical literature and machine technology.

21 The suggestion that machines let us move bodies in ways that they would not ordinarily move is a staple in the literature on the mechanics of machinery. Thus Andrew Gray (1804, p. 3) writes: 'bodies can be put in motion, or great weights raised by these powers when united, *which could not be raised by them individually*. Thus a power unable to produce the above purposes of itself, may effect its end by transferring a certain degree of the pressure upon the fulcrum or prop of a lever, or by the aid of pulleys, of the inclined plane, or of the screw' (my emphasis).

22 Florentius Schuly, in his Latin edition of Descartes's *l'Homme*, mentions a number of descriptions of automata. A detailed account of seventeenth-century automatism and mechanism is given in Price 1964, p. 342.

23 The most vocal proponent of the interpretation of Descartes's philosophy as relying on sensation for knowledge is MacKenzie (1989). See Garber 1978, pp. 15–16. Both Garber and MacKenzie regard the quest for certainty as so critical to Descartes's project that they are willing to abandon his avowed apriorism.

24 Michel Foucault (1973, pp. 54–6) rightly maintained that resemblance, or 'drawing things together,' was a fundamental category of knowledge for Renaissance thinkers.

25 See also Descartes's letter to Princess Elizabeth (28 June 1643), where he asserts that 'body (i.e., extension, shapes and motions) can likewise be known by the intellect alone, but much better by the intellect aided by the imagination ...' (*AT* III 692; *CSMK* 227).

26 Descartes submitted, in a letter to Mersenne (13 November 1639), that 'the imagination, which is the part of the mind that most helps mathematics, is more of a hindrance than a help in metaphysical speculation' (*AT* II 622; *CSMK* 141). This passage is offset by the sixth meditation, which asserts that we can have an understanding of a mathematical object without our being able to visualize it. Here Descartes remarks that the imaginary picture of a figure with one thousand sides is indistinguishable from another figure with one thousand and one sides. A mathematician, however, can almost certainly distinguish the two figures and prove appropriate theorems about the one and other. We can therefore have a purely intellectual understanding of a geometrical figure, quite different from a visualization.

27 Scholars often presume that pictorial significance is related to the resemblance between the picture and the objects that is depicted. Ernst Gombrich (1968) opposed this view by holding with Descartes that pictures do not really look like nature.

28 For the distinction between systematic and unsystematic theory, see Whewell 1897, vol. 2. Whewell invokes this distinction to address different kinds of resources in the history of botany, but it is just as applicable to the history of mechanism. He also refers to 'imaginary knowledge' or the 'earliest lore ... which we discover in the records of the past [which] consists of mythological legends, marvellous relations, and extraordinary ... qualities' (1897, vol. 2, p. 359). A not inconsiderable portion of seventeenth-century thought about machines was imaginary in the sense that it is steeped in hopes, fears, and love of the marvel of machine technology. It is important to recognize, however, that this lore about machines was intermingled with a great deal of information about the behaviour and properties of machines. Though this unsystematic theory was still a lore of sorts, it was not imaginary, in Whewell's sense of the word.

29 Newton submitted, in Proposition XVII, Theorem XVI, Book III of the *Principia*, that 'the axes of planets are less than the diameters drawn perpendicular to the axes.' Since the centrifugal force generated by the earth's rotation was in the opposite direction to that of the force of gravity, the pull of gravity would be lessened at the equator, causing the earth to bulge out. Newton's view was opposed by J.D. Cassini's contention that the equatorial diameter was shorter than the axis of revolution. This difference of opinion precipitated one of the most bitter disputes in the annals of science. Still, the shape of the earth was irrelevant so far as celestial mechanics was concerned.

4. Illustrating Chemistry

DAVID KNIGHT

1. VISUAL LANGUAGE IN CHEMISTRY?

Anybody who delights in the forms and colours of plants and animals will enjoy the pictures in works of natural history. In older books, they will look often curious to our eyes, as Dürer's rhinoceros does; and changes in the interests and theories of natural historians, as they mostly evolved into biologists, zoologists, and botanists, led to changes in the way they depicted species. In general, plates have become more austere from the layman's point of view: there is a tension between the needs of science and the demands of aesthetics, but at all times there have been artists for whom these constraints have been liberating, and who have produced immensely satisfying pictures. The same is true in medicine and surgery.

We do not expect to find this story among the physical sciences. There are electron micrographs, astronomical photographs, and pictures from X-ray crystallography which have a haunting beauty, but are hardly works of art; and in modern chemistry books, we find tables and diagrams, which with formulae flesh out the rather flat prose in which such works are written. But without the visual language, the material would be unintelligible; it is said that 'every language begins as poetry, and ends as algebra' (Myers 1992, p. 58); and in chemistry this evolution has become almost complete (Knight 1992b, pp. 176–7) as the overtones we value in ordinary language have been deliberately suppressed. The science has thus, curiously enough, become in a different way as arcane as alchemy was, cut off from the uninitiated by a rebarbative vocabulary.

Alchemy was richly illustrated, visual language being full of associations and ambiguities (Kaufmann 1993,) so that pictures of snakes biting their tails, of lions, the Sun, and the Moon all conveyed special meanings to initiates, and yet might form part of an aesthetically satisfying overall design. There are some very famous pictorial works of alchemy, notably the *Atalanta Fugiens* of Michael Maier (de Jong 1969); and such pictures lend themselves to iconographical, and even psychological, studies, as with Jung (1968). Alchemists also used complex systems of signs and sigils (Gettings 1981), which in the standard dictionary of them are indexed by strokes, in the way Chinese characters are; these are partly occult, partly shorthand. In alchemy, as in all sciences, the pictures did not stand by themselves: the visual message was meant to be seen with the associated text, and was complementary to it. We can thus in scientific illustration see the pictures in context and have a good idea of the artists' intentions.

By the later eighteenth century, the idea that chemistry was a mystery had been generally given up, and indeed the science was supposed to be accessible (Golinski 1992): in the event, research in chemistry soon became something for men of genius to perform with expensive apparatus, although its new language, devised by Lavoisier and his associates, was intended to be open, clear, and unambiguous (Guyton de Morveau 1787; Crosland 1978; and Brock 1993, ch. 3). If we follow the progress of chemistry through the nineteenth century, we see in its visual part a transition from pictures and illustrations through to tables and diagrams; and this is closely related to the growth of a chemical community, all trained in much the same way so that they could interpret these abstract forms of visual language, just as they could make sense of the brief and impersonal text which the illustrations accompanied. Looking at pictures in chemical publications will bring depth to our understanding of what it meant, and means, to practise chemistry.

2. FROM POETRY TO ALGEBRA: THE OUTWARD FACE OF THE SCIENCE

Alchemy was poetical, and often indeed written in poetry; and its illustrations were also meant to operate on more than one level. They were visual metaphors. Those in Ashmole's *Theatrum chemicum Britannicum* (1652), for example, do not show us what apparatus was like or how to use it, what chemical substances looked like, or how a laboratory should be organized; we do get some pictures of alchemists, but these are not straightforward portraits. None of the illustrations is simply descriptive. Like icons, they are part of the world of wonder and empathy that was undermined in the Scientific Revolution.

Lavoisier's great book of 1789 deliberately introduced the algebraic ideal for chemical language. To go with the new theory of combustion and acidity, *oxygen* rather than *phlogiston*, came the new vocabulary in which references to the appearance of things, to exotic languages, or the names of discoverers were omitted. In his novel *1984*, George Orwell described a new language, Newspeak, in which it would not be possible to think old thoughts; Lavoisier had anticipated him. Part of Lavoisier's project to get his readers to see things his way was to get them following his experiments; and his apparatus is carefully depicted at the back of the book.

We know what Lavoisier looked like from a number of portraits, but one of them (fig. 4.1) shows him in his laboratory with his wife (Donovan 1993); both sumptuously dressed, they are clearly not engaged in a real experiment, but there is apparatus around. There is another drawing, which shows her making notes at a side table during an experiment on respiration; but for most of the nineteenth century, chemistry like fishing was a male preserve, and husband and wife teams were rare.

Portraits do not often show the chemist at work in the laboratory; but just as those of landowners or admirals showed a great house or a ship in the background, so they usually show some tools of the trade. Sheridan Muspratt, chemist and industrialist, began his book with a series of fine engraved portraits of chemists, including himself; which are often details of bigger pictures, without this background. But in portraits of Humphry Davy (Knight 1992a; Muspratt 1860), we find, when he was young, electrical apparatus in the background of a portrait by Howard, done just (fig. 4.2) as he was making his name as lecturer and researcher in London, attracting large audiences to the Royal Institution. A later portrait by Thomas Lawrence (fig. 4.3) shows him as president of the Royal Society; a general, as he put it, in the army of science. Though it is not quite clear who was the enemy, military metaphors have always been popular in rhetoric about science and medicine; and Davy stares masterfully at us, while in the background is the safety lamp for coalminers which had made him unstoppable as a candidate for the presidency. Neither Howard nor (especially) Lawrence belonged to the 'warts and all' school of portrait painters; but, on the whole, in these pictures of chemists, we get an idea of what they

4.1 David's portrait of the Lavoisiers (Guyton de Morveau 1787).

4.2 Howard's portrait of Davy (J. Davy 1836).

4.3 Lawrence's portrait of Davy (J.A. Paris 1831).

looked like, and of what they did. Careers in Regency England were open to the talents, and the lives of Lawrence (who became president of the Royal Academy) and of Davy demonstrate social mobility – and sometimes its problems. Davy did his best to maintain the dignity of his predecessor, the wealthy landowner and voyager Sir Joseph Banks (Carter 1991; Banks 1994; Gascoigne 1994); and was anxious to make chemistry gentlemanly and attractive to the upper classes – philosophical rather than merely utilitarian.

There are also some less formal and flattering pictures of chemists, like Gillray's famous cartoon of Davy's respiration experiment, which contrasts with the decorum in Lavoisier's laboratory; here Davy is holding the bellows, Count Rumford stands by the door, and Thomas Garnett is administering the laughing gas; the audience includes women (fig. 4.4), as it generally did at the Royal Institution. This cartoon was funny but harmless, except for putting dignified men of science in a ridiculous light, but Priestley had earlier been venomously caricatured when his political and religious ideas (he was a democrat and Unitarian) had led him into enthusiastic support for the French Revolution (Fitzpatrick 1984). In general, chemists were proud of their manipulative skills, of interrogating nature with hands and mind. Chemistry could not be done in the armchair, or by the ham-handed. In the laboratory, chemists wore an apron like a butcher or a grocer (or a surgeon) rather than the modern white coat, as we can sometimes see when we have pictures of laboratories.

2.1 *Workplaces*

These indeed are our next category. Just as we have pictures (and often plans) of the interiors and exteriors of great houses and churches, so for the new laboratories and lecture theatres of the period around 1800 we find that there is a graphic record (Forgan 1986). Eighteenth-century chemistry had been carried on in kitchens, outhouses, or anywhere convenient for smelly, explosive activities, usually requiring a stove or furnace as well as some precision apparatus such as a balance, and some delicate glass and china ware. In 1804 we find an elegant linear drawing (fig. 4.5), with no attempt at perspective, in a Spanish book on applied chemistry (López Piñero 1987, p. 70). A British 'experimentalist's laboratory' of about 1820 (Mackenzie 1822, plate 1) has natural lighting coming both from the side and above, running water, bellows, and numerous shelves. The Royal Institution's

4.4 Gillray cartoon, laughing gas, 1801.

laboratory (Brande 1830, vol. 1, frontispiece) at about the same time had a more cavernous look – it was in the basement – and had gas lighting. Chemists on the move might take with them a travelling laboratory, a kind of chemistry set in a trunk, as Davy did when he went to France in 1813, to be set up somewhere convenient. By the 1870s, the famous survey ship *HMS Challenger* had a trim little laboratory constructed on board (fig. 4.6) for use during her scientific circumnavigation (Thomson 1887, vol. 1, p. 19).

All these are shown empty. Although Davy worked in public on some of his researches (Golinski 1992, p. 221), having a bank of seats fixed up to one side of the laboratory, this seems to have been very unusual; in general, experiments were done in public only for the purposes of demonstration, that is, in support of a lecture. The element of surprise involved in discovery was therefore absent; though a good deal of contrivance must have been necessary to ensure that the experiments worked when done before an audience.

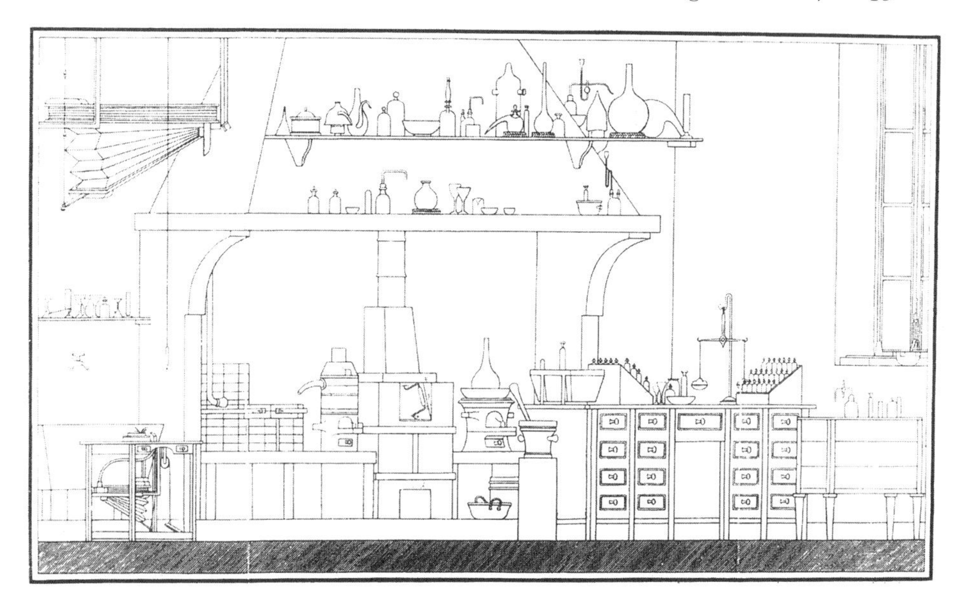

4.5 A laboratory, 1804 (J.M. López Piñero 1987).

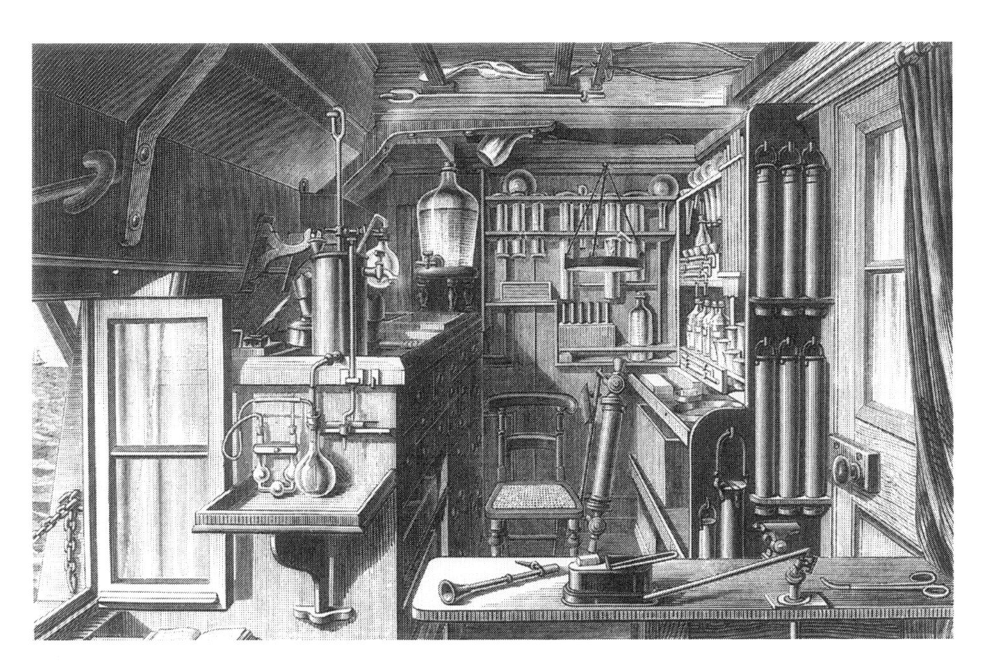

4.6 Laboratory aboard *HMS Challenger* (C.W. Thomson 1887).

Chemical lectures were given to large and enthusiastic audiences in the early years of the nineteenth century; for example, at the Royal Institution by Davy. Michael Faraday continued this tradition and began the famous series of Christmas Lectures for children. Their audiences were drawn from the upper ranks of society, who expected to be entertained as well as instructed. Knowledge of chemistry in its revolutionary decades was appropriate to an educated person; but such knowledge did not need to be detailed. Aristotle after all had written in his biological works that a gentleman should know the principles of sciences, not the details. But medical students increasingly needed to attend formal courses on chemistry, and pass examinations at the end of them. They formed a different kind of audience for a different kind of lecture.

In addition, they undertook courses of practical chemistry in the laboratory. Chemistry was the first science to be taught in this way. At first, such instruction was an optional extra, available, for example, in the University of Durham in the 1830s on payment of an additional fee. But following the example of Justus Liebig[1] at the previously obscure University of Gieseen, where he trained a great number of those who became professors of chemistry elsewhere in Germany (Farrar 1975), universities everywhere began to make laboratory training compulsory; for example, University College, London, had purpose-built teaching laboratories (fig. 4.7) in 1846.

2.2 *Printed Illustrations*

Textbooks for those involved in such courses were in general copiously illustrated; and the illustrations were often plagiarized. They show apparatus and indicate how it is to be used; one of the great features of chemistry during the nineteenth century was that the samples used became smaller and smaller as techniques became more sensitive and accurate. Chemical manipulation became neater; and chemists were trained in the making and use of their own apparatus. A classic work here was Faraday's *Chemical Manipulation* (1827); unlike most textbooks, this did not instruct students in the principles of chemistry and the properties of things, but only in how to perform experiments. Faraday was himself very skilful and enjoyed working with his hands; one of his great triumphs was the first isolation of benzene, by fractional distillation of whale oil in a glass tube bent into a zigzag. His careful instructions on weighing, grinding, or triturating, and on glass-working,

4.7 Teaching laboratory, University College, London (N. Harte and J. North 1978).

accompanied by illustrations, give a splendid route into the chemistry of his time. The book (Faraday 1827) is illustrated with little line drawings which help the reader understand the instructions – showing, for example, how to hold a finger over the top of a test-tube to block the escape of volatile substances – but to our eyes is prosy. Pictures were only used when description was not clear enough, making the book look solid rather than attractive – though the manipulative skills would still be useful to the chemist today.

Lavoisier's great book of 1789 (Lavoisier 1965) was illustrated with copperplate engravings. Here the picture is cut into the copper (in *intaglio*), and to print it the plate must be inked and then wiped; ink remains in the grooves, and under high pressure damped paper will pick up a very clear image. Engraving with a burin produces a hard line, and shading must be done with cross-hatching. It was a matter of great skill: high-quality full-page scientific engravings might cost twenty guineas each; an enormous sum in 1800. Contemporaries compared engraving from a drawing to translating from a rich language to a poorer one. An alternative was to etch the copper with acids under careful control. This, combined with processes such as aquatinting, could give a softer effect; and was somewhat easier to do. But copper was expensive, and also rather soft, so that hundreds of impressions rather than thousands could be expected; and it was printed differently from the metal type in use until very recent years (Twyman 1970a). The type stood up from the surface in relief; type and copper plates, hence, needed different kinds of printing presses, and ideally rather different paper. Engravings therefore were printed separately from type; illustrations and text were thus kept apart.

Engravings were gradually superseded in the early nineteenth century by lithographs, which were much cheaper; a drawing is made on suitable stone in wax crayon; the stone is wetted, and a greasy ink then applied, which sticks to the waxy but not the wet parts of the stone, so that a print can be taken off. Here the crayon is pulled across the stone, giving a much more flowing line than a pushed burin; and some artists proved capable of drawing directly on the stone, and thus cutting out the craftsman or 'translator' – this was especially important in natural history illustration (Rudwick 1992). It was more difficult to get the fine detail achieved by the best engravers, such as J. Basire, who worked for the Royal Society's *Philosophical Transactions*; but by the 1830s, copper engraving for scientific purposes was in decline. Lithographs again could not be printed with text; because the image is reversed, any writing on them must moreover be done in 'mirror' form.

Better for chemical textbooks were wood engravings done on the end-grain of boxwood, which is very hard; like type, the image stood up from the block, and these illustrations could be printed amidst the text to which they referred.[2] This technique was perfected by Thomas Bewick, who used it for natural history subjects and for witty tailpieces; and it rapidly proved valuable for chemical subjects too. Blocks, or copies of them, were often reused in other publications; metal castings from woodblocks, called *clichés*, were made for big editions, as pages of type were cast or *stereotyped* – these terms having entered our language.

Popular works on chemistry, chemical textbooks and journals all included illustrations of apparatus. Two handsome books on scientific illustration through the ages, from cavemen to the present, have recently come out (Ford 1992; and Robin 1992); but unfortunately they include hardly any chemical illustrations of the nineteenth century – that is one of the problems of dealing with millennia rather than genuine historical periods. When we do look at chemical textbooks and other compendia, we often find illustrations elegantly shaded to indicate three dimensions: this was no doubt partly for aesthetic reasons, but also reflects the fact that in the early years of the nineteenth century readers could not be expected to be familiar with the exact shapes of pieces of apparatus. A shaded drawing conveys more information to the uninitiated. By the end of the century, representations of apparatus had become diagrammatic in the way familiar to us. By then, readers knew what a conical flask, a funnel, or a test-tube was like; they did not need to be informed about its shape in visual language, where shading and realism would distract from the scientific information presented. Nor did they need to be shown how the different components were fixed together; that was now standard. A richer pictorial language had given way to something more like geometry.

Chemistry was not only a laboratory science (Brock 1993, ch. 8) but was also increasingly useful in industry. Pictures of chemical processes go back to the Renaissance, with mining and metallurgy; to us, and no doubt to contemporary readers, the illustrations are extremely helpful for understanding the text. By the nineteenth century, formal science was beginning to replace the organized common sense which had on the whole guided the pioneers of the Industrial Revolution. The new pattern is shown in Davy's safety lamp for coalminers (fig. 4.8); he had samples of the explosive gas found in mines sent to his laboratory, analysed them, found that the methane was only ignited at a high temperature, and devised various lamps (Davy 1839–40, vol. 6, plates 1 and 2) in which the flame was separated from the surrounding atmosphere by metallic tubes or gauze which would dissipate heat. A device evolved in the laboratory duly proved applicable down the mine: and in the light of this new lamp, the relationship between science and technology looked different. The idea of 'applied science' became increasingly popular as the century wore on; and the synthetic dye industry proved the economic importance of the latest chemistry.

Illustrated chemical books may include pictures of gasworks and of steam engines; the older tradition of science as the opposite of mere traditional practice ('rule of thumb') (Bud and Roberts 1984) continued to be extremely important in, for example, agriculture and in much of the 'heavy chemical' industry, producing sulphuric acid and alkalis: here organized common sense, and reasoning by analogy, were more use than new theory. There are many illustrations of chemical works (fig. 4.9) and the processes which went on in them that cast a

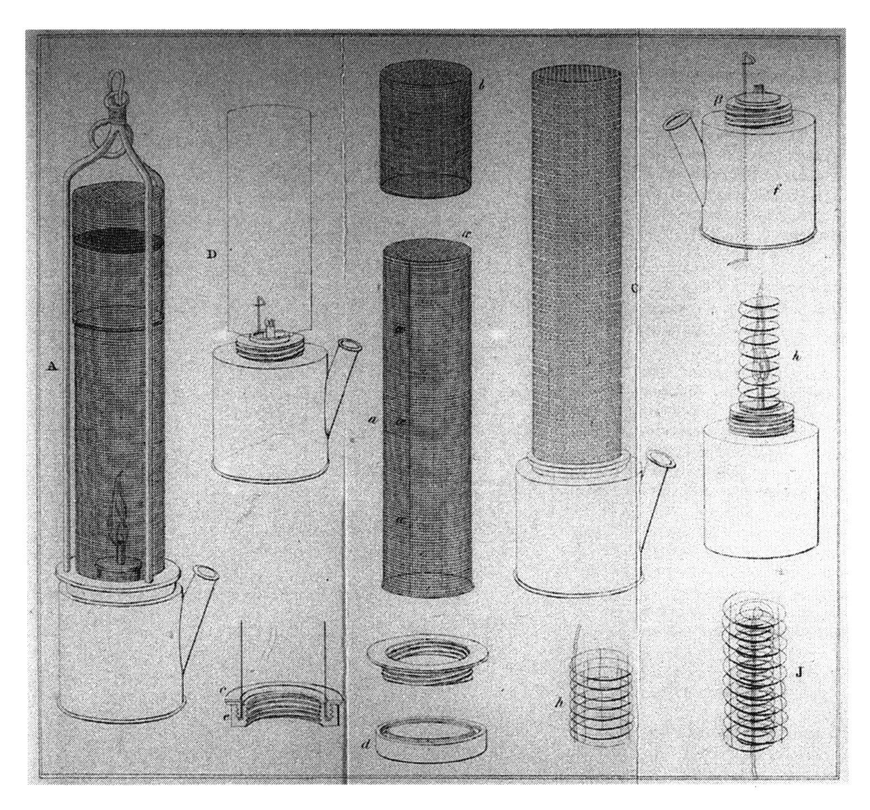

4.8 Davy's safety lamps (H. Davy 1839-40).

fascinating light upon their period; notably in encyclopaedic works (Mackenzie 1822; Vincent n.d.) dealing with chemistry. The nineteenth century's appetite for facts is always astonishing; and they were often well illustrated as well as described. We could call lithography a chemical process, depending on the fact that oil and water will not mix; but more up-to-date chemistry (see C. Roberts 1992, Schaaf 1992, and Rothermel 1993), involving Davy in some early experiments, lay behind photography.

2.3 *Photographs*

Photography was an important medium for illustration by the middle of the nineteenth century; and it found increasing use in portraits. In oils, these took many hours of sitting and were very expensive, so that

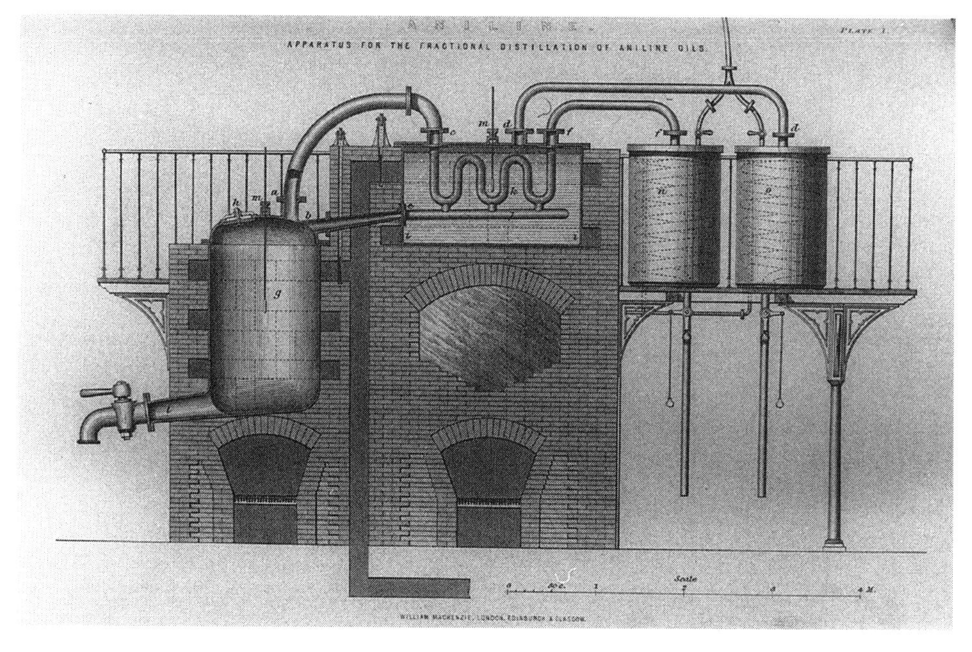

4.9 An aniline dye works, 1880s (C.W. Vincent n.d.).

only eminent people have had their portrait painted; an informal sketch, or a silhouette, was an inexpensive, or unshowy, alternative, but perhaps less revealing than a studied picture. Photographs were cheap, and thus we know much better what chemists of the later nineteenth century looked like. Long exposures were required, and the sitters always look rather stiffly posed by our standards; early photographs were also generally done in a studio, so that chemists look like anybody else: their photographs do not contain any pieces of apparatus, and they are wearing a sober attire. The exchange of photographs became a feature of correspondence between chemists; just as offprints of published articles were rapidly circulated around the chemical community, so were likenesses such as that of the great Italian, Stanislao Cannizzaro (fig. 4.10). The coming of railways and steamships meant that chemists from different countries met each other more; but letters (and photographs) remained important, and group photographs are an interesting feature of conferences and graduation days from the later years of the century. By then we also have some photographs of scientists in their laboratory.

4.10 Photograph of Cannizzaro, 1898 ('Memorial Lectures,' *Chemical Society* 1914).

Photography was not very much used for illustrating apparatus. Modern museum catalogues use it; but for standard apparatus, the same objections applied to its use as to carefully shaded engravings. The photograph displays a particular piece of apparatus in a particular setting; and for most scientific purposes something more diagrammatic is more useful. Unless they were worked up as wood engravings, it was difficult also to print photographs with type. Photography came into its own for the recording of some chemical results, notably, spectra. These represented the new phenomenon of chemists using physical data; spectroscopic observations seemed quite different from the traditional methods of analysis, involving test-tubes and reagents – spectroscopy could be done without getting the hands dirty, and the spectrum when photographed could be studied at leisure.

New techniques were a feature of the chemistry of this period as they are generally in experimental science; and we can follow the changes most easily in pictures. The coming of Bunsen's burner made spectroscopy possible, or at any rate convenient, because the chemist now had a source of intense heat readily available; before, there had been furnaces or rather weak spirit lamps. Standard configurations of glassware also came in: for example 'Kipps' apparatus' involving three connected vessels mounted on top of each other, with a tap from which a flow of hydrogen sulphide could be easily obtained for use in analyses. There had been few safety measures in use in laboratories early in the century, and as a result many eminent (and no doubt ordinary) chemists suffered injuries; for chemistry was a science of bangs as well as stinks. It was also a science of poisons; and with safety goggles there also came in fume cupboards, with a flue behind and a sash window in front: work with hydrogen sulphide, chlorine, and other noxious substances could be carried on behind the almost-shut window. Most dangers were discovered the hard way, by trial and error: gradually, chemists came to treat substances with more respect as the science moved from the heroic early days into its classical period as a mature science by the mid-nineteenth century.

Depictions of chemical experiments towards the end of the century were often in the striking form of white-on-black illustrations, done by a process akin to the blueprints used by engineers (Baynes and Pugh 1981). We see these, for example, in William Crookes's early papers on the cathode rays. Some of these (fig. 4.11) appeared in his *Chemical News* (from 1860) or in Norman Lockyer's *Nature* (from 1869), journals in magazine format appearing frequently and keeping a large circle of

Radiant Matter and "Radiant Electrode Matter."

In recording my investigations on the subject of radiant matter and the state of gaseous residues in high vacua under electrical strain, I must refer to certain attacks on the views I have propounded. The most important of

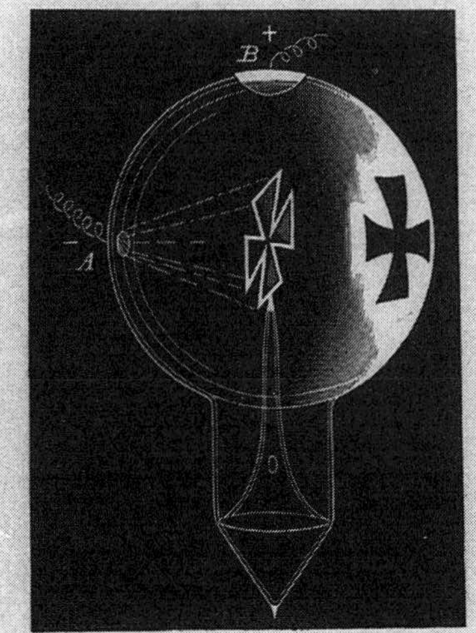

Fig. 20.—P. = 0·00068 m.m., or 0·9 M.

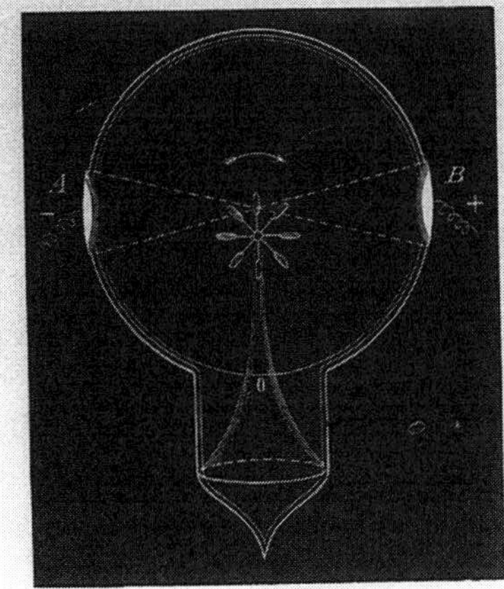

Fig. 21.—P. = 0·001 m.m., or 1·3 M.

these questionings are contained in a volume of "Physical Memoirs," selected and translated from foreign sources under the direction of the Physical Society (vol. i., Part 2). This volume contains two memoirs, one by Hittorff on the "Conduction of Electricity in Gases," and the other by Puluj on "Radiant Electrode Matter and the So-called Fourth State." Dr. Puluj's paper concerns me most, as the author has set himself vigorously to the task of

Fig. 22. P. = 0·000076 m.m., or 0·1 M.

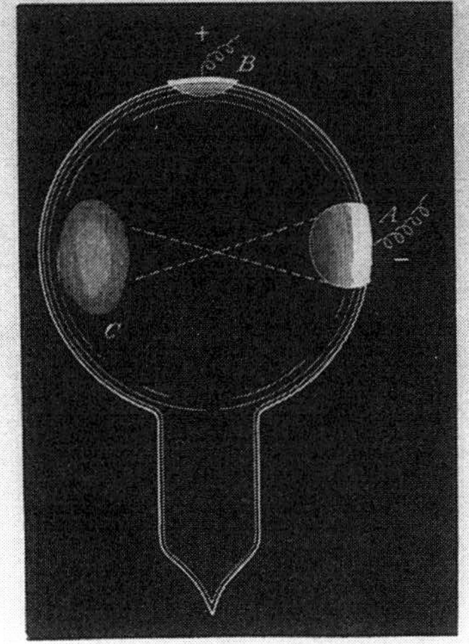

Fig. 23.—P. = 0·00068 m.m., or 0·9 M.

opposing my conclusions. Apart from my desire to keep controversial matter out of an address of this sort, time would not permit me to discuss the points raised by my

4.11 Cathode rays illustrated (W. Crookes 1891).

readers right up to date. These differed from the heavyweight journals, mostly published by scientific societies or academies, in which research papers would appear only after careful scrutiny by a referee who was supposed to repeat any experiments and check calculations. In these publications, addressed to the *cognoscenti*, illustrations were schematic; they might even be intended to be occult (Ford 1992, p. 97), in that they would be incomprehensible to the uninitiated reader, and thus help bond the chemical community together.

3. CHEMICAL THEORY IN VISUAL LANGUAGE

3.1 The Copperplate Era

We have looked at the outward face of chemical illustrations, how they were reproduced and what impression of chemistry they might convey to outsiders: about the status of chemists, about their workplaces, and the tools of their trade. The interest of such readers might be desultory; their enthusiasm needed to be aroused if it was to become serious. The time has now come to turn inwards, as it were, and examine in more detail how changes in chemical knowledge affected what was represented, as the science moved from poetry to algebra by way of arithmetic in Lavoisier's (1965) calculations of weights, and geometry in the straightforward depiction of people, places, and apparatus. Some of Lavoisier's plates show complicated pieces of equipment, such as his ice calorimeter for experiments on heat changes; but others have unrelated objects squeezed in to get as many pictures on the page as possible. We get a strong feeling of the form of the vessels, some of which are cut away to show what is inside. Around the edge are depicted some important parts magnified, as in a botanical illustration. We can also follow in the plates the crucial experiments which confirmed the role of oxygen, rather than the hypothetical phlogiston, in burning. The illustrations, with every tap and joint displayed, gave enough information for Lavoisier's experiments to be repeated exactly; and the idea was simply that anybody doing so would get the same results and come to the same conclusions. The visual language, accompanying a text in plain prose, is thus a part of making chemistry public knowledge.

In the fifth year of the Republic, 1796, two years after Lavoisier's death, the third edition of J.A. Chaptal's *Elemens de chimie* was published; in contrast to Lavoisier's book, this was unillustrated. The book contains a mass of information, in three stout volumes; it was a textbook, by a professor at Montpellier interested in the chemical industry but writing for medical students; it was not meant to be controversial. The pages of unrelieved text make, and must have made, it rather unattractive. William Nicholson's *First Principles of Chemistry* (1796) also had a third edition in that year; this has a single illustration in the form of a fold-out frontispiece showing apparatus and furnaces, some cut away. Nicholson was careful to eschew theory as far as possible; the plate makes the descriptions easier to understand, but makes no theoretical point. By the early years of the nineteenth century, illustration came to be more prominent in chemical books, making them more agreeable.

Fredrick Accum had close connections with the Royal Institution in London; his *System of Theoretical and Practical Chemistry*, of which the second edition came out in 1807, was said on its title-page to be 'with plates' – and indeed, it has seven, all of them showing several pieces of apparatus. It also has a list of 'Apparatus and Instruments sold by Messrs. Accum and Co.,' which ends with: 'and every other instrument made use of in Experimental Chemistry.' The book, dedicated to the managers of the Royal Institution, was intended to teach practical chemistry as well as to promote the author's apparatus-selling business. Some of the plates show disembodied hands holding instruments; clamps and stands are also evident. With the book as guide, anybody could assemble complex arrangements of apparatus. He published another little book, *Chemical Reagents or Tests* (1828), for those analysing waters, earths, soils, ores, or alloys; this was revised by W. Maugham in 1828, and its frontispiece shows a balance, a pestle and mortar, and standard pieces of chemical apparatus.

In Accum's writings, theory is not prominent; but like Lavoisier, he wanted his readers to know exactly how to carry out the experiments he described. He was also a great promoter of the new coal-gas industry, and his *Practical Treatise on Gas-Light* (1815) is a splendid piece of promotion, with fine hand-coloured plates. Similarly Colin Mackenzie (1822) has a magnificent coloured frontispiece to his *Thousand Experiments in Chemistry*, showing heroic figures manning the coke ovens: artists were not always made to think of Hell, or of dark Satanic mills, when confronted by innovation in chemical trades. Mackenzie's other plates show not only chemical apparatus, but also an early steam locomotive and some industrial plants.

The surgeon James Parkinson, from whose observations Parkinson's disease is named, published *The Chemical Pocket-Book* (1801) for the 'professional student,' that is, a medical practitioner, or for anyone who wanted a general knowledge of chemistry. His frontispiece shows apparatus on stands, elegantly framed by pillars and a curtain; but it also shows the chemical symbols of the Frenchmen Hassenfranz and Adet. Their zigzags, curves, circles, squares, and triangles never came into general use – not even within Parkinson's book – but they, like the tabulated chemical properties at the back of the book, show an intention to go beyond straightforward depiction. The industrial chemist Samuel Parkes's *Chemical Catechism*, which reached its third edition in 1808, also boasts a 'theatrical' engraving on its title-page, but this shows coal gas being prepared and used to light a stage: that is, a chemical process being carried on. His frontispiece in this edition (others showed a laboratory) was made by etching glass with fluoric acid, and is thus itself a chemical curiosity; but it shows the usual collection of apparatus, looking rather more delicate and linear than in the engravings in other books.

Parkes's book was an elementary one, though it has much information in its extremely copious notes forming at least half of every page; it was directed to practical men. Those whose interest was more intellectual, and presumably women readers, bought Jane Marcet's contemporary *Conversations on Chemistry* (1825), which turned the young Faraday towards experimental science: like Parkes's, this was an extremely successful work, going through many editions. It has a series of plates spread through the two volumes; in which two very bright girls, Emily and Caroline, rapidly learn the science from their governess, Mrs B. The illustrations came to include safety lamps and steam engines as well as what we normally think of as chemical subjects: the boundaries of sciences after all are socially constructed, and vary over time. The illustrations are linear, and not nearly as shaded as those of Accum, which gives less feeling of solidity and less indication of what materials things are made of; there are sometimes hands holding things. The plates, like the book, emphasize chemistry as an intellectual accomplishment rather than a professional activity; and must have made it seem attractive. Jane Marcet, the wife of a prominent doctor, hoped that it would be used in conjunction with lectures such as those Davy gave at the Royal Institution.

Davy's own first book, on nitrous oxide (1839–40, vol. 3), had a frontispiece showing his gas-holder (a small ancestor of those now holding natural gas for towns) and breathing apparatus, in the design of which James Watt had been involved. Much of the book is indeed taken up with reports, Davy's own especially vivid, of the subjective effects of laughing gas; but there are also analyses in the tradition of Joseph Priestley, and Lavoisier, which by 1800 did not need illustration because the procedures were familiar to experts. The success of these researches took him to London, where some of his lecture courses were on agricultural chemistry; these were published (Davy 1839–40, vols. 7–8) in 1813 as a handsome quarto volume. As well as showing apparatus for soil analysis, this volume reproduced engravings of sections through trees which were copied from Nehemiah Grew's *Anatomy of Plants* of 1682; microscopes had been improved so little in the intervening years that the plates were not out of date. It also contains a fold-out plate of an imaginary landscape, illustrating geological formations, which is an unusual feature of a chemistry book. Davy's *Elements of Chemical Philosophy* of 1812 (see Davy 1839–40, vol. 4), by way of contrast, had plates of chemical apparatus comfortably spaced on the page and in a clear but unshaded style.

With John Dalton's *New System of Chemical Philosophy* (1808–11), we find illustrations of hypothetical atomic arrangements, also as copper-plate engravings (fig. 4.12). Because Dalton had nothing to guide him to the absolute numbers of atoms composing different compounds, he could only rely upon the assumption that they will have the simplest formulae: water, for example, will have one atom of hydrogen and one of oxygen in its molecule. We have thus come a long way from pictures of something into a kind of abstract geometry. Although Dalton knew that atomic structures must be three-dimensional, he drew them out as though they were flat; and it was to be many years before theoretical chemists took three dimensions seriously. When they did, they needed models (commercially available by 1867) to play with in addition to pencils and paper.

Some of Dalton's contemporaries tried to understand crystal structure in terms of arrangements of particles. In the journal Lavoisier founded, *Annales de chimie* (Crosland 1994), Rene Haüy (1793) expounded his view that crystals were composed of 'unit cells'; this enabled him to classify them and understand their planes of cleavage, along which they easily break. His illustrations look more like geometry than chemistry. William Hyde Wollaston (1813) invented an optical instrument, the goniometer, for measuring the angles of crystals; and tried building crystal forms up from atoms which were prolate or oblate spheroids – slightly elongated or flattened: his models still survive in the Science Museum in London, and his illustration in the Royal Society's *Philosophical Trans-*

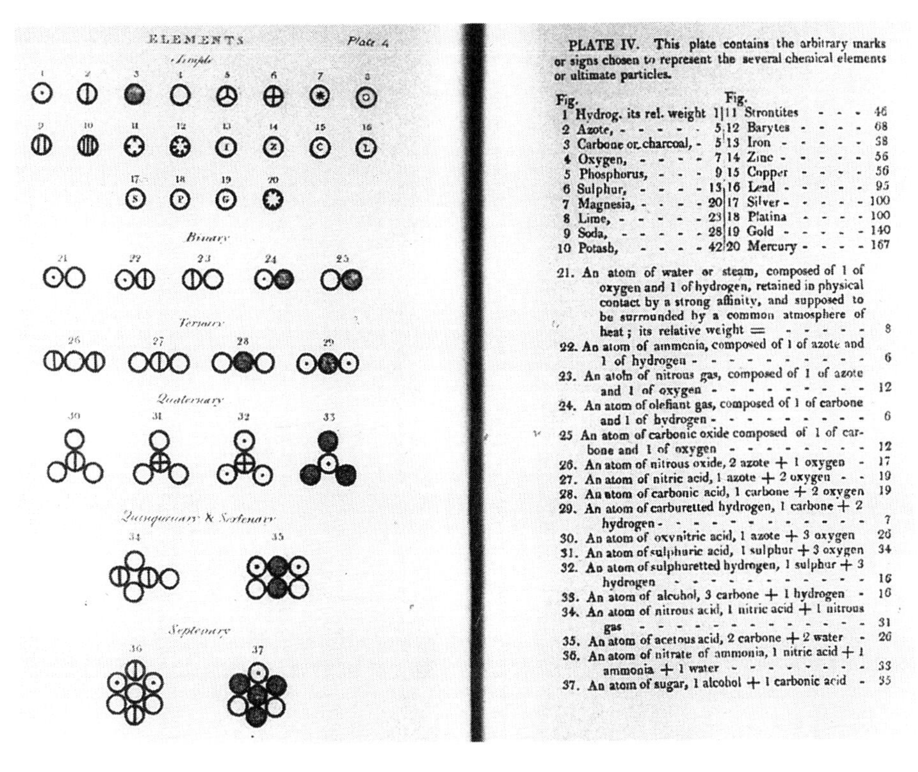

PLATE IV. This plate contains the arbitrary marks or signs chosen to represent the several chemical elements or ultimate particles.

Fig.			Fig.		
1	Hydrog. its rel. weight	1	11	Strontites - - -	46
2	Azote, - - - - -	5	12	Barytes - - - -	68
3	Carbone or charcoal, -	5	13	Iron - - - - -	38
4	Oxygen, - - - -	7	14	Zinc - - - - -	56
5	Phosphorus, - - -	9	15	Copper - - - -	56
6	Sulphur, - - - -	13	16	Lead - - - - -	95
7	Magnesia, - - - -	20	17	Silver - - - - -	100
8	Lime, - - - - -	23	18	Platina - - - -	100
9	Soda, - - - - -	28	19	Gold - - - - -	140
10	Potash, - - - - -	42	20	Mercury - - - -	167

21. An atom of water or steam, composed of 1 of oxygen and 1 of hydrogen, retained in physical contact by a strong affinity, and supposed to be surrounded by a common atmosphere of heat; its relative weight = - - - - - - 8
22. An atom of ammonia, composed of 1 of azote and 1 of hydrogen - - - - - - - - - - - 6
23. An atom of nitrous gas, composed of 1 of azote and 1 of oxygen - - - - - - - - - - 12
24. An atom of olefiant gas, composed of 1 of carbone and 1 of hydrogen - - - - - - - - - - 6
25 An atom of carbonic oxide composed of 1 of carbone and 1 of oxygen - - - - - - - - - 12
26. An atom of nitrous oxide, 2 azote + 1 oxygen - 17
27. An atom of nitric acid, 1 azote + 2 oxygen - - 19
28. An atom of carbonic acid, 1 carbone + 2 oxygen 19
29. An atom of carburetted hydrogen, 1 carbone + 2 hydrogen - - - - - - - - - - - - - 7
30. An atom of oxynitric acid, 1 azote + 3 oxygen 26
31. An atom of sulphuric acid, 1 sulphur + 3 oxygen 34
32. An atom of sulphuretted hydrogen, 1 sulphur + 3 hydrogen - - - - - - - - - - - - - 16
33. An atom of alcohol, 3 carbone + 1 hydrogen - 16
34. An atom of nitrous acid, 1 nitric acid + 1 nitrous gas - - - - - - - - - - - - - - 31
35. An atom of acetous acid, 2 carbone + 2 water - 26
36. An atom of nitrate of ammonia, 1 nitric acid + 1 ammonia + 1 water - - - - - - - - - 33
37. An atom of sugar, 1 alcohol + 1 carbonic acid - 35

4.12 Dalton's atomic symbols (J. Dalton 1808-11).

actions shows them, rather than something hypothetical. But crystallography took a more abstract direction, following ideas of the polymath William Whewell (Fisch and Schaffer 1991, p. 100), and in W.H. Miller's *Treatise* of 1839 we are back with a highly abstract geometrical analysis, in terms of what have become known as 'Miller Indices.' By 1858, in Greg and Lettsom's text, we have both geometry and chemistry (fig. 4.13).

Plates were occasionally supplemented or replaced by actual specimens. This happened in botany, where a pressed flower or other plant could be better than an illustration. In the journal *Records of General Science*, the eminent chemist Thomas Thomson wrote in 1835 a long article on calico-printing. This was before the first synthetic dyes. To make his points, small pieces of cloth (about 8 cm by 4 cm) dyed with the various colouring matters, in bright and attractive patterns, were stuck onto the pages where their preparation is described. They

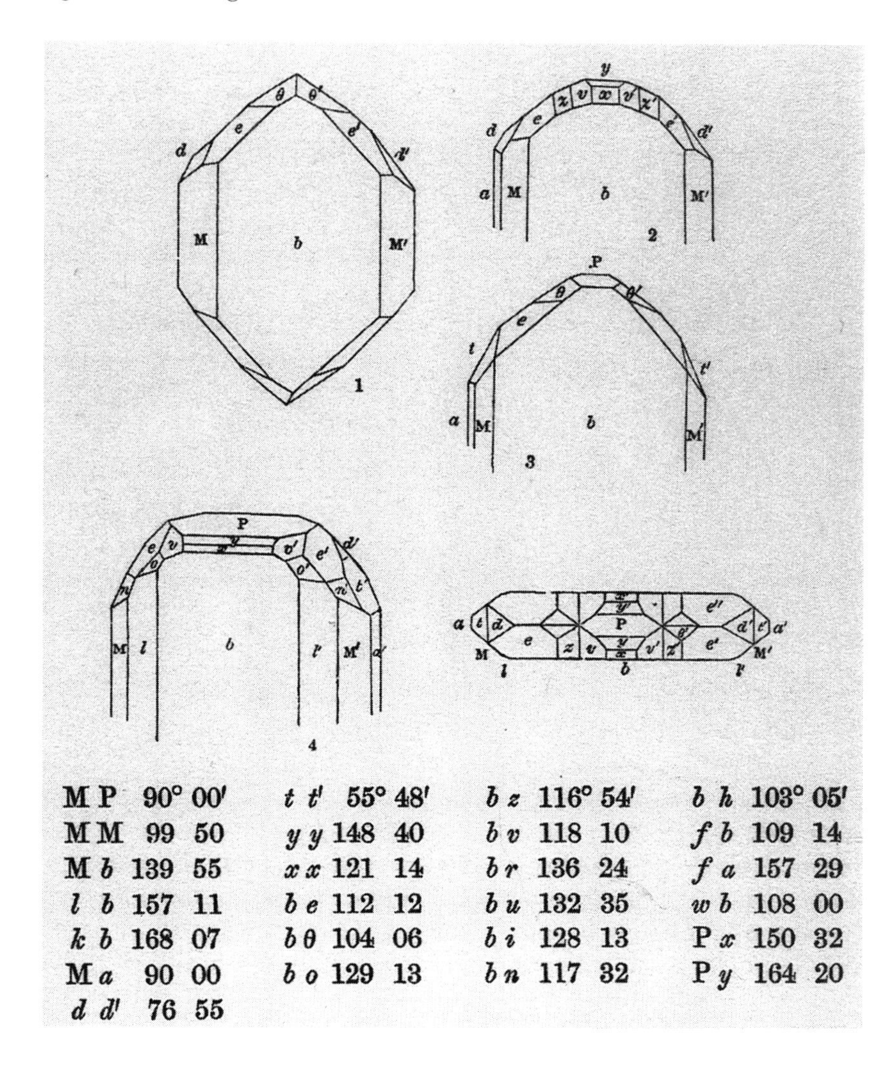

M P	90° 00′	*t t*′	55° 48′	*b z*	116° 54′	*b h*	103° 05′
M M	99 50	*y y*	148 40	*b v*	118 10	*f b*	109 14
M *b*	139 55	*x x*	121 14	*b r*	136 24	*f a*	157 29
l b	157 11	*b e*	112 12	*b u*	132 35	*w b*	108 00
k b	168 07	*b θ*	104 06	*b i*	128 13	P *x*	150 32
M *a*	90 00	*b o*	129 13	*b n*	117 32	P *y*	164 20
d d′	76 55						

4.13 Crystal forms and chemical formulae (R.P. Greg and W.G. Lettsom 1858).

remain brilliant enough for the most part to make us realize that vivid colours were available to our ancestors at the beginning of Queen Victoria's reign; and that chemistry and dyeing were connected long before W.H. Perkin and his successors began their work on coal tar and its derivatives.

3.2 Text Illustrations

Greg and Lettsom's illustrations, and Thomson's scraps of cloth, were scattered liberally through the text, much more conveniently than engravings on copper could be. This practice had been taken up in the next generation of textbooks. One of the most handsome of these was W.T. Brande's *Manual of Chemistry* (1830) with numerous figures in the text. He was Davy's successor, and Faraday's predecessor, at the Royal Institution; and while undistinguished in research, he taught medical students, who from 1815 were required to take formal classes in chemistry. His book was clear and has many tables as well as pictures: it also has some diagrams illustrating chemical affinity, and chemical reactions; thus on page 24 we find barium nitrate and sodium sulphate interacting thus:

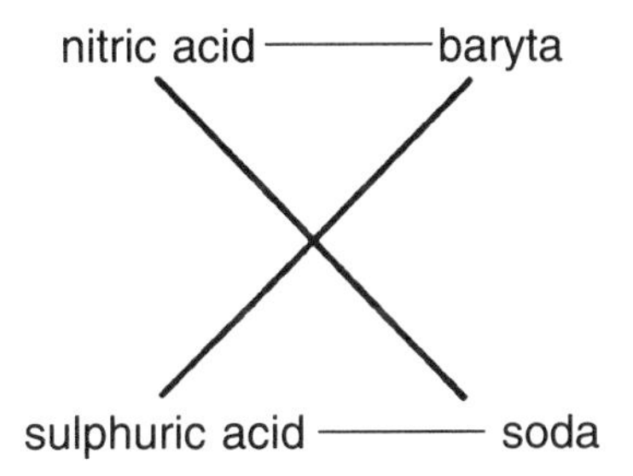

This indicates the course of the reaction, and that Brande (like his contemporaries) saw salts as composed of two halves, an acidic and basic component, rather than as a unified molecule. In this particular reaction, a double decomposition, the acids change partners. The reaction could also be set out with a different geometry, the different compounds being arranged along the sides of a square. These were not original with Brande, but he drew attention to such notations.

Other textbooks of the day are not unlike Brande's in their use of illustrations, though they differ in their emphases upon different branches of chemistry. Thus Edward Turner of University College,

London, was an analytical chemist, whereas his colleague at King's College, J.F. Daniell, was a disciple of Faraday, whose interests were in electrochemistry. Their books reflect this (Turner 1831; Daniell 1839); but the illustrations, in which Daniell's is much richer, are descriptive. Thomas Thomson's very successful textbook (1831; which had even had the accolade of a French translation) was rather skimpy in its illustration; while part of the success of the Irishman Robert Kane's text was his use of illustrations and tables. His book also used (fig. 4.14) the more algebraic symbols for the elements (C for an atom of carbon, H for hydrogen, Fe for iron, etc.), which had been proposed earlier by J.J. Berzelius and gradually came into general use. Though it was not until after 1860 that there was general agreement on the formulae even of compounds such as water, which might be HO or H_2O, if an author used symbols consistently his reasoning was easier to follow, and it was economical of space. Kane was nevertheless chary of actually writing chemical equations using the symbols. It was not until 1860 and the Karlsruhe Conference of chemists that agreed formulae became possible, and chemistry books began to have their modern appearance.

Copious use of equations had nevertheless already begun with Auguste Laurent and his book *Chemical Method* (1855); and he also has some hexagons which give a look of great modernity to the book, but it is spurious, for they do not stand for benzene rings. Because Laurent believed that the structures of molecules could never be inferred inductively from analyses, he proposed a deductive approach to the science: chemists should think of a structure, work out its consequences, and test them experimentally. This was the approach followed by Auguste Kekulé and his school, and it led through benzene rings to the three-dimensional structures of J.H. van't Hoff, notably in his book *The Arrangement of Atoms in Space.* He believed that round each carbon atom four other atoms or groups would be arranged in tetrahedral configuration; perceived that a consequence of this would be that some molecules would be asymmetrical; and found that it was so. They formed two sorts of crystals which were mirror images of one another, like our left and right hands; and in solution, they rotated the plane of polarized light to left or to right. To show three-dimensional structures required the use of quite complicated geometrical drawings; which in turn led to more understanding of how molecules actually react with one another. We can see some of this in the syntheses described by Carl Schorlemmer (1894, pp. 252–63) in his book on organic chemistry.

Van't Hoff also modified the way chemical equations were written, by introducing a double arrow in place of an equal sign for those reactions

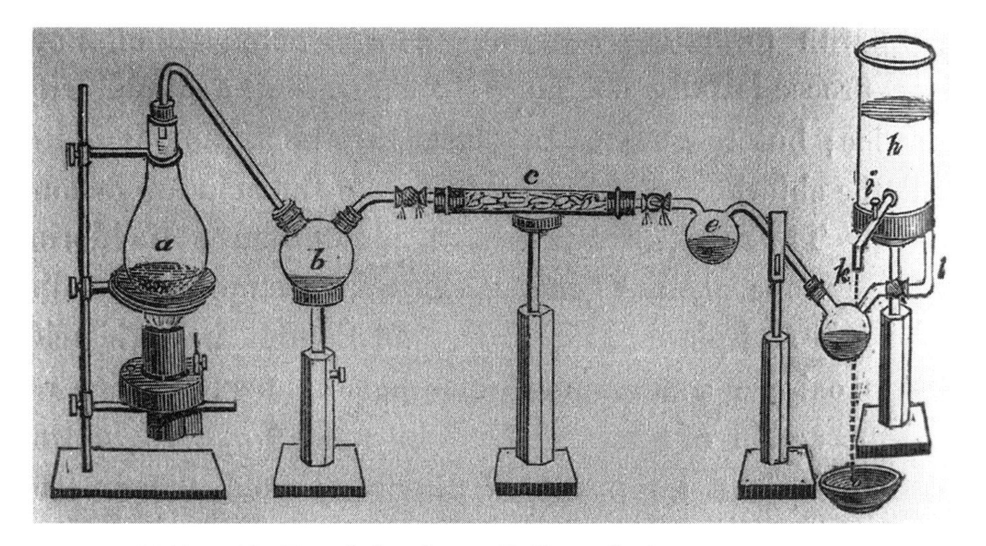

4.14 Making chlorides of phosphorus (R. Kane 1842).

which are reversible – under different conditions, they go either way, and at a particular temperature and pressure there is probably a dynamical equilibrium. He was a founder of 'physical chemistry'; and among contemporary chemists such as Thomas Andrews (1889, p. 465) we find, as well as rather splendidly detailed plates of apparatus, graphs being used to demonstrate what is happening in some processes. This had been practised in physics and other sciences for a long time (Tufte 1983), but was rather new in chemistry.

Tables of affinities or of elements had a long history in chemical writings, but the Periodic Table of Dmitri Mendeléeff (1897, p. xvi), in which the chemical elements were set out in their families and groups, transformed the science. Versions of this table have decorated chemistry lecture rooms for over a century; and it proved a wonderful way of condensing an enormous amount of information. Elements like sodium and potassium, which come in the same column, behave very similarly; and we can infer that others, like rubidium, in their column or family will be very like them. As well as vertical relationships, horizontal and diagonal neighbours display some similarities. Mendeléeff perceived that there were some 'gaps' in the table, and made detailed predictions of the properties of these unknown elements: most of which turned out to be accurate, propelling him into great prominence as one of the greatest scientists of his day. Intelligent use of the Periodic Table spared the student from having to memorize a vast number of brute facts.

At the end of the nineteenth century, then, we still find some purely descriptive illustration showing instruments and apparatus; but as new equipment became more familiar, it could be depicted more schematically, and by 1900 apparatus was not usually shown in carefully shaded pictures indicating three dimensions. We also find tables and graphs; equations and molecular structures; and spectra, sometimes reproduced in colour because these were now an important part of chemical analysis. Chemistry was by then the science most taught: its central position made it important for engineers and doctors as well as pharmacists and dye-makers. There were many textbooks, all much alike in appearance, expounding a fairly standard syllabus.

In the twentieth century, the Periodic Table was explained in terms of the nuclear atom of Ernest Rutherford and Niels Bohr, while the older electrochemistry was replaced by an electronic understanding of how atoms combine, notably with G.N. Lewis, Christopher Ingold, and Linus Pauling. This can be portrayed as a reduction of chemistry to physics (Knight 1992b, pp. 157–70), but to some chemists at least, their science remains as autonomous as architecture. The chemist is creating molecules, many of which do not exist in nature, and must obey the rules of electron sharing; but these allow scope for different building techniques, and for intuitive and inventive approaches. There is an attractive and strikingly illustrated recent account of synthetic chemistry, *The Name Game* (Nickon and Silversmith 1987), in which the making and naming of molecules of striking shapes is described. Some of the reactions look a bit like Japanese paper-folding; they result in molecules which have been given informal names like 'housone' and 'churchane' for architectural-looking structures, and others which look like sandwiches or bagels. Here the symbols for the elements in the various rings are omitted, so that it would be difficult for anybody without a chemistry degree to understand the pictures, except as abstract geometry like Miller's: but for the well-informed they are full of compressed meaning. Not exactly algebra, which the equations in effect are, they are nevertheless mathematical-looking symbols of definite power; but tamed by having been given funny informal names.

In this book, and in Mary Jo Nye's *From Chemical Philosophy to Theoretical Chemistry* (1993), we even find the artists who drew the diagrams given credit: they are Leanne M. Nickon and Glenn Dryhurst, respectively. Nye (1993, p. 260) points out that different ways of writing a formula may be appropriate in different chemical contexts. Sometimes a constitutional formula like $H_2C{=}CH{-}CH{=}CH_2$ may be what is needed, but for other purposes a fuller version showing all the bonds, or one indicating floating valence, a set of ionic or electronic formulae, a bond-angle formula, or an electron-cloud formula will be appropriate. The chemist will choose the most suitable for a particular context: some, like the carefully shaded flasks, have distractingly more content than is needed on most occasions. Often it is easiest to draw the benzene ring C_6H_6 as a plain hexagon, ⬡, for example.

We can conclude with a book (Hoffmann and Torrence 1993), really the catalogue of an exhibition, that was devised to make chemistry accessible through both text and illustrations in our own day: *Chemistry Imagined.* The main illustrations are coloured prints from collages, using fragments of chemical illustrations of all sorts chiefly from the eighteenth and nineteenth centuries. Almost all of them show at least one person: chemistry is a human activity; and most of them contain a good deal of empty space – perhaps Newton's vast ocean of undiscovered truth. There are also formulae, structures, and diagrams in the text, some of which is poetry. Here again, different ways of expressing the structure of a molecule are emphasized (pp. 24–6); camphor can be portrayed as a bare and elegant hexagonal structure that chemists would see usually in journals; it could have each of the atoms labelled as C or H; it could be three-dimensional; or like balls and rods; or a mass of interacting spheres producing a blobby three-dimensional space-filling appearance. Context will determine which is 'right,' giving the reader what is wanted.

In works on chemistry, then, we shall not find many illustrations that are a joy forever; but there is much of interest, particularly perhaps in older works which were written with a general readership in mind: for what is the use, as Alice wondered, of a book without pictures or conversations? Jane Marcet's famous text for girls, *Conversations on Chemistry* (1825), had both; and so in later editions did Davy's last book, *Consolations in Travel.* But even in serious works addressed to students or to experts, and becoming increasingly austere, the illustrations are an essential part of the message, a visual language to be read with the verbal language. For different reasons, chemistry needs illustrating as much as natural history does.

NOTES

1 W.H. Brock is writing a biography of Liebig.
2 More fully discussed in Knight 1977, chapter 1.

5. Representations of the Natural System in the Nineteenth Century[1]

ROBERT J. O'HARA

Naturalists try to arrange the species, genera, and families in each class, on what is called the Natural System. But what is meant by this system?

Charles Darwin (1859), p. 413

1. INTRODUCTION

The Natural System – the idea of the order in living diversity – is one of the great theoretical conceptions in the history of science. Although systematists – those who study the Natural System – have not always been able to agree upon 'what is meant' by this conception, they generally have agreed that the results of systematic research are best presented diagrammatically. In proposing his 'map-making' approach to systematics, for example, the British naturalist Hugh Edwin Strickland (1811–53) observed that

> the true order of affinities can only be exhibited (if at all) by a pictorial representation on a *surface*, and the time may come when our works on natural history may all be illustrated by a series of *maps* on the plan of those rude sketches which are here exhibited. (Strickland 1841, p. 192; italics in the original)

Alfred Russel Wallace (1823–1913), a promoter of Strickland's methods, also wished

> that in every systematic work each tribe and family should be illustrated by some such diagram, without which it is often impossible to tell whether two

families follow each other because the author thinks them allied, or merely because the exigencies of a consecutive series compels him so to place them. (Wallace 1856, p. 207)

And indeed the only illustration in Darwin's *Origin* was his well-known diagram of an evolutionary tree, illustrating the theoretical structure of the Natural System. 'The accompanying diagram,' he wrote, 'will aid us in understanding this rather perplexing subject' (1859, p. 116).

In a previous paper (O'Hara 1988b), I defined three periods in the history of nineteenth-century systematics. The first of these, the quinarian period (1819–40), was embodied in the writings of William Sharpe Macleay (1792–1865), Nicholas Aylward Vigors (1787–1840), and William Swainson (1789–1855). Quinarian systematists believed that two sorts of relationship – affinity and analogy – obtained among taxa, that taxa existed in natural groups of five, that circular chains of affinity connected taxa within each group of five, and that relationships of analogy obtained among taxa occupying corresponding positions in different circles of affinity. During the subsequent map-making period (1840–59), Strickland and Wallace reacted against the quinarians and argued for the exclusion of analogy and symmetry from the domain of systematics. They promoted an empirical approach to systematics which compared the reconstruction of the Natural System to the geographical surveying of an unknown territory. Finally, during the evolutionary period (1859–1901), a variety of systematists explored, to varying levels of depth, the implications that the new doctrine of evolution held for their discipline. In the present essay, I will discuss diagrams from all three of these periods, but I will focus less on their temporal development (as I did in my previous paper), and more on the representational elements which vary among them. As Mayr has rightly said (1982, p. 144), 'the most important aspect of the history of systematics is that it is, like the history of evolutionary biology, a history of concepts rather than facts.' I hope that this survey will encourage others to investigate these issues in greater detail, and that it will alter the mind of any who may still believe that the history of systematics is a history of classifications and nomenclatural technicalities.

I have selected ten diagrams to analyse here and have arranged them chronologically as figures 5.1–5.10; all of these diagrams are ornithological, but none of them appeared in my previous paper, and most are being reproduced here for the first time since the nineteenth century. Other studies which have examined systematic diagrams (most of them non-ornithological) include Wilson and Doner (1937), Voss (1952), Greene (1959), Barrett (1960), Stresemann (1975), Winsor (1976), Nelson and Platnick (1981), Stevens (1983, 1984), Reif (1983), and Gaffney (1984).

2. ELEMENTS OF THE NATURAL SYSTEM

The elements of the Natural System that I wish to consider are affinity, analogy, continuity, 'directedness' in its various forms, symmetry and predictivity, reticulation, branching, and dimensionality. The methods by which these elements of the Natural System were recognized or discovered by investigators are a fascinating but entirely separate matter, and are beyond the scope of this survey.

Affinity

Affinity is in many ways the core concept underlying the idea of the Natural System, and in the pre-evolutionary systematic literature the term 'affinity' denoted a relationship based on some sort of essential similarity. While the later and somewhat related concept of homology was rarely discussed in the purely systematic literature (homology was a relation that obtained among characters, in contrast to affinity, which obtained among taxa), discussions of affinity pervaded that literature. Vigors titled his quinarian study of bird systematics, from which figure 5.1 is taken, 'Observations on the Natural Affinities That Connect the Orders and Families of Birds' (Vigors 1824); Strickland published 'Observations upon the Affinities and Analogies of Organized Beings' (1840), and even included a 'Scale of Degrees of Generic Affinity' on his diagram of kingfisher systematics (Strickland 1841), reproduced in my previous paper (O'Hara 1988b, p. 2751). It is easy to understand how Strickland was able to compare the Natural System to a map, because the language of affinity was almost always spatial language: taxa were spoken of as being 'close' to one another, or 'far apart,' or as 'approaching' other taxa.

In the evolutionary period, affinity came to be viewed by some authors as genealogical relationship on a tree, but the traditional spatial language continued to be used in many cases, and map-like illustrations of the Natural System, depicting affinity in the old sense, often existed side-by-side with tree-like illustrations depicting genealogical affinity (compare figs. 5.8 and 5.9, in which the map-like view is represented as a cross-section of the tree). Although it is not as popular

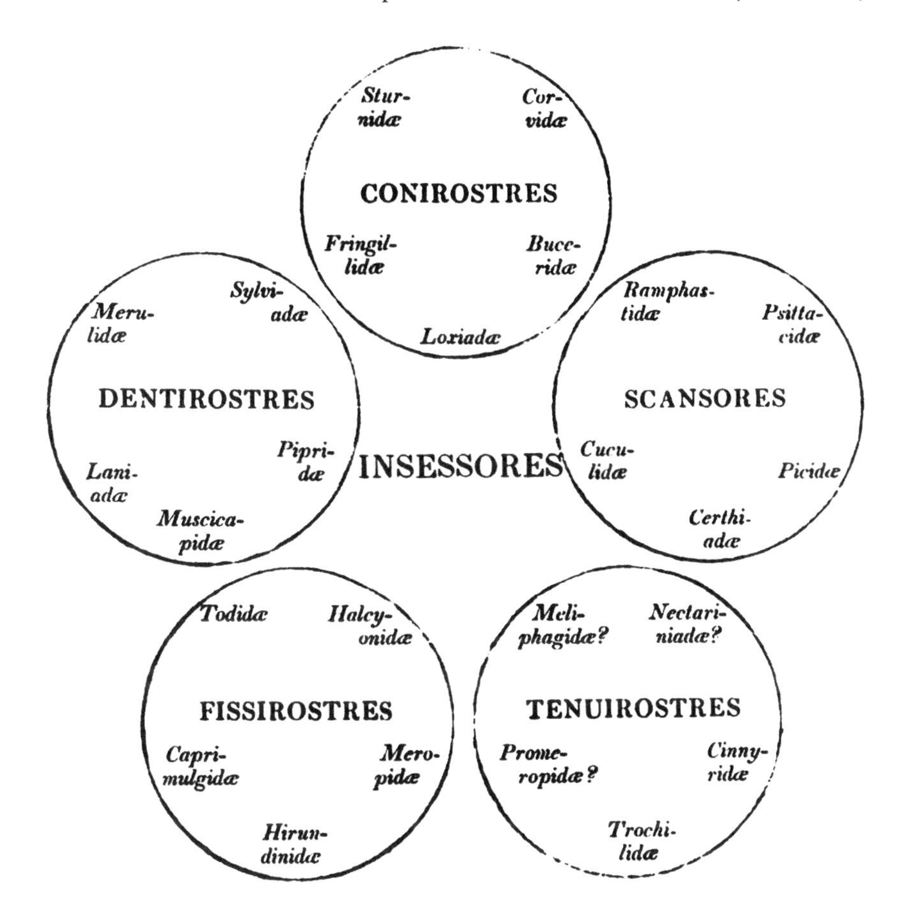

5.1 The circular affinities of the insessorial order of birds (N.A. Vigors 1824).

today as it was in the nineteenth century, the term 'affinity' is still used by some contemporary systematists (for example, Harrison 1969; McGowan 1982; Olson 1987).

Analogy

Affinity was not the only aspect of the Natural System for many nineteenth-century systematists, however. William Swainson and the other members of the quinarian school, followers of the entomologist Macleay, considered analogy to be equally important:

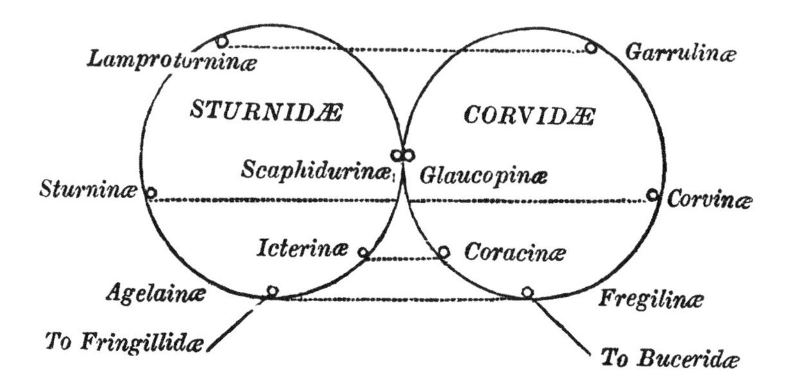

5.2 Analogies between the starling and crow families (W. Swainson 1837). According to Swainson's 'law of representation,' the same five 'primary types' are analogically represented in each circle of affinity.

> ... we shall consider that to be *a natural system* which *endeavours* to explain the multifarious relations which one object bears to another, not simply in their direct affinity, by which they follow each other like the links of a vast chain, but in their more remote relations [analogies], whereby they typify or represent other objects totally distinct in structure and organization from themselves. (Swainson 1835, p. 197; italics in the original)

Figure 5.2 (from Swainson 1837) illustrates both the circular affinities of the starling and crow families, and also the analogies between them, analogies which connected every circle of affinity in the quinarian system.

Strickland and his followers in the map-making period explicitly denied that analogy had any place in the Natural System (Strickland 1840, 1841) and did not depict it in any of their systematic maps. Similarities among taxa showing little affinity to one another were undeniable, however, and the acceptance of evolution allowed systematists to in some measure reintroduce the depiction of analogy (under the name of evolutionary convergence) into their systematic diagrams. The hoatzin, for example, a South American bird of the family Opisthocomidae, has features in common with both the galliform and cuculiform birds, and Maximilian Fürbringer (1846–1920) could depict this evolutionarily in 1888 by showing the branch of the Opisthocomidae emerge from the galliform section of his tree, but then continue upward, cross into the top section of the diagram, and end near the cuckoos (fig. 5.8, upper right). The representation of evolutionary convergence in this manner has extended well into the twentieth century (see for example Mayr 1969, p. 227).

Continuity and Directedness

Continuity and 'directedness' in its various forms (progress, advancement, time, evolution) are among the most complex notions connected with the Natural System, and they run into many of the larger currents of Western thought. Lovejoy's classic work *The Great Chain of Being* (1936) treats extensively of the notion of continuity in the period preceding the one considered here and provides important philosophical background, because the systematic diagrams of the nineteenth century are in a very real sense the wrack of the Chain of Being (see, particularly, Stevens 1983).

In the quinarian period there was a clear belief in the Natural System's continuity. Vigors complained that 'by an oversight of the printer's, the circles in [fig. 5.1] were not made to touch each other ... and they thus seem to convey an erroneous idea of the series of affinity being incontinuous' (1824, p. 509). And Swainson, in his *Preliminary Discourse* (1834, pp. 228–35), discussed at length the 'law of continuity.' For the evolutionists, beginning with Wallace, continuity was transformed from an abstract philosophical principle into a reflection of the real and physical connections of evolutionary genealogy, and it was manifest not only among the living taxa of the present but, more importantly, between living taxa and their ancestors. And in this sense, continuity remains a central element of the Natural System for evolutionary biologists: 'relationship' is defined today, at least by most cladistic systematists, as the relative recency of genetic continuity among taxa which are now reproductively isolated.

I have argued (1988a) that belief in any sort of directedness in the Natural System, apart from that of time itself, is mistaken, and the product of an inappropriately narrative way of viewing the world. Metatemporal directedness was, however, and continues to be, an important element in many systematic representations. Pre-evolutionary quinarian diagrams by virtue of their circular nature do not exhibit strong directionality, although the arrangement of taxa into upper and lower circles is not likely to have been accidental. Wallace (1856) referred to a 'main line' of affinities in his text, and intended the central axis of

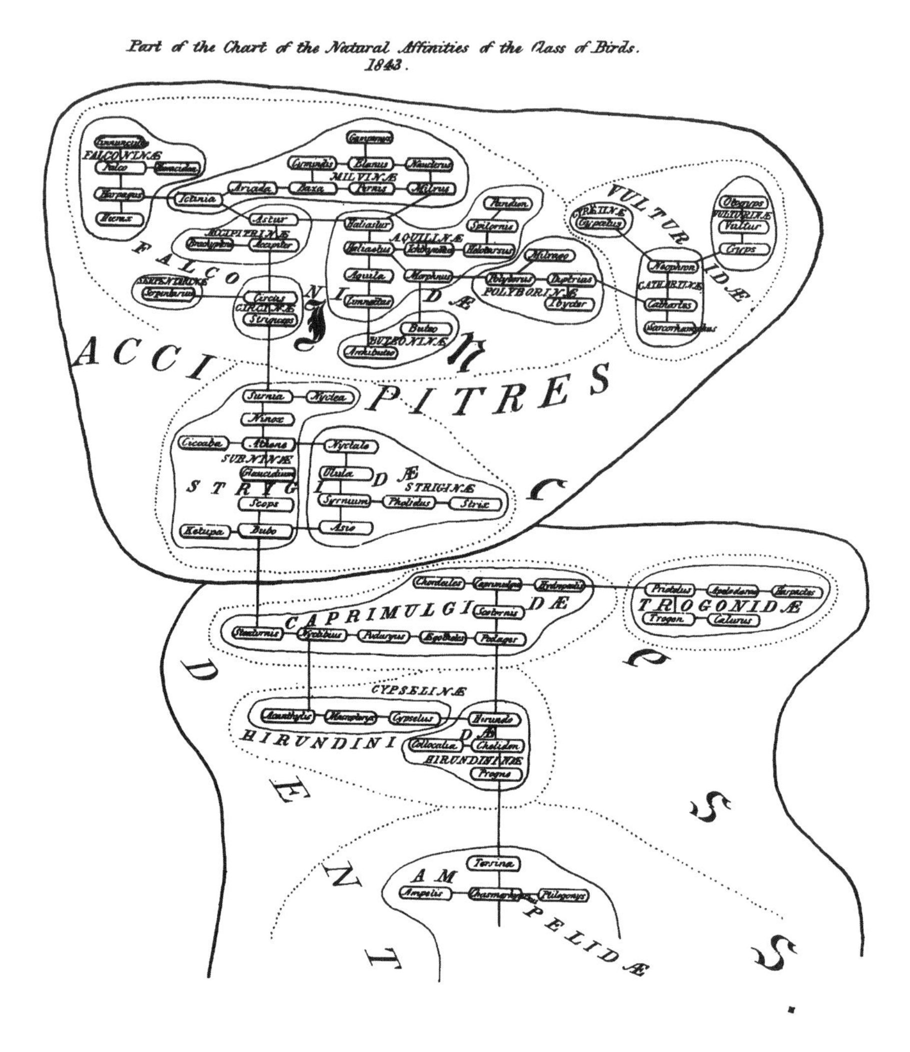

5.3 A portion of Strickland's 'Chart of the Natural Affinities of the Class of Birds,' displayed at the 1843 meeting of the British Association for the Advancement of Science and published after Strickland's death (Sir William Jardine 1858).

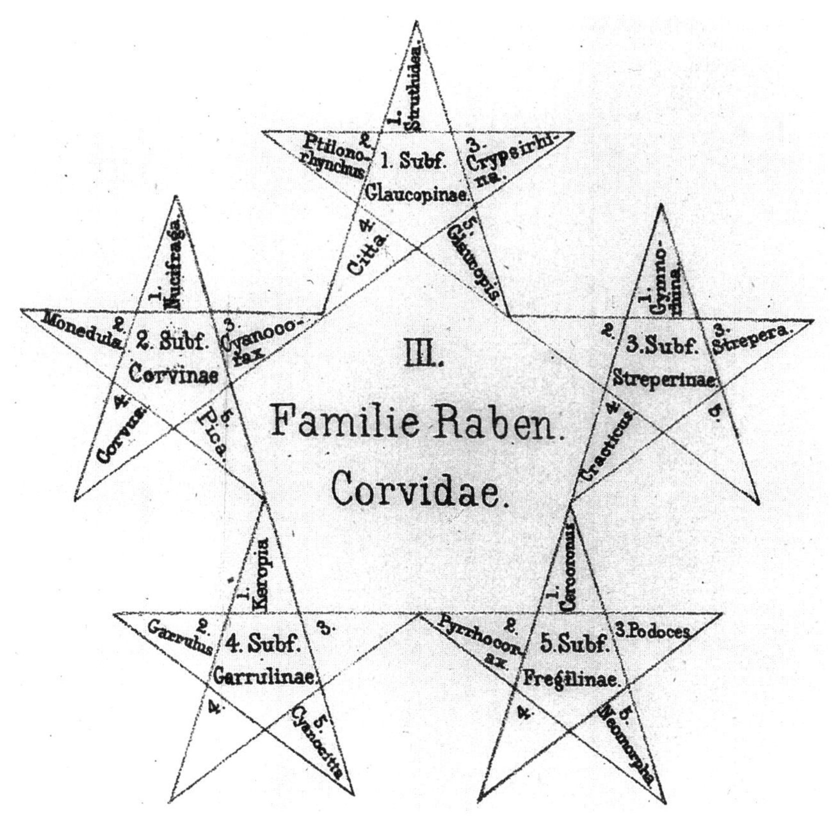

5.4 'Hexenfuss' of the crow family (J.J. Kaup 1854). Blank triangles stand for taxa not yet discovered. Compare the arrangement of the sub-families in this diagram to their arrangement in the corvid circle of Swainson (fig. 5.2).

figure 5.5 to represent that main line. It was not until later in the evolutionary period that direction regained some of the prominence it had lost in the partial collapse of the Chain of Being. Evolutionary trees almost invariably were drawn extending up to a crown (figs. 5.6, 5.7, and 5.8), and even when they were not, as in the avian tree (fig. 5.10) drawn by Richard Bowdler Sharpe (1847–1909), direction was communicated by the left-to-right sequence of the branches. Remnants of the sequence in figure 5.10 can be found today in the ordering of taxa in any popular field guide to birds, as well as in many technical handbooks and checklists. The Chain of Being has by no means been unlinked in its entirety.

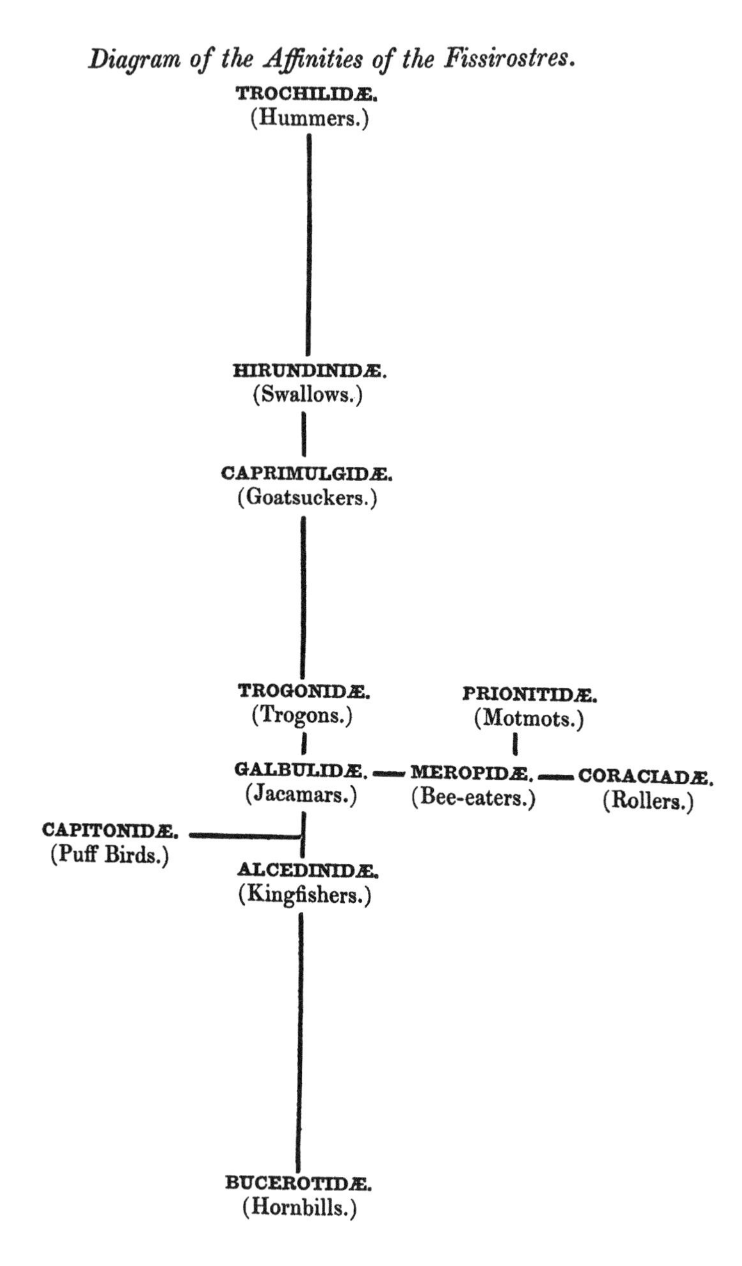

5.5 The affinities of the fissirostral birds, one of two diagrams published by A.R. Wallace (1856). Note the empty node between the Alcedinidae and Galbulidae.

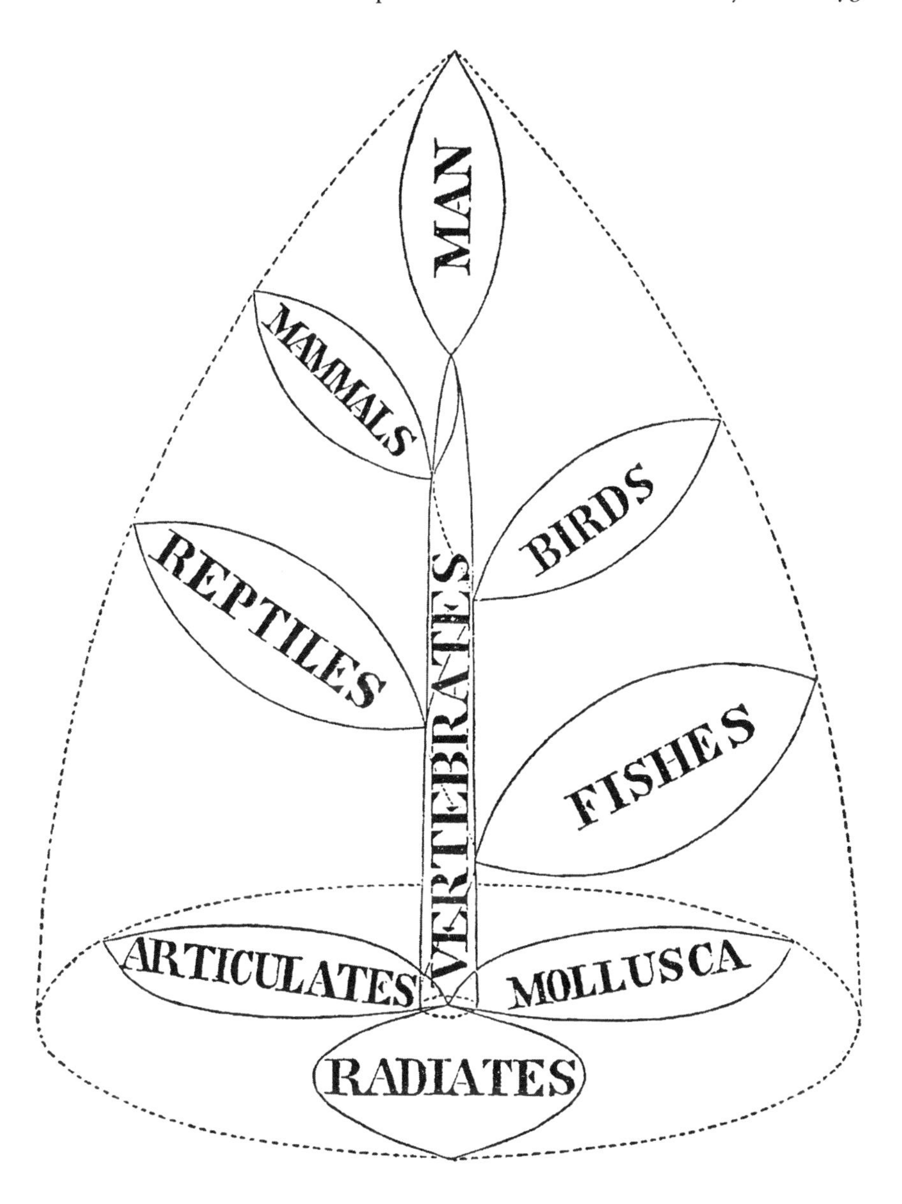

5.6 The animal kingdom (G. Lewis 1866). 'Trifiling as it may seem, the rising of the germ to meet the warm bosom of the mother, in reality marks the whole distance from the lowest Radiate to the Warm-blooded animals.'

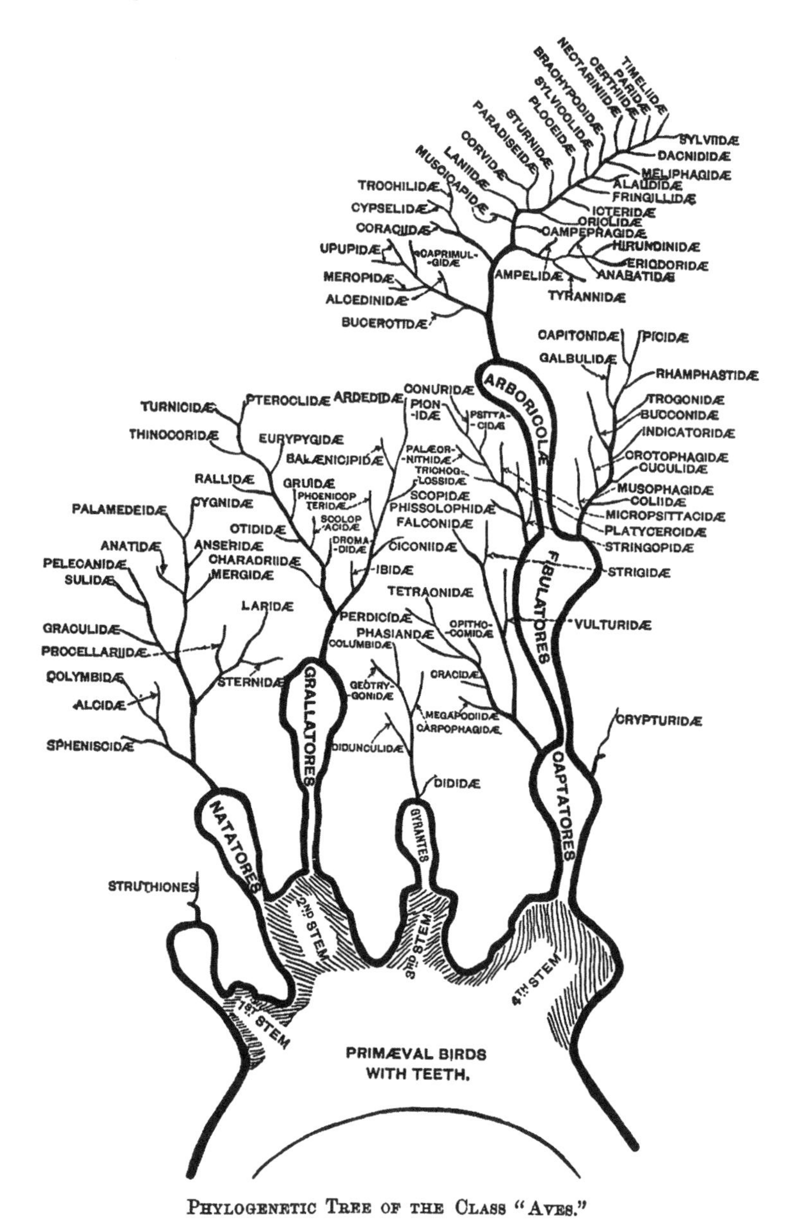

PHYLOGENETIC TREE OF THE CLASS "AVES."

5.7 Reichenow's 'Stammbaum' of the class Aves (1882) as redrawn by Sharpe (1891). Reichenow's original uses German vernacular names and also numbers keyed to his text, and so acts as an evolutionary table of contents to his book.

Symmetry

The quinarians believed that the Natural System was symmetrical and numerically regular (figs. 5.1, 5.2, and 5.4), and for this belief they were widely criticized. They insisted that this numerical regularity was a simple fact of nature, and not a product of their own preconceptions, but their critics always found these claims difficult to believe (Strickland 1841; Wallace 1856). One might expect that symmetrical and regular views of the Natural System would be completely incompatible with the views of evolutionists (they were certainly incompatible with the views of Wallace, for example), but this was not always the case (*contra* Ghiselin 1969, p. 104). At least one of the quinarians' contemporaries objected to their work because he thought it sounded *too* evolutionary:

> We are told, for example, that 'the nearest approach of the mammalia to the birds exists, according to Macleay, among the glires, which *make several attempts*, as it were, *to attain the structure of the feathered class*,' as plain, strong, and precise terms, as Darwin [Erasmus Darwin!] or Lamarck himself could have used in talking of a jerboa (*Dypus*, GMELIN) trying to convert its legs into wings, or a porcupine (*Hystrix*, BRISSON) endeavouring to barb its quills with feathers. (Rennie, 1833, p. xli; italics in the original)

A further example of the compatibility of symmetrical views of the Natural System with evolution can been seen in figure 5.6, the tree of the animal kingdom published by Graceanna Lewis (1821–1912). Lewis accepted evolution, but she was also heavily influenced by Lorenz Oken's *Naturphilosophie* and opened her study of the natural history of birds (1866) with Oken's declaration that 'the animal system is a multifariously constructed temple, with its nave, choir, chapels and towers.' Lewis's views of evolution were little shared by her contemporaries, however (Warner 1979), and later evolutionists did indeed abandon all notions of a symmetrical Natural System. A remarkable example of how completely the notion of symmetry did disappear can be seen in figure 5.7, a phylogeny of birds published in 1882 by Anton Reichenow (1847–1941).

Predictivity

If the Natural System was symmetrical and numerical as the quinarians believed it was, then it could also be predictive: whenever we find taxa which appear to exhibit an incorrect number of subgroups, or inappropriate analogical relationships, we know that there must be other taxa in that group which have not yet been discovered. Thus figure 5.4, which shows the relationships of the crow family according to Johann Jakob Kaup (1803–73), was drawn with several empty triangles for taxa which were believed to exist, but which had not yet been found. According to Swainson (1835, pp. 225ff), these 'gaps' could be caused either by undiscovered living taxa, or by extinct taxa.

One might expect that the acceptance of evolution would cause the problem of predictivity to disappear, at least for those evolutionists who rejected systematic symmetry and numerical regularity. But evolution is in fact highly predictive with regard to the structure of the Natural System, because it takes the matter of continuity to an extreme: as noted above, evolution converts continuity into a physical, genetic phenomenon. If evolution is true, then *every* 'gap' in the Natural System must be filled by extinct taxa, which may yet be discovered. Wallace, in discussing the earliest of the evolutionary diagrams reproduced here, declares it to be

> an article of our zoological faith, that all gaps between species, genera, or larger groups are the result of extinction of species during former epochs of the world's history ... Thus if the space between the Kingfishers and Hornbills [in fig. 5.5] has been filled up by a natural succession of families, we can see that the change must have been to heavier, larger, and larger-billed-birds, and we see such a change begun already from the Jacamars to the Kingfishers. (Wallace 1856, p. 206)

In the *Origin*, Darwin was at pains to show how incomplete the geological record was precisely because of the predictions evolutionary theory made about the structure of the Natural System.

Reticulation and Branching

A key element in the quinarian view of systematics, seen in both figures 5.1 and 5.2, was that chains of affinity were circular: they returned on themselves. Strickland rejected the numerical regularity of the quinarian system, but he did not necessarily reject circular reticulation (Strickland 1841, 1844), and two circular chains of affinity are visible in the upper part of figure 5.3, one connecting the six central genera of the Milvinae, the other adjoining the first and connecting the lower genera of the Milvinae with one genus in the Accipitrinae and one genus in the Aquilinae. Reticulation was abandoned by the evolutionists, beginning with Wallace, but in special cases, namely in those taxa in which hybridization or symbiosis are believed to have played an important evolutionary role, a reticulate Natural System is accepted again today.

Dimensionality

Although printed on a two-dimensional page, many systematic diagrams attempted to represent three-dimensional structure. Figures 5.6, 5.8, and 5.9 illustrate such three-dimensional systems, and figure 5.10 was also, like figure 5.8, part of a double view, showing the Natural System not only from the side but also from above (see O'Hara 1988b, p. 2758, for the top view). In a remarkable passage I have quoted previously (1988b, p. 2750), Strickland even wondered whether the ramifications of the Natural System might exist in more than three dimensions:

> Whether they are so simple as to admit of being correctly depicted on a plane surface, or whether, as is more probable, they assume the form of an irregular solid, it is premature to decide. They may even be of so complicated a nature that they cannot be correctly expressed by terms of space, but are like those algebraical formulae which are beyond the powers of the geometrician to depict. (Strickland 1841)

In the evolutionary period, when affinity could be taken to mean branching genealogical relationship, an interesting conflict was set up between the depiction of the topological connections of the branches in what may be called a 'graph space' (like the space of a subway system, within which a traveller must follow the branching of the tracks in order to arrive at a destination) and the deployment of those branches in a two- or three-dimensional Euclidean space (like the space of the city itself, under which the subway runs) in which the traditional spatial language of affinity could be used. This conflict was manifest in the many attempts to illustrate both branching trees and map-like cross-sections through trees, and is particularly apparent in the work of Sharpe, who constructed his map-like views first, and then 'tested' those views by suspending evolutionary trees below them (Sharpe 1891; O'Hara 1988b). The conflict between Euclidean spatial thinking and what I have called 'tree thinking' (O'Hara 1988a) is far from resolved today.

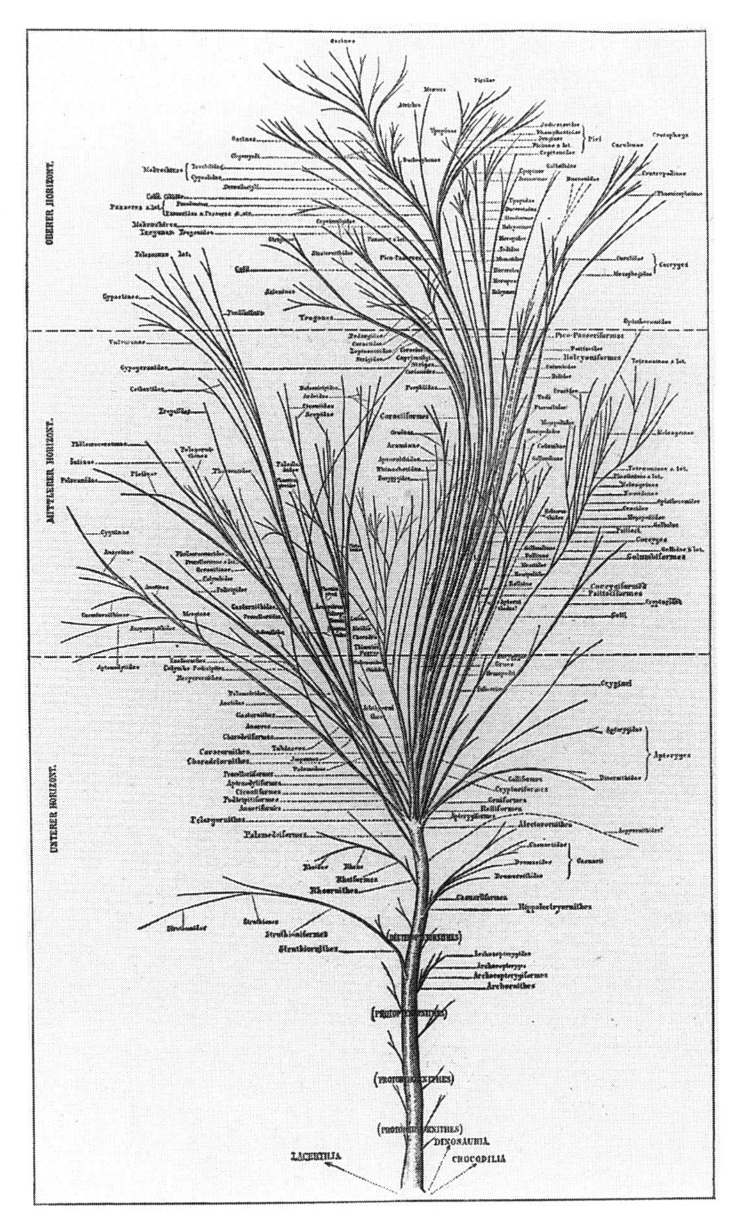

5.8 'Verticale Ansicht' of the evolutionary tree of birds (M. Fürbringer 1888) as reprinted by Sharpe (1891). Fürbringer also published a view of this tree from the opposite side, so that the branches on the back of the tree could be seen more clearly.

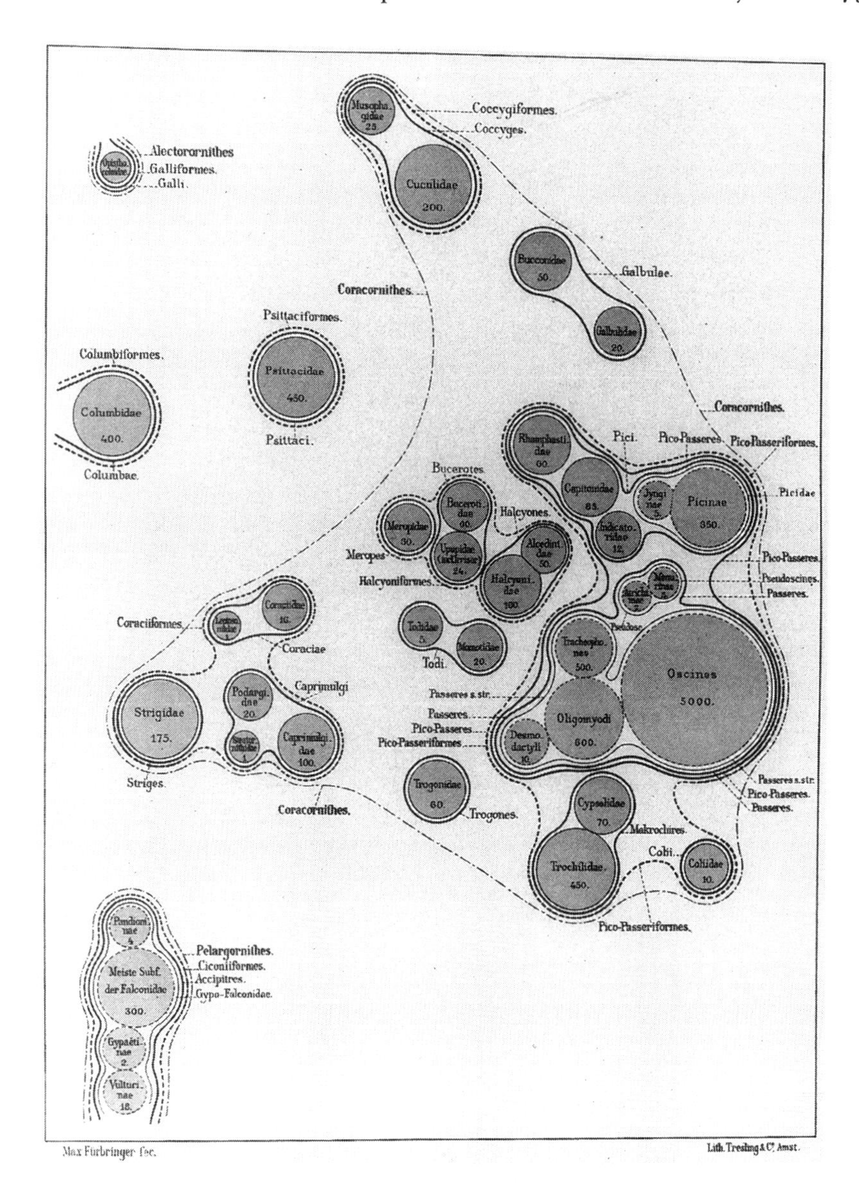

5.9 'Horizontale (Planimetrische) Projection' of Fürbringer's evolutionary tree (1888) at the upper horizon. The diameter of each circle corresponds to the number of species contained in it. Fürbringer also published cross-sections at the lower and middle horizons.

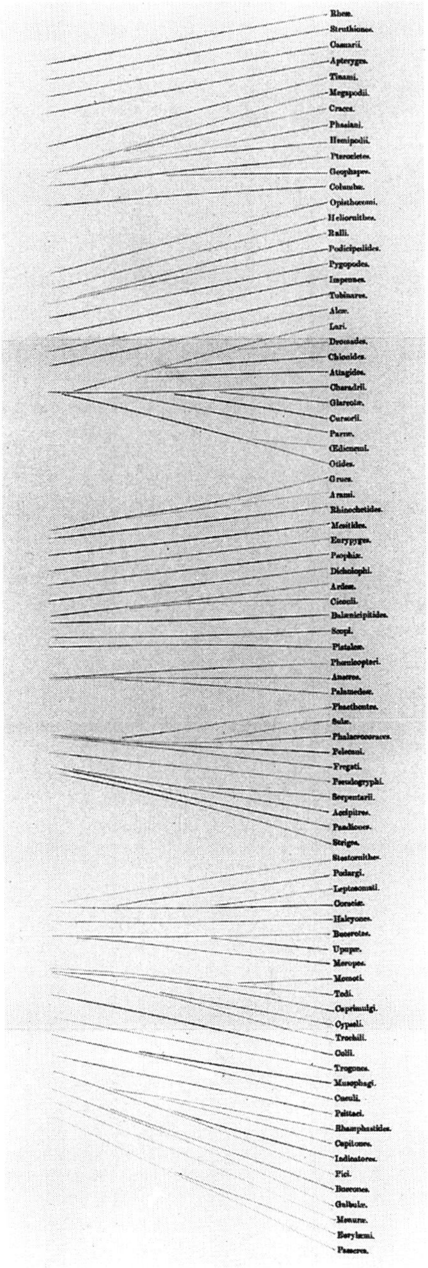

5.10 The phylogenetic tree of birds (R.B. Sharpe 1891). The original folds out horizontally; it is shown here on its side. Note how all of the branches come up to the same level. Sharpe also drew a top view of this tree, which is reproduced in O'Hara 1988b, p. 2758.

3. THE PROBLEM OF INTERPRETATION

While many aspects of the diagrams I have examined here are easy to interpret, others aspects may always exceed our hermeneutic abilities. Because we can no longer directly question the authors of these diagrams, we cannot determine in all cases whether a particular element of a diagram was intended by its author to carry meaning, or whether it was simply an arbitrary illustrative or printing feature. In Vigors's original of figure 5.1, for example, the circles are printed in brown ink. Was this an attempt to contrast the real nature of the taxa themselves with the abstract nature of the affinities which connect them, or was it (as I suspect) simply an illustrative accident? In Strickland's diagrams, what is the significance of the positions of the taxa on the page in relation to the lines connecting them? If all the lines in figure 5.3 were erased, but the positions of the taxa on the page were preserved, would Strickland say that the diagram still conveyed the same information? Probably not, considering his apparently conscious inclusion of circular chains of affinity in figure 5.3. What if the topological connections were maintained, but the positions of the taxa on the page were altered? This question is perhaps impossible to answer. T.H. Huxley published two evolutionary trees of birds in 1868; one of them (1868a) pointed up, as most of the trees shown here do, but the other (1868b) pointed down, showing 'descent' rather than ascent. What, if anything, did Huxley intend to communicate by this difference? From his text one cannot tell. There is a point at which a work moves from the interactive and manipulative domain of its creation, where it can be directly challenged and where it can be explained and revised by its author, into the comparative and observational domain of history and hermeneutics, where an understanding of the work can only be built up with the tools that its author left behind. All the works discussed here have long since passed into the domain of history.

Yet even in the interactive domain of science and philosophy, meaning is teased out of works only to the extent that they *are* challenged and questioned, and this suggests an interesting project for some contemporary philosopher of science: take a collection of recently published systematic diagrams and interview both their authors as well as a variety of other systematists about what precisely the diagrams communicate. Can branches be moved without changing meaning? If so, in what ways? Do left-to-right sequences convey meaning? If not (or if so), do both authors and readers understand this? Such an inquiry would not only be a fascinating study in scientific communication and its difficulties, and of the structure of a scientific community, but it might also be a substantive contribution to systematics, to the extent that it would point out areas where improved communication is needed.

4. CONCLUSION

The representational richness I have outlined in this essay disappeared around 1900 as interest in the large-scale structure of the Natural System declined, and as scepticism about the possibility of reconstructing phylogeny grew in the experimental atmosphere of the early 1900s (Coleman 1971; O'Hara 1988b; Bowler 1989). Darwin's question of what exactly is meant by the Natural System, and along with it the question of what exactly is the purpose of systematics, remained unresolved. These problems smouldered through the Synthesis era of the thirties, forties, and fifties, and in the systematics controversies of the last thirty years we have seen their re-emergence, a re-emergence which has taken place with only a superficial understanding of their nineteenth-century history.

Some see the recent systematics controversies as an attempt to free systematics from its entanglement with evolution, and return it to a more empirical, pre-Darwinian form (Brady 1985). Perhaps the sought-for empiricism is that of Vigors, the author of figure 5.1:

> Devoted to no school of natural science, and carried away by the dictates of no authority however high, no reputation however imposing, I have come to the investigation of my subject, – and I trust I may here be allowed to know myself, – unseduced by the fascinations of theory, and unfettered by the trammels of system. (Vigors 1824, p. 513)

Perhaps it is the empiricism of Swainson, the author of figure 5.2:

> ... science is founded upon facts, and upon a cautious process of inductive and analogical reasoning drawn from those facts: it has nothing to do with speculative opinion or metaphysical reasoning. (Swainson and Richardson 1831, pp. xlv–xlvi)

Or perhaps it is the empiricism of their opponent, Strickland, the author of figure 5.3, who declared *his* approach to systematics to be the truest to nature:

Being a purely inductive process, the details of any branch of natural history may be in this way worked out and depicted without reference to any theoretical assumptions. (Strickland, 1844)

The philosophically inclined student of scientific diagrams might well ask today's empirical systematists whether figures 5.1, 5.2, or 5.3 could be published in a systematic work today.

I do not share the views of those who would create a theory-free systematics. Indeed, I believe that such is impossible, because in systematics – in any discipline – observation and theory are inextricably intertwined. Meta-systematic beliefs always influence systematists, as the diagrams in this paper show; likewise the notions of systematists act as meta-influences on those in other fields. Far from showing a need to free systematics from evolution, the controversies of the last thirty years illustrate to me that systematics still contains a great many pre-evolutionary concepts and structures, concepts and structures which ought now to be purged. We have only just begun to understand the truly evolutionary answer to Darwin's question of what is meant by the Natural System. We are only now coming to realize that the Natural System is in fact the branching chronicle of events in evolutionary time, and that the analogy of systematics to *classification* is mistaken. The task of a systematist in the evolutionary world is not the construction of classes, but the reconstruction of evolutionary history (de Queiroz 1988; O'Hara 1988a), and diagrams of the Natural System today are not information retrieval devices, illustrated classifications, or summaries of character distributions: they are representations of history.

NOTE

1 This paper was prepared for the symposium 'Making Sense of Science Making Diagrams,' held at the History, Philosophy, and Social Studies of Biology meeting, London, Ontario, 1989. I am grateful to Peter Taylor for inviting me to take part in the symposium, and also to P. Ericson, G.C. Mayer, M. de Pinna, M. Ruse, P. Taylor, and R.L. Zusi for their comments on various drafts of the manuscript. P.F. Stevens has long encouraged my studies of the history of systematics, and I extend my thanks to him as well. This paper is reprinted with permission from *Biology and Philosophy* 6 (1991): 255–74.

6. Visual Representation in Archaeology: Depicting the Missing-Link in Human Origins[1]

STEPHANIE MOSER

1. INTRODUCTION

Archaeology is an explicitly visual science. As with disciplines such as geology and palaeontology, prehistoric archaeology has from its very beginnings developed a distinctive visual language that it has used to communicate theories, technical principles, and data (Moser 1992, 1993). In this paper, I would like to show how one type of archaeological illustration functions within the discipline, and in doing so outline some aspects of how ideas are visually represented in archaeology. The type of illustration I will discuss is the pictorial reconstruction of prehistoric life. This type of visual display, in which our hominid ancestors are seen engaged in particular activities in a landscape, plays a crucial role in archaeology because it fulfils our desire to know what our distant ancestors looked like and how they behaved. The paper examines ways in which these representations make arguments in a distinctly visual manner, in a way that verbal text cannot. My analysis focuses on an examination of some key images depicting what the very first hominid ancestors looked like and how they lived. These images of the 'missing-link' or hominid species known as the australopithecines are particularly important because they attempt to reconstruct what the most unimaginable of our ancestors looked like. Moreover, they demonstrate how visual imagery is central to debates about human origins.

Pictures of our ape-like ancestors that have appeared in illustrated newspapers and books throughout the century represent attempts to comprehend that which is seen to have been not quite human and yet not completely ape. The goal of research into human origins has been to establish the point at which our ape-like ancestors started to acquire human-like behaviours, and this has involved the specification of what it is that makes us uniquely human. For instance, was it the ability to make tools, hunting and meat-eating, food sharing, or the ability to stand upright? By emphasizing certain attributes and conveying human qualities, the pictures of 'missing-links' aim to define the boundary between apes and humans. The ancestor that is thought to have represented the first transitional stage in the evolution from apes to humans is especially hard to conceive since it is an ancestor for which we have no contemporary analogue. While palaeontology has long depicted extinct animal forms and creatures for which there is no contemporary analogue (see Rudwick 1992), the archaeological depictions of extinct hominids constitute something entirely different because they make reference to, and are assessed in terms of, our own human status.

In this sense, the pictures of ancestral 'missing-links' embody a constant tension between the desire to characterize the fossil remains as human-like and thus close to us, and to characterize them as ape-like and thus distant. The illustrations are critical to the aim of settling the question of whether particular fossil hominids were part of or excluded from the human lineage. Reconstruction drawings of the appearance of certain ancestors have been enormously influential in conferring human or non-human status upon the fossil specimens in question. This is precisely why archaeologists and other evolutionary specialists have enlisted scientific illustrators to flesh out the bones that they have found in ancient deposits. It is my contention that such illustrations are distinct from and not simply derived from verbal statements in the accompanying text. Put simply, it is in the illustrations that arguments are made about what constitutes humanity.

While the more abstract pictorial reconstructions, such as depictions of stratigraphic layers, plans of archaeological sites and their settlement patterns, and charts of the distribution of artifacts, have been described as representations of ideas (Addington 1986; Adkins and Adkins 1989), little attention has been paid to the role of more naturalistic images. Illustrations rendered realistically are generally thought to be peripheral to the substance of archaeological arguments, not simply because they are seen as being purely hypothetical, but also because they are presented in a relatively universal language, instead of a specialized and professional one. Most archaeologists see reconstruction drawings as being entirely separate from the more abstract images and diagrams.

Yet reconstruction drawings are very much part of the tradition of archaeological illustration, and are just as integral to explanation as are the more abstract forms of illustration.

Histories of archaeology, physical anthropology, and palaeoanthropology have paid much attention to how early evolutionary scientists widely advertised their findings by presenting their ideas in the major scientific journals and illustrated newspapers and science magazines of the day. However, little or no mention is made of the way in which images functioned in these efforts to promote new arguments about the hominid lineage. It is important to look at how new hominids were introduced to the professional and non-specialist audience in the form of illustrations, and to establish how the images were used to place hominid ancestors in the scheme of human evolution. It can be shown that contemporary images of our ancestors have set the theoretical views and informed the interpretation of new fossil data.

2. VISUAL REPRESENTATION IN SCIENCE

In a recent book on the subject of representation in the sciences, it is claimed that interest in the history of scientific illustration has arisen as a result of greater use by historians and philosophers of science of sociological explanations to account for the acceptance of theories and theory change, the emergence of specialities, and the resolution of controversies (Lynch and Woolgar 1990, p. 3). In introducing the papers in *Representation in Scientific Practice*, Lynch and Woolgar refer to works by Fleck, Kuhn, Polanyi, Hanson, Horton, Lakatos, and Feyerabend, all of which discuss efforts made by scientists to enlist 'agreement' through persuasive appeals. Ranked high on the list of such appeals are visual displays. For Lynch and Woolgar (1990, p. 7), the analysis of visual representation is about exposing some of the devices or 'tricks' that scientists employ to ensure the success of their theories.

Lynch and Woolgar challenge the privilege assigned to verbal statements and call for the examination of the variety of representational devices used in science, such as graphs, diagrams, equations, models, photographs, and computer programs. Of the range of representational devices that scientists employ, pictorial forms are shown to be a most effective means of communicating ideas, and as Bastide (1990, p. 200) points out in her contribution to the volume, visualization is a highly seductive method of argument that is often subtle and unconscious. Furthermore, Latour's (1990) contribution argues for the centrality of pictures to the crafting of knowledge, using several important case-studies of the development of pictorial languages. Here and in an earlier paper, Latour (1986) argues that images effectively escape our attention because they are so practical, so modest, so pervasive – so close to the hands and the eyes. He emphasizes the way in which groups of people argue with one another using illustrations and argues that the importance of images is the unique advantage they give in the rhetorical or polemical situation – 'you doubt what I say? I'll show you.' By showing illustrations, the effect is to make an argument more visible and thus more believable. As Hacking (1991, p. 980) argues in his review of the book, the issue is whether the point of the representations is to convey information at all, or rather to convince us that this is solid stuff, not to be challenged, and not challengeable. While one can make one's work appear solid by supplementing it with finished diagrams and illustrations, such work may be solid because it meets or exceeds current standards for the presentation of arguments. However, it need not be at all convincing.

The subject of visual representation has also been explored in terms of the emergence and professionalization of particular scientific disciplines. In his seminal paper on the visual language of geology, Martin Rudwick (1976) highlighted the impact that visual modes of communication had on the formation of geology as an autonomous discipline. He demonstrated the vast potential of the subject of scientific illustration, and has continued to explore the implications of visual imagery for the development of the earth sciences in the nineteenth century (Rudwick 1989). More recently, Rudwick has traced the origins and emergence of a major illustrative genre in palaeontology. In *Scenes from Deep Time: Early Pictorial Representations of the Prehistoric World* (1992), Rudwick examines how the reconstructions of prehistoric life dealt with the problem of conveying a truly deep or geologically ancient earth history (see Moser 1993b). Another recent study of how illustrations assisted in the creation of scientific disciplines is Winkler and Van Helden's (1992) analysis of the place of visual communication in the development of astronomy. These authors see naturalistic representations, or pictures of the heavens, as a unique category of evidence in early astronomy, which assisted the delineation of astronomy as a visual science.

3. VISUAL REPRESENTATION IN PREHISTORIC ARCHAEOLOGY

Pictorial reconstructions of past human and hominid lifeways have a special place in archaeology because they represent theories that have been advanced by archaeologists to account for the discovery of particular fossil data (Moser 1992). Early images of our cave-dwelling ancestors hunting hairy mammoths and fighting off sabre-toothed tigers emerged alongside the excavation and interpretation of cave deposits in Europe from the mid-nineteenth century. By the turn of the century, such images became a primary means of presenting new and extremely ancient hominid fossil discoveries to the wider public. When skeletal remains of what appeared to be our ape-like ancestors began to be found in Asia, Africa, and Europe around this time, the community of scholars who were fast defining themselves as evolutionary specialists began to draw heavily on the skills of 'expert' or scientific illustrators to flesh out the specimens and bring their ideas to life. The reconstruction drawings that were produced made sense of the often meagre fragments of bone and stone. Visual representation in the field of human origins not only stems from the fragmentary nature of the data and the desire to see specimens fleshed out 'as they once were,' but also from a flourishing visual tradition of producing scenes from deep time that pre-existed in the field.

When archaeological scenes of deep time first appeared in books and museum displays, a diverse range of images were used to communicate ideas and arguments about human evolution. For instance, depictions of early humans either in caves or the jungles of Asia accompanied pictures of contemporary landscapes, flint implements, skeletal remains, primate and human anatomy, and ethnographic peoples from around the world. Such illustrations were part of the fast emerging discourse on human antiquity in the mid-1800s. Huxley (1863), Lubbock (1865), and Dawkins (1874) used visual materials to convey principles and arguments central to their new science. The visual aids that were employed by geologists and natural historians to accompany their discussions of tumuli, Danish peat bogs, Swiss lake habitations, bone caves, river drift gravels, and the 'customs and manners of savages' enforced various theoretical viewpoints, ranging from the principles of uniformitarian geology to biogenetic law ('ontogeny recapitulates phylogeny'). Pictures of 'savages' were used as templates for making inferences about the lifeways of ancient humans.

In palaeontology much has been written about the history and practice of reconstructing fossil animals (for example, Desmond 1974, 1975, 1979; Rudwick 1992). Recently, archaeologists have begun to address the issue of how pictorial reconstructions are embedded with meanings far beyond their claimed intention of summarizing findings from archaeological sites (Conkey 1991; Gamble 1992; Gifford-Gonzalez 1993; Moser 1992, 1993a). In two important new books on the place of Neanderthals in human ancestry (Stringer and Gamble 1993; Trinkaus and Shipman 1993), special attention has been given to the production and persistent recycling of images of the species, which the authors suggest have had a key role in the history of Neanderthal research.

What precisely is it that makes images of early hominids so special and so compelling? My contention is that the images show how archaeologists have grappled with the problem of defining what it is to be human, and that this appears to be a highly visual problem in itself. The images have a vocabulary of their own. They use symbols and icons that are impossible to replicate in a verbal format. They are part of a visual tradition of 'talking' about or, more precisely, 'seeing' and thus understanding human origins. A firmly established genre (to use Rudwick's term) of pictorial reconstructions of human origins now exists as a result of the fact that illustrators and archaeologists have been working together to produce such images for over a century. By genre I mean that certain canons of representation have been standardized, and that there is an assemblage of visual icons that are repeatedly used. That such a genre exists is attested to also by the fact that professional reconstruction artists have been employed by various museums and institutions around the world, and that their work is defined and often referred to in the scientific literature (see, for example, Rensberger 1981). A brief analysis of a set of important images reveals some characteristics of this representational genre.

A good starting point is the illustrations of the first hominid ancestors of modern humans, the so-called 'missing-link.' Four major species of australopithecines have been identified – *Australopithecus afarensis, Australopithecus boisei, Australopithecus africanus,* and *Australopithecus robustus.* For this analysis, I have selected some of the major images of australopithecines produced in association with the major fossil discoveries and archaeological analyses of the species. The aim is to examine how the images have been used to define these ancestors and accord them a place in the human lineage. The pictures reveal how australopithecines were conceptualized as the first ape-like ancestors who exhibited some traits of supposedly human behaviour, but more fundamentally they are crucial in understanding the difficulties that archaeologists and evolutionary specialists experienced in defining the extent to which these ancestors were 'human.' The images not only embody a traditional resistance to accepting the genus as part of the hominid lineage, they were crucial to the argument that the australopithecines were merely fossil apes. Put simply, the images made their own distinctive contribution to the dialogue on whether the ancestor in question was 'one of us' or not.

4. INTRODUCING THE NEW ANCESTOR

The first depiction of an australopithecine appeared in the *Illustrated London News* in 1925 (Elliott Smith 1925; see fig. 6.1). This picture is a reconstruction based on Raymond Dart's discovery of a fossil cranium in Taungs, South Africa (Dart 1925). Published by Grafton Elliott Smith, the image presents Dart's australopithecine on the left, standing next to the 'Rhodesian Man' representing a skull found in a cave at Broken Hill, Northern Rhodesia, in 1921 and thought to be the remains of the later hominid species *Homo erectus.* This image is a direct attack on Dart's view that the remains from Taungs were of an extinct race of apes intermediate between living anthropoids and humans. Dart had argued that the skull from Taungs represented the real 'missing-link' between apes and humans. His view was regarded as extremely controversial for a number of reasons. Among them was the belief that the 'missing-link' had come from East Asia (an 'ape-man' named *Pithecanthropus* was found in Java by Eugene Dubois in 1891). Because Asia was seen as the cradle of humankind, there was resistance to the idea that the new African remains represented the beginnings of humanity. Furthermore, Dart was battling for a place in the small community of evolutionary specialists; he was not accepted as part of the newly defined group of professionals who were considered authorities on human evolution. By confidently announcing to the world that he had found the missing-link, Dart upset senior anatomists who considered it their task to bestow such status on prehistoric remains.

Elliott Smith had been Dart's professor at University College in London, and he promptly attacked Dart for labelling the fossil as the missing-link. Elliott Smith's picture of the australopithecine from Taungs very effectively achieves what it sets out to do. The image directly challenges Dart's claim that the Taungs specimen represents the missing-link between humans and apes, and conveys a reluctance by the academic community to accept the australopithecines as a member of the hominidae. The picture does not simply suggest that Dart's ancestor was not 'part of the family,' it asserts that the specimen

6.1 Grafton Elliott Smith's australopithecine from Taungs, South Africa, 1925.

was nothing more than a fossil chimpanzee. In his *Nature* article, Dart (1925) argued that the brain and teeth of the specimen were very different to those of the apes, and that from these two features it was possible to infer that the species walked upright, with hands free for offence and defence. Dart labelled this alleged intermediary between humans and apes *Australopithecus africanus*. There was a huge media response to his find, with claims that this missing-link now replaced the other contenders, such as the 'Piltdown Man' and 'Java ape-man.' However, soon after Dart presented the new species to the international community, evolutionary experts intervened. Like Elliott Smith, Arthur Keith (1925) described Taungs as a young anthropoid ape and promptly rejected its missing-link status.

Elliott Smith's picture shows how the australopithecine from Taungs was not accepted as a human ancestor by the scientific establishment. It is noteworthy that a senior evolutionary authority chose to enter into a visual dialogue and produce a picture to exclude the ancestor from the hominid lineage. His argument that Taungs represented nothing more than an anthropoid ape could not be more effectively made than it is here in visual terms. It was not so much the case that he lacked the evidence to make the case verbally, but rather that such arguments are always more convincing when made visually. In this picture, the author and illustrator have created a visual way of signifying what it means to be human. By placing the two ancestors – the australopithecine and the 'Rhodesian Man' – in the same picture, Elliott Smith was showing how one of these is like us and the other is not. Not only is the three-foot chimp-like australopithecine from Taungs towered over by the six-foot Broken Hill ancestor, but he passively holds a couple of stones in his hands, while the Broken Hill ancestor assertively holds a long or spear-like stick. While the australopithecine is hairy and ape-like, without clothes, hunched, and has bow legs and chimp-like feet, the Broken Hill ancestor has body hair like a human, wears a loincloth, stands fully erect, and has feet just like us. These are some of the visual symbols that are used to denote the human and non-human status of the fossil hominids. In essence the picture argues that the australopithecines were not part of the hominid lineage because they did not possess enough familiar human-like traits.

Elliott Smith was not the first to use such representational devices. He was taking advantage of a visual tradition of communication that had already been used earlier in the century. The creation of a visual language that served to characterize human ancestors according to whether they possessed ape-like or human-like features had been used to debate the evolutionary status of the Neanderthals. And it was in association with the production of knowledge on this species that reconstruction drawings were shown to be an extremely useful way of dealing with fossils whose claim to human ancestry was contested. For example, a visual language that had the power to 'make or break' an ancestor had been created when Marcellin Boule and Arthur Keith employed pictorial reconstructions to debate the Neanderthal's place in human evolution (Moser 1992). Similarly, in Elliott Smith's image of the australopithecine from Taungs, we see the use of a fast emerging iconography in which human-like and ape-like attributes were compiled and juxtaposed in order to denote whether the species in question was entitled to be labelled a 'human' ancestor.

5. THE PREDATORY ORIGINS OF HUMANITY

The next two important images of the australopithecines selected for discussion here were published by Raymond Dart in 1959 in the *Illustrated London News* (Dart 1959; figs. 6.2 and 6.3). The first image features a 'family group' of australopithecines in the process of making and using tools of bone and horn, and the second depicts a group of australopithecines fending off hyenas and a vulture from an animal kill. Both pictures are presented under the title 'The Ape-Men Tool-Makers of a Million Years Ago: South African Australopithecus – His Life, Habits and Skills.' Based on Dart's analysis of the faunal remains from a site at Makapansgat in South Africa, these images communicate Dart's hypothesis (1948, 1949, 1953, 1955, 1957, 1959) that the tool-using and predatory behaviour of the australopithecines had a key role in human evolution. This interpretation was based on the recovery of baboon skulls together with australopithecine remains. There was damage on the baboon skulls that led Dart to conclude that they had suffered a blow to the head with a hard object. Dart thought that weapons in the form of large bones of the antelope were used to kill baboons and other animals. In the article that features these reconstructions, he demonstrates 'the effectiveness of an ox scapula (shoulder blade) as a weapon' (Dart 1959, p. 798). In the drawings, the australopithecines were distinguished as being human-like because they had developed a technology to facilitate their predatory habits.

The first picture (fig. 6.2) presents Dart's (1957) theory that the australopithecines had their own bone tool culture, which he called the

6.2 Raymond Dart's depiction of a South African ape-man family making and using tools of bone and horn.

6.3 Raymond Dart's depiction of ape-men contesting a lion's kill with hyenas and a vulture.

'osteodontokeratic' culture. Here humanness or human-like behaviour is signified by the visual icon of tools, since nothing suggests human ingenuity as much as a picture of someone who – like us – could sit down and craft special tools. In the centre of the picture, a male figure is using one bone to strike another in order to break off or splinter sharp-edged tools that could be used for processing meat. On the left, a female figure is using such a tool to scrape meat from a hide. There are two youths also participating in the tool-making process. Littered around their feet are the products of their labour, all evidence of their bone tool culture. The depiction of this family-like scene with the four figures seated together at the mouth of a cave serves to further enhance the human quality of these ancestors. It is clear that Dart really pushed for the acceptance of the species with these two images, where tools, weapons, and other attributes are used to emphasize their attainment of human-like behaviour.

The second picture (fig. 6.3) shows Dart's bone tool culture in action. In this scene, two male figures with sharpened sticks and bone implements are featured warding off their competition. As the title of the picture claims, this was the 'struggle for existence a million years ago,' where 'ape-men contest a lion's kill with hyenas and a vulture.' Equipped with their superior technology, the australopithecines have the distinct advantage – the human advantage – over the other scavengers on the African savannah. While the picture does not explicitly typecast the species as mighty hunters, Dart went on to infer that the tools would have fostered the development of hunting behaviour. The significance of Dart's images is that they crystallize the 'picture' of a meat-eating way of life that was at the heart of Dart's theory that predatory behaviour separated our early ancestors from the apes. These are the first images to really place the australopithecines in a behavioural context, or, more precisely, the first images that accord the species with a distinctive way of life. At the heart of these pictures is the development of the hunting hypothesis, a hypothesis that was to be articulated more fully in the archaeological literature in the 1960s. The pictures are highly significant because they served to prepare both the professional and wider community for the verbal articulation of the thesis that hunting was pivotal in human evolution (Lee and De Vore 1968; Pfeiffer 1969; Washburn and Lancaster 1968), and in this sense they serve to set the scene, or pave the way, for the acceptance of the theory. Consequently, when people came to read about the role and significance of hunting in human origins, they 'knew' it already from the striking pictures that Dart had produced. While some discussion of the role of hunting was raised in the literature (Washburn and Avis 1958), Dart's theory about the predatory origins of humanity attracted much attention because it was expressed in such a graphic way.

Dart had been greatly assisted in making his case about the predatory origins of humanity by a spate of research carried out in South Africa during the 1930s and 1940s. Since his discovery of the Taungs specimen, fieldwork in South Africa had resulted in the recovery of many more australopithecine fossils. Of particular importance was the work of Robert Broom (1925), who had been supportive of Dart's claims for the status of Taungs, although it was not until he started to find skulls at Sterkfontein in the 1930s that he became convinced that the australopithecines were hominids. Based on the recovery of a skull in 1936, Broom (1936) reported in *Nature* and the *Illustrated London News* that he had found a new ancestral link between humans and apes that he called *Australopithecus transvaalensis.* Broom (1950, p. 55) later described how he too experienced great resistance to his ideas about the australopithecines, but that he continued to report his findings (Broom 1938, 1942, 1943). Broom and Schepers's 1946 monograph, in which they argue that the australopithecines were related to humans, was central to leading evolutionary specialists' acceptance of the australopithecines as members of the hominid lineage (for example, Le Gros Clark 1947, 1948). Furthermore, it was in association with Broom's work that Arthur Keith (1948), in his synthesis of human evolution, described the species as being closer to humans than was once thought.

What is generally neglected in the accounts of Broom's work is that he introduced a new species of the australopithecines – *Paranthropus robustus* – via the medium of reconstruction drawing. Broom presented this new species, based on the recovery of a skull from Kromdraai, in the *Illustrated London News* in 1938 (Broom 1938b; see fig. 6.4). This picture featured members of the new species 'repelling an attack,' in a manner which suggests that they had a more erect posture than apes. The focus of the picture is the posture of the species, and this was based on the position of the brain, which Broom thought indicated the species walked upright. Hence the image that illustrates the 'fighting methods' of the new ancestors is designed to show how the acquisition of erect posture was closely connected to, or facilitated by, their needing their hands free to throw objects. At the centre of the picture are two male figures standing erect with their hands raised, about to

6.4 Robert Broom's australopithecines — *Paranthropus robustus* from South Africa, 1938.

throw large rocks at the invader. Two other figures with infants are crouching and retreating back into what appears to be a cave. This picture is significantly different from Dart's set of images. It does not characterize the species as having an advanced technology, and the ancestors appear to be far less human. In the caption that accompanies the image, entitled 'A Step Nearer to the Missing Link? A Fossil Ape with "Human" Teeth – "*Paranthropus robustus,*"' Broom claims that the 'discovery leaves no doubt that in Pleistocene times there were a number of forms of large non forest-living anthropoid apes, which in structure were much nearer to man than either the chimpanzee or gorilla' (Broom 1938b, p. 310). However, Broom, unlike Dart, did not make a strong visual argument about the acquisition of human-like behaviour. The group of figures are not depicted making and using things outside a large cave opening, as in Dart's picture, which features a setting that is more suggestive of being a 'home.' In Broom's picture, the figures stand in front of what appears to be a rock crevasse. It is not evident that they actually inhabit it. There is no material culture to suggest that Broom's ancestors made tools and lived in the cave. Furthermore, while these ancestors used unmodified rocks to defend themselves, Dart's ancestors were equipped with sophisticated bone weaponry. Finally, the figures in Broom's picture appear somewhat defenceless. They do not possess the aggressive and more distinctively human quality that Dart's figures do.

With Broom's work, Dart had more evidence to fill out his picture of the lifestyle of the australopithecines. When Dart re-entered the missing-link debate in the late 1950s with his striking pictures of the australopithecines, he went far beyond the skeletal morphology to infer substantial information about the actual behaviour of the species. He had a clear vision of their lifestyle that struck a chord in the imagination of writers and film-makers. For instance, Dart's hypothesis is at the centre of Ardrey's (1961, 1976) best-selling books on the place of aggression, sex, and tool-making in human origins, and in the opening scene of Stanley Kubrick's film *2001: A Space Odyssey*. Dart's vision was reproduced and recycled in a mass of pictures of the australopithecines and their meat-eating and tool-using way of life (for example: Augusta and Burian 1960, plate 1; Binford 1984, p. 2; Cornwall 1960, p. 40; Scheele 1957, p. 74; Wood 1977, pp. 58–9). One example by Maurice Wilson, also reproduced many times (Andrews and Stringer 1989; see fig. 6.5), features an australopithecine in the foreground with a raised arm and rock in his hand. While this figure is running, others in the

6.5 A depiction of *Australopithecus africanus* by Maurice Wilson, ca. 1950.

background are waving sticks as they chase hyenas away from an animal kill. The bones strewn on the ground are further evidence of their meat-eating pursuits. Pictures such as this spread the idea that the species was characterized by an aggressive meat-eating way of life. Furthermore, the depiction of the australopithecines waving or holding sticks and hurling stones (see also Cornwall 1960, p. 40; Scheele 1957, p. 74; Wood 1977, pp. 58–9) reflects Dart's view that erect posture was facilitated by carnivorous habits. While this line of reasoning can be traced to the nineteenth century (Balkwill 1893, p. 180; Darwin 1871), it was not until Dart produced his images that it became a key evolutionary icon. Another classic example of this type of picture is the 'pack hunting model' of human origins that was recently reproduced in Lewis Binford's (1984; see fig. 6.6) book on the faunal remains from a site in Africa. This picture features a group of early hominids who are so adept that they can hunt down large antelope.

The point to make here is that by reproducing lots of images with groups of australopithecines chasing, throwing, meat-eating, making tools, and living in caves, archaeologists and palaeoanthropologists ensured that the australopithecines were continually being seen as hunters. Even if they were not shown bringing down a giant mammoth, the pictures suggested they were engaged in hunting behaviour. The pictures that appeared in books, magazines, and newspapers had a range of visual icons in common, and it was the repetition and recycling of these visual clues that was crucial. Myers (1990), in his study of the illustrations in E. O. Wilson's *Sociobiology: The New Synthesis* (1975), argues that even when there are signs to the contrary, certain images are still believed to show what really is. A drawing in Wilson's *Sociobiology*, for example, has early humans fighting off rival carnivores and eating a dead mammoth, and although the picture is labelled as speculative, the effect of the drawing is to convey that early humans were evidently carnivores (Myers 1990, p. 263).

6. THE POWER OF TECHNOLOGY

A new development in the construction of visual icons for debating human origins occurred in association with the recovery of a major australopithecine fossil at Olduvai Gorge in East Africa in 1959. Two illustrations were produced by Louis and Mary Leakey for the *National Geographic* to accompany the introduction of a new species of australopithecine – *Zinjanthropus boisei* – to the world (L. Leakey 1959, 1960, 1961; see figs. 6.7 and 6.8). According to Louis Leakey, the remains of 'Zinjanthropus' indicated that another species more robust than Dart's *Australopithecus africanus* and Broom's *Australopithecus* or *Paranthropus robustus* existed. Louis and Mary Leakey had been working in East Africa since the 1930s, searching for bones of hominid ancestors. While they had found and detailed the remains of stone tool culture, they were not successful in finding skeletal remains until 1959, when they unearthed what appeared to be a very robust australopithecine ancestor. The remains consisted of an upper jaw and some simple chopper tools. While Dart had promoted the idea that the australopithecines had a bone tool-making culture, it was the Leakeys who sealed the 'man the tool-maker' vision of human evolution with their discovery of 'Zinjanthropus.'

6.6 Early hominids featured in a pack-hunting model of human evolution.

Announcing their new discovery in *Nature* in 1959 (L. Leakey 1959), Louis Leakey then went on to produce a portrait of 'Zinjanthropus' for *National Geographic* and the *Illustrated London News* in 1960 (L. Leakey 1960a, 1960b). The striking portrait of the new species in the *Illustrated London News* is the basis for the even more striking colour picture in the *National Geographic*, which is the focus for discussion here (fig. 6.7). This picture features a close-up of the face of a male figure engaging the eye of the viewer. The species is portrayed as being distinctively human, in the sense that it has a beard and moustache, a fine nose, fine eyebrows, and thin lips. While the distribution of neatly trimmed hair on the face of 'Zinjanthropus' gives him a human appearance, it is, above all, the distinctively human glance of the figure that makes him seem human. This is a new attribute that has been added to the repertoire of visual symbols used to denote the level of humanity reached in an ancestor. The expression on the face, or the way that 'Zinjanthropus' is depicted looking out at the viewer, suggests that the species is fully self-aware or self-conscious. It makes the species seem far less ape-like than the figures that feature in the pictures by Broom and Dart. Louis Leakey was familiar with the strategy of illustrating arguments because he had already produced a visual reconstruction of

6.7 Louis and Mary Leakey's australopithecines — *Zinjanthropus boisei* from Olduvai Gorge, East Africa, 1960.

the later ancestor *Homo erectus* ambushing baboons at the site of Olorgesailie in East Africa (L. Leakey 1946).

Louis Leakey believed that it was the use and manufacture of stone tools that differentiated this new ancestor from other 'missing-links.' In his discussion of why 'Zinjanthropus' was the first real human, Louis Leakey made reference to the problem of establishing the boundary between 'man and near-man.' For instance, he argued that

> Zinjanthropus represents a stage of evolution nearer to man as we know him today than to the near-man of South Africa. The dividing line between man and near-man in that dim past is certainly a hard one to draw, but arbitrarily we set it when man began to make tools for his own use ... It is precisely by his manufacture of the first known pattern of implements that I believe Zinjanthropus can claim the title of earliest man. (L. Leakey 1960a, p. 435)

6.8 Louis and Mary Leakey's australopithecines — *Zinjanthropus boisei* and its successors at Olduvai Gorge, East Africa, 1961.

While he makes explicit reference to the use of tools in the text, it is in the picture that Leakey really 'draws' the dividing line between the more human-like and more ape-like ancestors. This suggests that while the discussion about the use of tools proves that 'Zinjanthropus' was part of the human lineage, it is the picture of a very human-looking figure that convinces us of its status. The picture of the dignified and assertive-looking 'Zinjanthropus' does far more in the way of telling us that he was human, than does the statement about his tool-making abilities.

The second picture produced in association with the recovery of 'Zinjanthropus' is a reconstruction of the lifestyle of 'Zinjanthropus' in relation to its hominid successors (L. Leakey 1961; see fig. 6.8). In a sequence of panels that make up the picture, 'Zinjanthropus' features in the second from the left. In the caption adjoining the text, Louis Leakey (1961, p. 571) writes that 'wooden clubs may have helped Zinjanthropus, a true man in the tool-making sense, fell larger game, like this prehistoric zebra colt. In lean times Zinj ate rats and mice, as did all Olduvai's dwellers through the ages.' In the successive images, the more evolved hominids are shown hunting larger animals with the aid of spears. The key inference is that the australopithecines were the first ancestors to use tools or weapons in order to maintain a hunting way of life. The cracked bones found at the site were seen to confirm the earlier belief that Zinjanthropus lived largely by killing young animals. Louis Leakey (1960a, p. 433) claimed that 'we know from broken bones strewn on the "living floor" – that is, the actual site where Zinj made his rude home – that he ate small animals, the young of the giant beasts, he could not hope to kill as adults.' The difference in the conditions of the hominid skull (found subsequent to the jaw bone) and the splintered animal bones was seen to 'indicate clearly that this skull represents one of the hominids who occupied the living site; who made and used the tools and who ate the animals' (L. Leakey 1959, p. 491). Despite the fact that no wooden remains were found at the site, the image of 'Zinjanthropus' swinging a club is clearly meant to suggest that this species had acquired skills that were designated as human. The sequence of hunting scenes that make up the picture also serve to reinforce the human status of 'Zinjanthropus.'

Highly significant in the case of 'Zinjanthropus' was the application of potassium argon dating to the site, which saw 'Zinj' become the first securely dated hominid fossil (Leakey et al. 1961). The 1.75 million-year-old date attributed to 'Zinj' had fundamental importance for the place of this ancestor in the hominid lineage. As Noel Boaz (1982, p. 243) argued, while discoveries in South Africa 'wrought a profound change in attitudes towards the australopithecines,' the species was still not accepted as a hominid ancestor until the Leakeys recovered 'Zinjanthropus' at Olduvai Gorge. For instance, despite the fact that a number of authors such as Broom and Schepers (1946), Gregory (1945, 1949), Gregory and Hellman (1938), and Le Gros Clark (1947, 1948) accorded the australopithecines hominid status, there was still a strong reluctance to accept the genus as a member of the hominidae prior to 'Zinjanthropus' (see Ashton and Zuckerman 1950). With the recovery and dating of 'Zinjanthropus,' the hominid status of the genus was finally accepted by the wider professional community.

The other crucial feature that contributed to the acceptance of 'Zinj' was that the fossil was found in what was interpreted as an ancient living floor. Mary Leakey (1984, p. 121) claimed her husband was disappointed when she found 'Zinj' because it was not the remains of

Homo, which was thought to be the first 'real' hominid. However, Louis Leakey was compensated when stone tools and bone fragments were found in the same layer as 'Zinj.' The tools were seen by him as indicating that 'Zinj' was the earliest known human; a contention that was, to some extent, based on Oakley's (1949, 1956) idea that tool-making signified humanity. The discovery of bone fragments led to the image of 'Zinj' chasing animals at Olduvai Gorge and reiterated Dart's characterization of the australopithecines as the 'earliest known hunters' (Leakey 1961, p. 570). Louis and Mary Leakey also claimed that these first ancestors had left 'living floors' or campsites for archaeologists to investigate. The dense clusters of splintered bones and stone artifacts found with the remains from 'Zinj' at Olduvai Gorge were interpreted as being the remains of meals and foraging activities. Mary Leakey (1966, p. 463) characterized the 'FLK Zinj' site at Olduvai Gorge as a densely concentrated mass of flakes, chips, and small flake tools with smashed animal bones, and labelled it an 'undisturbed living floor where occupational remains were found *in situ* sealed in by subsequent deposits' (M. Leakey 1966, p. 463). The 'Zinjanthropus' site effectively became the type-site of a Plio-Pleistocene hominid camp and has since been the subject of much reinterpretation in the debates about whether such sites are the remains of hunting or scavenging activities, whether they represent home-bases or butchering sites, and whether they were created by hominids or animals (Binford 1981, 1983, 1988; Bunn 1981, 1986; Bunn and Kroll 1986, 1988; Isaac 1976, 1978, 1984; Potts 1983, 1984, 1987, 1988; Potts and Shipman 1981; Shipman 1983, 1986).

Despite the wider acceptance of this ancestor, Louis Leakey's crusade to have 'Zinj' accepted as the first 'man' was short-lived. When the remains of *Homo habilis* with its bigger brain were later found on the same 'living floor' as 'Zinjanthropus,' the latter was promptly stripped of its status as being 'man the tool-maker.' 'Zinj' went from being a 'near-man' to being designated as the victim of the larger brained hominid, who was now seen as being the one responsible for crafting the tools. The initial argument that 'Zinj' was a tool-maker was renounced by Louis Leakey (1963, p. 453), who admitted his earlier claims 'cannot be regarded as more than the expression of a pious hope'; he went on to argue that the distribution of the bones and stones 'makes it clear that "Zinj" was on the outskirts, possibly the remains of a meal.' This conclusion led to the construction of another famous image of the australopithecines.

7. COEXISTENCE AND COMPETITION

Besides the Leakeys' depictions of life at Olduvai Gorge, an image of the australopithecines that was also important in the visual tradition of depicting missing-links appears in Clark Howell's (1965) book *Early Man*.[2] Featuring a clash between *Australopithecus boisei* (or Louis Leakey's 'Zinjanthropus') and *Homo habilis* at Olduvai Gorge, the illustration precipitated the loss of the australopithecines' human status. The newly found *Homo habilis* became the first 'real' human ancestor instead (L. Leakey 1961). Based on the recovery of bones of *Homo habilis* on the same 'living floor' as 'Zinjanthropus,' this picture is again making an argument about human phylogeny in visual terms. By depicting two species fighting one another, the image suggests that the successor is the ancestor who possesses the greatest range of human traits.

This picture makes a strong case about the place of australopithecines by introducing the theme of competition between the species. On one side, down the slope of the hill, are the hairy and now ape-like and somewhat brutish australopithecines, who are holding unworked rocks. On the other side of the picture, on top of the hill, are the human-like *Homo habilis*, who confront them with their sharp-edged crafted tools. By positioning these two groups of ancestors in such a way, the picture effectively distinguishes the species with more human traits from the species that is clearly not as evolved. The result is that the superior group is included in the hominid lineage and the less evolved group is excluded. This theme of competition between hominid species is a new visual element in the iconography of human origins, and, together with other visual symbols, was used to define the boundary between what was considered uniquely human and what was not. By placing the older, more archaic ancestors in battle against the more modern and recent ancestors, the image conveys the human fight for survival, in which those that are seen to be winning are those that are more like us. Like the other icons that were introduced into the visual vocabulary of human origins, this theme of competition was subsequently reproduced and recycled in various other formats (for example, Clapham 1976, pp. 61–2; Waechter 1976, p. 80).

This picture presents also the idea that the two different hominid species coexisted in the same place. Furthermore, it conveys the inference that if two hominid species coexisted, it was likely that the more advanced of the two would be predatory towards, or at least attack, the other. This idea was based on Mary Leakey's (1966, p. 466) claim that it was highly improbable that 'Zinj' and *Homo habilis* lived together at the same 'camp,' and that 'it must be assumed that one represents the occupant and the other the intruder or victim.' By showing *Homo habilis* as the species responsible for making the variety of cutting tools and 'weapons' found at the site, the picture suggests that items of material culture were a key feature in evolutionary success.

8. ROLE-REVERSAL AND THE NEW VICTIMS

The final characterization of the australopithecines singled out for discussion is the picture of the species produced by C.K. Brain in 1968 (fig. 6.9). Here Dart's vision of the australopithecines as carnivorous aggressors is finally replaced by the image of them as the helpless victims of large cats. The image features a leopard sitting in a tree above a sinkhole. The leopard has taken an australopithecine up into the tree and is in the process of devouring it. Standing on the ground below are two other australopithecines holding sticks, who appear to be challenging the leopard. This picture constituted a radical revision of the entrenched image or stereotype of the species as hunters and toolmakers, and was reproduced in a slightly different form by Brain (1981) in his major monograph on the Swartkrans sites. In this second image (Brain 1981, p. 268), the leopard is still up in a tree over a sinkhole, but, in place of the two other australopithecines, two hyenas are standing at the base of the tree. The text explains that an australopithecine is 'being consumed in a *Celtis* tree overhanging the entrance to a subterranean cave, while hyenas of several species wait hopefully below' (Brain 1981, p. 268).

Brain's pictures communicate the hypothesis that leopards, and not australopithecines, were responsible for the accumulation of hominid and baboon remains in the deposits at Swartkrans. Based on an investigation of the site and its faunal remains, Brain conducted a major study in taphonomy, or the process by which archaeological sites are formed. While questions regarding the taphonomy of cave sites in South Africa had been raised earlier (Washburn 1957), it was Brain (1981) who systematically tested the hypothesis that other agents may have been responsible for the accumulation of material in these deposits. The significance of the reconstruction is that it presents a sudden role reversal in the depiction of our early hominid ancestors. Brain characterized early hominids as being 'the hunted' rather than

6.9 C.K. Brain's australopithecines — *Australopithecus africanus* from Swartkrans, South Africa, 1968.

'the hunters.' The dramatic image of a leopard devouring the figure in a tree more effectively makes this case than any array of words.

Brain was not the first to suggest that our ancestors were victims rather than predators. In 1894 Worthington Smith (p. 56) argued that 'primeval man is commonly described as a hunter of the great hairy mammoth, of the bear and the lion, but it is in the highest degree improbable that the human savage ever hunted animals much larger than the hare, the rabbit, and the rat. Man was probably the hunted rather than the hunter.' What is significant in Brain's contribution is that the idea, when translated into this visual form, has a far more profound impact. This does not mean that pictures necessarily say more than words, but rather that they are a powerful component of arguments, just like words. As Gooding (1989) argues in his analysis of the role of representation in research on the magnetic field, images that might start out as rough sketches can go on to acquire great theoretical and practical significance. Such is the case with Brain's illustration, where the idea that early humans were victims rather than predatory killers takes on greater meaning as a result of being translated into an image.

The notion that the australopithecines were no different from other hunted animals of the African plains is now firmly entrenched in our consciousness through the use of particular images. With these images, archaeologists and palaeontologists made an explicit attempt to move away from the tendency to humanize our early ancestors, and tried to see them as part of the animal world. The growing concern to apply ecological principles to the problem of reconstructing past hominid species (see Foley 1984; Kinzey 1987) also represents a challenge to the assumption that remains found in a stratum with fossils are effects of hominid behaviour; it makes credible the idea that animals often have a key role in the formation of archaeological sites. Brain's reconstruction reflects the concern with understanding site formation processes, and particularly the ways in which animals' bones ended up in archaeological sites (Binford 1981, 1983; Behrensmeyer 1975; Behrensmeyer and Hill 1980; Brain 1967, 1968, 1981; Gifford 1978, 1980, 1981; Hill 1976; Schaller and Lowther 1969; Schiffer 1983, 1987). These arguments were a major challenge to the practice of projecting human-like behaviours back into the past. By casting the species in a completely new role, Brain's picture captures a major shift in disciplinary thinking.

The impact of Brain's image is indicated by the fact that it was reproduced in other books and magazines (for example, Johanson and

Edey 1981, p. 67; Weaver 1985, p. 607). Of particular significance is the rendition of the thesis that appeared in *National Geographic* some seventeen years later (Weaver 1985, p. 607). This image attempts to show how the leopards would have held the head of their australopithecine victims (by the jaw), based on the discovery of puncture marks in the skulls of our early ancestors. Furthermore, the image of a leopard eating an australopithecine has been further entrenched into popular consciousness with the construction of museum displays, as for example the diorama in the gallery on human evolution at the Australian Museum (The 'Tracks through Time' exhibit; see Moser in press).

9. RECENT AND OTHER VISIONS OF THE AUSTRALOPITHECINES

There are many more images of australopithecines besides the illustrations presented here that have had an important place in the perception of the species. One particularly important example is the reconstruction, published in 1979, by Mary Leakey in *National Geographic* (1979, p. 446).[3] This reconstruction introduced another new species known as *Australopithecus afarensis*, and was based on the preservation of a series of footprints in the volcanic ash at Laetoli. It features a 'family group' of australopithecines walking across the ash-covered plains at Laetoli. Its numerous reproductions in texts on human origins (for example, Fagan 1985, p. 299; Johanson and Edey 1981) stems from the striking effect that it has. As Roe (1980, p. 108) claims, Mary Leakey's picture 'can hardly be surpassed for sheer visual and human impact.' Two other reconstructions are worth mentioning because they suggest that different lifestyles were maintained by two of the species of the australopithecines. One of the images that appear in Howell's (1965) *Early Man* shows the species known as *Australopithecus robustus* foraging for plants (pp. 64–5), while the other shows the species known as *Australopithecus africanus* fighting with hyenas over an animal kill (pp. 66–7). While the latter conveys Dart's views of the species, the first picture reflects the views of Robinson (1954), who argued that *Australopithecus africanus* and *Australopithecus robustus* were distinguished by their different diets. For instance, the morphological difference between the two species was seen as resulting from the former being meat-eaters and the later being vegetarian.

The more recent pictures of the australopithecines produced throughout the 1980s reflect an awareness of the tendency to humanize the ancestors and emphasize their dietary habits. An important example is the set of images that appeared in *National Geographic* in 1985 (Weaver 1985).[4] These pictures generally portray the species as living more like primates than humans. In the image that represents the lifestyle of *Australopithecus afarensis* (Weaver 1985, p. 595), the species is shown as living in bigger social groups, as not being fully erect all the time, and as living in close association with trees and thick vegetation, eating plant foods. The figures are not using tools, nor are they in a home-camp or a cave setting, and they are not holding sticks. Their identity is reinforced as much by what they do not have as by what they do. Similarly, in the image of *Australopithecus africanus* (Weaver 1985, p. 598), the focus is on a female figure digging for tubers. This image stands in stark contrast to earlier images of this species chasing after hyenas, waving sticks, and using bone implements. It is clear from this picture that the species is now seen as being fundamentally different from what it was once thought to have been like. This is not only conveyed in the way that individuals are shown foraging for plant foods, rather than hunting or scavenging, but it is also conveyed in the way that sexual differentiation is depicted. No longer are we presented with the stereotypical picture of the aggressive male dominating the landscape. Now we are being told that female members of the species played an active part in the evolutionary process.

In another recent image from Roger Lewin's book *In the Age of Mankind* (1988, p. 77), members of the species are shown as being tentative but fully self-aware.[5] The artist employs the same visual strategy that was introduced by Louis Leakey in his reconstruction of 'Zinjanthropus,' whereby the australopithecine figures are shown confronting the viewer with a direct gaze. The look that Lewin's australopithecines possess is distinctly confrontational, yet there is a degree of apprehension. It is as if we by chance have come upon them whilst walking across the grassy plains. Their gaze is compelling and suggests that they too are investigating us and are just as curious about us as we are about them. Much has recently been written about the way in which the subjects in pictures are shown looking or gazing (for example, Byars 1988; Lutz and Collins 1992; Mulvey 1975; Rodowick 1982). In light of the arguments made about the power and significance of 'the gaze,' it appears that we have finally encountered the new fully self-conscious missing-link. The knowing look that has been conveyed in their eyes suggests that even if we do not know what their place in evolution is, they certainly do.

10. VISUAL REPRESENTATION AS A FORM OF REASONING IN ARCHAEOLOGY

Some time ago scholars such as Gombrich (1968) and Arnheim (1969) gave a new dimension to the study of visual imagery by using principles from psychological and linguistic theory. Both authors developed the argument that images constitute a form of reasoning that written text cannot replicate. In this context, visual reconstructions of prehistoric life constitute an important form of reasoning that archaeologists employ, and one of the most fascinating aspects about them is the epistemological significance that they have. For instance, these pictures achieve much in the way of convincing us that they are a reasonable explanation of the data, because they make use of a range of icons and symbols that draw on our own human experience. They are fundamentally different from other types of archaeological illustration – such as stratigraphic sections, models, or diagrams – in the sense they are presented in a naturalistic format that is a highly familiar form of representation. They are full of what Myers (1990, p. 235) has described as 'gratuitous details,' the elements in the picture that do not seem relevant to the actual claim that is being made. Trees in the background of a reconstruction drawing are gratuitous in the sense that they are not central to the argument, yet they serve to make the picture continuous with our world. While the focus in a reconstruction may be on a figure hunting or making a stone tool, the cave setting, the scatter of artifacts and debris, and the details of the environment serve to reinforce the validity of the claim. These elements become more than just circumstantial evidence; they become integral to the behavioural interpretation that is being made in the reconstruction.

Furthermore, in these reconstruction drawings new ideas were often expressed with reference to previously published images. While many iconographic elements were reproduced and recycled, new elements were enlisted to make an even stronger case about the status of the ancestor in question. Unlike some sciences, such as chemistry, that have moved towards using more and more abstract forms of illustration that rely on the ability of a specialized audience to understand their meaning, archaeology continues to use these realistic types of illustration at the same time as developing a more abstract visual language. While contemporary archaeologists are more reluctant than the early professionals to utilize the realism of reconstruction drawings to present their arguments, they continue to participate in the reconstruction enterprise. This is partially in response to the pressure from publishers and major magazine organizations to present such images, but it is also a result of the interest that we still have in picturing what the past looked like. The discipline has developed in such a way that these pictures remain necessary to our professional discourse. For instance, without them it is difficult to reason about the human and non-human status of our fossil ancestors. The illustrations of our hominid ancestors are an integral part of the dialogue about the hominid lineage, and we will continue to use them as long as the debates about human phylogeny are carried out. In considering the place of these pictures in archaeology, it is clear that they are not only a powerful form for reinforcing arguments, but, moreover, that the practice of illustrating ideas is tantamount to reasoning in the discipline.

NOTES

1 I would like to thank Meg Conkey and Lori Hager for inviting me to participate in the symposium 'Envisioning the Past – Visual Forms and the Structuring of Interpretations' held at the American Anthropological Association Meetings in San Francisco in 1992, where I first presented this paper. I would also like to thank Clive Gamble and Brian Molyneaux for inviting me to participate in their session 'Visual Information and the Shape of Meaning' at the Theoretical Archaeology Group Conference held in Southampton in December 1992, where a version of this paper was presented. Finally, I am very grateful to Alexander Zahar, Penelope Allison, Brian Baigrie, and Roland Fletcher for useful comments on the text.

2 Unfortunately, it was not possible to obtain permission to reproduce this publication. However, because it is a very important reconstruction, I have elected to keep the discussion of it in the text.

3 A section in this paper was devoted to the significance and place of this image in the history of research on the australopithecines; however, because the artist refused to give permission to reproduce the image, I have omitted this discussion from the text.

4 The artist refused permission to reproduce any of these images.

5 The author was unable to reproduce this image.

7. Towards an Epistemology of Scientific Illustration

DAVID TOPPER

Scientific illustrations are not frills or summaries; they are foci for modes of thought.

Stephen Jay Gould (1991), p. 171

1. INTRODUCTION

For several decades, art historians, psychologists, philosophers, and other theorists have been directing much effort towards understanding the nature of visual imagery. Nevertheless, a reading of this literature reveals that little has been directed towards the study of scientific illustration. As the art historian Samuel Y. Edgerton, Jr (1985, p. 168) puts it, 'few scholars have ever sensed that it [i.e., scientific illustration] has any historical interest. Most art historians have disdained it'; except, of course, when an illustration comes from the hand of a genius, such as Leonardo da Vinci. 'Historians of science have shown a little more curiosity, but they too tend to treat scientific pictures only as after-images of verbal ideas.' Yet Edgerton contends that scientific illustrations comprise 'a unique form of pictorial language,' which by 'symbols and conventions' convey information (1985, p. 168).

Similarly, in a pioneering article on visual imagery in geology, Martin Rudwick chastises contemporary historians of science for ignoring 'the strong visual component of the original source-materials.' Even when historians do reproduce geological illustrations in an article, notes Rudwick (1976, p. 149), the pictures usually have a mere decorative role – 'they are rarely integrated with the text.' He, like Edgerton and others, sees this as a reflection of the more general phenomenon regarding visual matters in 'the hierarchy of our educational institution,' where 'visual thinking is simply not valued as highly as verbal or mathematical dexterity' (1976, p. 150). This provides Rudwick with a rationale for his paper: 'A study of the conceptual uses of visual images in an early nineteenth century science may help in a small way to counter the common but intellectually arrogant assumption that visual modes of communication are either a sop to the less intelligent or a way of pandering to a generation soaked in television' (Rudwick 1976, p. 150).

As well, in a provocative article on the 'nonverbal' component of technology, Eugene S. Ferguson (1977, p. 835) states that visual imagery has been ignored by historians of technology 'because its origins lie in art and not in science.' In his discussion of illustration in technology, he shows the key role visualization has played in the thought processes of inventors. Ending with pedagogical matters, Ferguson deplores the demise of mechanical drawing in engineering curricula (perhaps his impetus for writing the article?).

Although the articles by Edgerton, Rudwick, and Ferguson (to be discussed in some detail later) are relatively isolated cases, they (along with the other articles discussed below) at least provide a starting point for delving into a subject that has been neglected far too long. This paper, accordingly, has a major twofold purpose: to provide a much needed overview of the literature and, in the course of this perusal, to lay out some of the historical and philosophical issues at the heart of the matter. Another issue, which runs throughout this paper as a minor theme, is the question of the demarcation between art and science. If this exposition clarifies some problems and stimulates some projects, it will have served its purpose.

2. A FRAMEWORK

Anyone who has read a popular account of the theory of relativity will remember that, at the start, the reader usually is eased into the principle of relativity through imagery of a familiar sort – specifically, the experience of riding in a train. Often this is made explicit with a picture: an illustration of a train, an embankment, and some necessary observers. Einstein's own popular account, *Relativity*, contains such an illustration (1961 reprint, p. 25), although of an abstract variety (see fig. 7.1).

IX

THE RELATIVITY OF SIMULTANEITY

UP to now our considerations have been referred to a particular body of reference, which we have styled a "railway embankment." We suppose a very long train travelling along the rails with the constant velocity *v* and in the direction indicated in Fig. 1. People travelling in this train will with advantage use the train as a rigid reference-body (co-ordinate system); they regard all events in

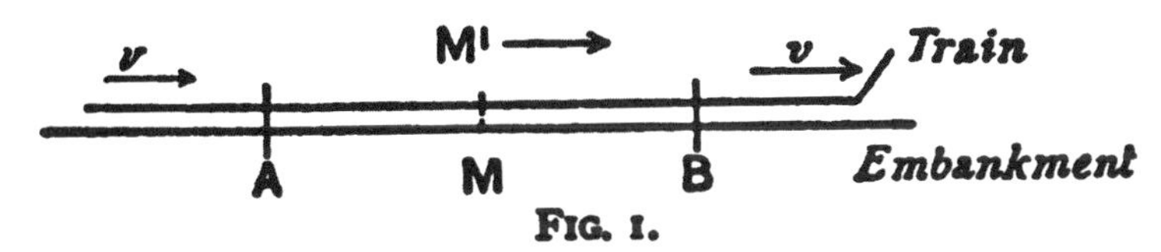

Fig. 1.

reference to the train. Then every event which takes place along the line also takes place at a particular point of the train. Also the definition of simultaneity can be given relative to the train in exactly the same way as with respect to the embankment. As a natural consequence, however, the following question arises:

Are two events (*e.g.* the two strokes of lightning *A* and *B*) which are simultaneous *with reference to the railway embankment* also simultaneous *relatively to the train*? We shall show directly that the answer must be in the negative.

When we say that the lightning strokes *A* and *B* are

25

7.1 Illustration of train and embankment (Albert Einstein 1961).

What role do such diagrams play, beyond the obvious one of providing a prosaic and visual context for grasping the concept of the relativity of motion? One answer is that eventually the illustrations become superfluous. The train and embankment are replaced by two coordinate systems, and ultimately everything is embodied in the Lorentz transformation. Since the essence of the problem is encompassed by the equations, the illustrations, at best, play merely a pedagogical role. Accordingly, they have no epistemological value.

The mode of thinking entailed in this answer is, I believe, symptomatic of the common belief that in human consciousness there is a cognitive hierarchy: from visual perception, which is at the 'bottom,' to language (and this would include mathematics and, today, computer 'language'), which is at the 'top.' The process of moving up the hierarchy is one of abstraction – at once a retreat from the sensual and an advance towards the intellectual. A variation of this viewpoint is the idea that human thinking itself takes place in words, and only in words; without language, therefore, there is no thought and hence no perception. An extreme version of this position – what has been called 'linguistic determinism' – is the tenet that humans visually perceive only what they have words for. Although containing the truth that language and perception are often interactive, taken literally this is certainly an untenable position – contradicted by the most trivial of perceptual experiences.

Nevertheless, the idea of such a hierarchy – from sensual knowledge to 'pure' thought – forms a thread that runs throughout so-called Western thought, from Pythagoras to the present. Indeed, I would say that it lies at the core of why, until recently, the subject of scientific illustration has hardly been studied as a topic unto itself, especially by historians of ideas (including historians of science).

A recent work that, from its title, would seem to rectify this is Barbara Marie Stafford's, *Voyage into Substance: Art, Science, Nature, and the Illustrated Travel Account, 1760–1840* (1984). This profusely illustrated book is a comprehensive study of travel accounts of naturalists and explorers, most of whom also produced drawings of flora and fauna in the regions explored. In Stafford's words (1984, p. xx), these accounts combined 'literature of fact and pictorial statement' such that 'descriptive word [was] wedded to accurate image.' The seventeenth-century scientific attitude itself formed an intellectual underpinning to the voyages. As she writes: 'The Baconian imperative to get to the bottom of physical things ... is mirrored in a leitmotif of the travel account; the profoundly expressed need to penetrate the inward substance of natural particulars.' Voyaging was an inductive process, the goal being to 'unmask nature' (Stafford 1984, p. 284). This belief formed the basis of both the travellers' attitude towards the written descriptions and their visual depictions of what they saw. There thus seems to have been no inkling of an epistemological problem either in the relationship between the written word and the visual image or between the accounts themselves and the 'world' being described (or nature being 'penetrated'). The connections were made easy by eliminating metaphorical language and writing in a 'plain' and 'masculine' style. The pictures then supplemented the text. 'Since the empiricists postulated that knowledge was conveyed in the form of sensible images, painting could help to dispel any lingering obscurity still clinging to words' (Stafford 1984, pp. 51–2). Of course, none of this is really surprising, given the historical context of these travel accounts.

It is surprising, however, to find that Stafford herself apparently also believes this story. At least I could find no evidence to the contrary. Although there are 270 illustrations in her book, she rarely refers to them or discusses them in any detail. Instead, the pictures sort of tag along behind the written text – as if they speak for themselves, visually complementing the words. Even though Stafford analyses the historical context of the *written* accounts (pointing to personal biases, ideological factors, and so forth; what is often called 'theory-ladenness'), it does not seem to have occurred to her that this analysis could apply to the *pictures*, too. Only the text, it seems, is of epistemological interest. Thus, despite being profusely illustrated, Stafford's book, at least tacitly, reinforces the hierarchical hypothesis.

A cursory perusal of the pictures in her book reveals, however, that the images are fraught with elements of the 'picturesque' style. Popular at the time, this style (based heavily upon the art of Claude Lorrain and various Dutch artists) entailed depictions of tranquil landscape scenes, frequented with Roman ruins or windmills, often containing a requisite wayfarer or peasant wandering down a lonely path, usually framed by the bending boughs of strategically placed trees, and bathed in the golden light of dusk or dawn. In the eighteenth century, some wealthy estate owners in England attempted to create similar scenes in a real landscape; to make, in other words, an English garden look like a painting by Claude!

What is therefore quite remarkable about the illustrations in Stafford's book, ostensibly depicting people and places around the non-

European world, is how many of them are picturesque vistas, containing too many European motifs to be straightforward depictions of what the artists 'saw.' Rather, these facts point towards the theory-ladenness of *pictures.* Perhaps today, more than ever, theorists should be more conscious of this, especially with the demise of formalism and structuralism (in both literary and visual art studies). In this postmodern world, art works (pictures or texts) are less viewed as hermetically sealed artifacts – isolated from cultural, political, psychological, and other environs. Rather, they are 'theory-laden,' so that pictures – like words – must be 'read.'

Acknowledging the framework of a language of art implies cognitive factors in pictorial perception. This, in turn, threatens the hierarchical hypothesis by bringing the faculty of mental processing 'down' to the perceptual level. One key theorist in the field, Rudolf Arnheim, coined the phrase 'visual thinking' (used above by Rudwick) to delineate this act. In fact, Arnheim (1969; cf. Root-Bernstein 1985) has gone so far as to speak of 'perception as cognition,' thus postulating a theoretical position antithetic to the hierarchical hypothesis and/or linguistic determinism.[1]

3. BACKGROUND

The modern history of scientific illustration may properly be said to begin in the sixteenth century with the introduction of scientific texts accompanied by printed illustrations. Most surely, there is a prehistory of the subject, beginning in antiquity when Egyptians illustrated papyrus scrolls. The Greeks of the fifth century B.C. also illustrated literary works (e.g., Homer); they probably illustrated scientific works as well. Scholars believe, for example, that Dioscorides' *De materia medica* (ca. A.D. 65) – the most important herbal for over a millennium – was illustrated by the author. The original manuscript, of course, is no longer extant; the oldest copy is in the *Anicia codex* (ca. 512), an extensively illustrated manuscript (Riddle 1985, pp. 176–217; Anderson 1977, pp. 3–10).

Illuminated manuscripts thus remained the major artifacts of all illustrated texts (not only scientific ones) until the invention of print making in the Renaissance. Prints constituted a key turning point in the history of imagery; they made possible the communication to a wide audience of identical visual information, or, in the apt phrase of William M. Ivins (1953), of 'exactly repeatable pictorial statements.'

Thus, in the fifteenth century, when Europeans began producing repeatable words (that is, using the invention of the printing press with moveable metal type), techniques for making woodcuts, engravings, and etchings were also invented. Later, in the sixteenth century, books with text and pictures appeared (Ivins 1953), and thus arose what may be called the scientific illustration in the usual sense of the term.

Except for some of the famous profusely illustrated books (several mentioned later in this paper), imagery in many textbooks was often confined to the frontispiece, perhaps because engraving was an expensive process. Such print-making was a three-step progression – from the artist's drawing, through the engraver's plate (artist and engraver were not always the same person), to the making of the print. This process was simplified with the invention of lithography in the last decade of the eighteenth century, since it eliminated the middle-person; however, this medium was not widely used for scientific illustration until the 1820s. About the same time, wood engraving was revived and widely used; this medium was particularly important since it did not require special paper, so that both the illustration and the text could appear on the same page (Rudwick 1976, p. 157). Eventually photography eclipsed virtually all other forms of illustration, minus some key exceptions to be noted later (Twyman 1970b, p. 225).

4. NATURAL HISTORY AND THE LIFE SCIENCES

This brief historical survey of some the salient points in the material history of illustration thus sets the stage for the following discussions of specific areas of scientific illustration. A paragraph from David Knight's article 'Scientific Theory and Visual Language' (1985, pp. 106–7) is a good starting point:

> Scientific illustrations are pictures designed not to stand on their own, but to accompany a text: they are in partnership with prose (or, occasionally, as in Erasmus Darwin, verse) intended to convey knowledge old or new. They may be diagrams: showing apparatus with, or in our century without, an elegantly cuffed hand holding it; illustrating mathematical propositions, like the epicycles in Copernicus, or psychological ones, as in works of physiognomy or phrenology; or, like the only illustration in the *Origin of Species*, showing hypothetical divergence and extinction over time. Some are thus more theory-laden than others, but even diagrams of apparatus (especially without hands or supports) make sense only to those who have learned the conventions. They are concise

visual languages which must be taught before they can be read, and sooner or later get out of date. In scientific illustration, the artist's intentions can be illuminated by the text; they are pictures with a clear context.

Knight's paragraph constitutes less a definition of 'scientific illustration' and more an exemplification of what may be construed as legitimate subject matter – and, as such, approaches a delineation of the boundary conditions.

Knight's key idea is contained in both the first and last phrases of the above quotation: that scientific illustrations 'are pictures designed not to stand on their own' but are 'pictures with a clear context.' The context – the scientific theory contained in the text – is then a source of the theory-ladenness of the pictures; and, indeed, most of Knight's paper is taken up with excellent examples of the theory-ladenness of scientific illustration drawn from eighteenth- and nineteenth-century natural history works. As well, he notes that the consequent phenomenon of image 'reading' has a historical dimension – something usually taken for granted for the written text, but not for the pictures. He also reminds us that the text may illuminate the artist's intentions; in other words, a text may supplement a picture, rather than vice versa (as usually assumed).[2]

In a superb article on natural history, 'Taking It on Trust: Form and Meaning in Naturalistic Representation' (1990), Martin Kemp also stresses the theory-ladenness of scientific illustrations. Although 'naturalistic representation in natural history appears to be quite straightforward,' he contends that the artist and viewer are dependent upon 'a complex interaction of prior knowledge, automatic expectation, illustrative technique, emotional context and the given framework of verbal information' (1990, pp. 127–8). Accordingly, Kemp provides many examples of this interaction, focusing on two topics: the rise of naturalism in Renaissance natural history illustration, and the depiction of the motif of the jungle in the eighteenth and nineteenth centuries.

The complexity of the early history of naturalism may be seen in the case of the representation of two salamanders in Conrad Gesner's *Historia animalium, II* (1554), an important treatise in this history (fig. 7.2). Here Gesner juxtaposed a 'stylized' salamander (clearly a fanciful copy based upon numerous other copies) with a 'real' one (ostensibly drawn from life), thus making explicit the difference between copied pictures and depictions of, so to speak, the real thing. Today's viewer

De Salam. A. Lib. II.

7.2 Salamander (Conrad Gesner 1554).

may applaud Gesner's 'real' salamander, but one's judgment is quickly tempered by the realization that his treatise also contains mythological creatures such as the 'man-ape' (fig. 7.3).[3]

In his discussion of the motif of the jungle, Kemp introduces an important perceptual fact when he notes that the influence of Romanticism (particularly that found among some artists) 'played an essential role in the recognition of the dynamic drama of nature which came to feature so prominently in the writings of the nineteenth century naturalists' (1990, p. 133). Thus, in contrast to his discussion of Renaissance art, where he argues that theory-ladenness may limit naturalistic depiction, here Kemp implies that the Romantic sensibility played a positive role in alerting naturalist-artists to 'the sublime savagery of nature' (1990, p. 134). But he is aware, nevertheless, that this sensibility may be regressive, as in the cases of anecdotal anthropomorphizing or extremely contrived settings for pictures (1990, pp. 135–6). In short, theory-ladenness in depiction – as in all facets of perception – is a double-edged sword: on the one hand, concepts aid us in seeing what may otherwise be missed; on the other hand, they also can impede us in recognizing something that does not fit the given categories but which may, in fact, be sitting in front of our noses. In the

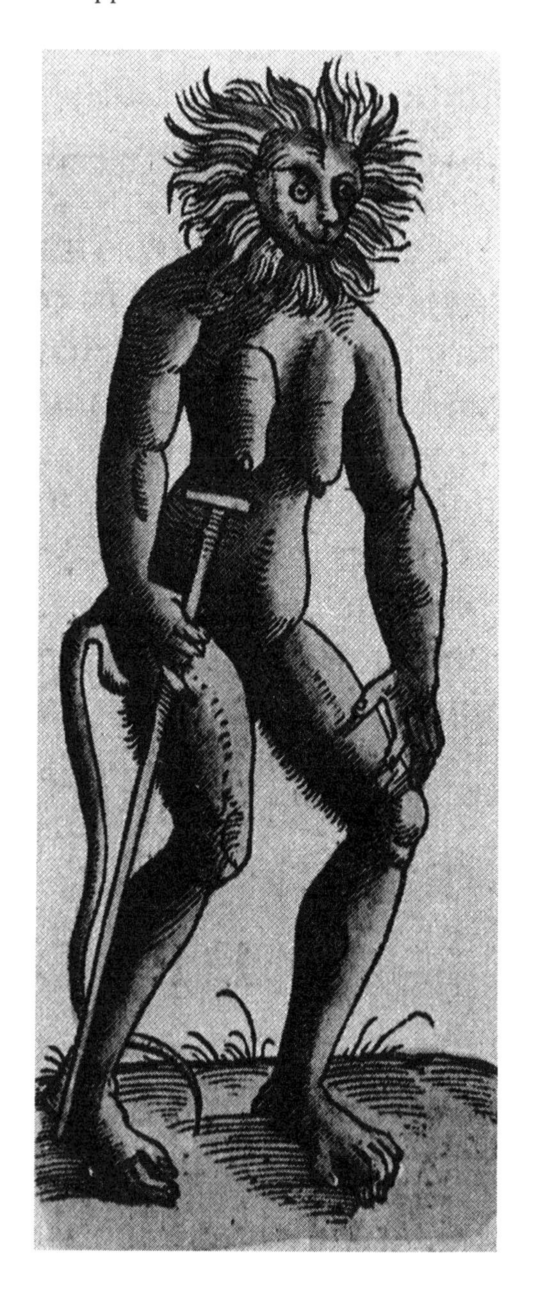

7.3 Man-ape (Conrad Gesner 1551).

end, Kemp returns to the truism that all images 'exist within a framework of belief,' and, therefore, 'to be read properly such images require a highly developed ability to know what the illustrator is requiring of us' (1990, p. 137). Thus, both Kemp and Knight emphasize that scientific illustrations exist within a given context, and therefore the theory-ladenness of scientific illustration is acknowledged by these contemporary authors on natural history illustration.

It is instructive to recall that earlier scholarly work on the subject took place within the framework of the rationalism/empiricism polarity. It was the art historian Erwin Panofsky, who, in a now classic 1952 lecture, 'Artist, Scientist, Genius: Notes on the "Renaissance Dämmerung,"' proposed that the rise of empiricism in the Scientific Revolution of the seventeenth century had its roots in the art of the fifteenth and sixteenth centuries. In particular, Panofsky was challenging a thesis of the eminent historian of science George Sarton, who had categorized the Renaissance (from a 'scientific point of view') as merely 'an anticlimax between the two peaks' (the fourteenth and seventeenth centuries) – a position Sarton later modified. Panofsky, in contrast, claimed that 'some of the achievements of the arts' made 'vital contributions to the progress of the sciences' (Panofsky 1962, pp. 127–8). As he wrote:

> ... the rise of those particular branches of natural science which may be called observational or descriptive – zoology, botany, paleontology, several aspects of physics and, first and foremost, anatomy – was so directly predicated upon the rise of the representational techniques that we should think twice before admitting that the Renaissance achieved great things in art while contributing little to the progress of science. (Panofsky 1962, p. 140; this article is an expansion of the original 1952 lecture)

The same point is made in another pioneering (although frustratingly cryptic) article, 'The Role of Art in the Scientific Renaissance,' by the science historian Giorgio De Santillana. In the Renaissance, De Santillana (1969, p. 33) claimed, there were 'the direct contributions of art in the rendering of observed reality.' This point was echoed by A.R. Hall (1960, p. 369) when he wrote that 'it was the artist, not the man of learning, who returned to the natural model,' so that 'naturalism is older in art than in science' (p. 29).

This theme has been repeated in some recent work. James S. Ackerman, in two articles, 'The Involvement of Artists in Renaissance Science' (1985a) and 'Early Renaissance "Naturalism" and Scientific

Illustration' (1985b), argues that verisimilitude in imagery (or naturalism in art) was a key factor in the return to nature and hence in the rise of empiricism in the Renaissance and in post-Renaissance Europe. Ackerman includes in this framework the geometric underpinnings of art, such as perspective and foreshortening and the study of human proportions (e.g., Albrecht Dürer's books on the subjects). This material, also mentioned by Panofsky, was utilized by artists who illustrated books in natural history, anatomy, and other scientific subjects.

In the second article (1985b), Ackerman specifies another way in which Renaissance art influenced science. When early naturalistic drawings first broke stylistically from their medieval antecedents, they tended to be depictions of individual specimens; the artists, so to speak, drew literally what they saw – warts and all. But later a shift took place towards depicting types, ideals, or species. This is a facet of natural history illustration mentioned by several authors; for example, Kemp (1990, p. 129) speaks of it in terms of depicting 'archetypical' images instead of 'particulars' or 'accidents' of the specimen in question.

One might thus think that once naturalism was achieved – and hence the idealized scientific illustration as an artistic genre unto itself was born – then the rest of the story would consist merely of lists of treatises illustrated by competent naturalist-artists depicting what they saw. But, in fact, this is far from the truth. Despite what may be called the heyday of scientific illustration (approximately mid- to late sixteenth century, which saw, for example, such masterpieces as Vesalius's *De fabrica* of 1543), there followed a period of extensive copying from other illustrators (recall Gesner's 'stylized' salamander) rather than from nature! Indeed, most illustrations were copies of other illustrations, some going back through medical and botanical manuscripts to classical texts, such as those of Theophrastos, Dioscorides, Galen, and Pliny (Arber 1953; Pächt 1950, p. 26; Reeds 1976, p. 520). As Marie Boas Hall (1962, p. 52) aptly put it, 'the illustrations illustrated the text, not nature – a peculiar view, no doubt ...' Or, in the words of Wolfgang Harms (1985, p. 80): 'Libraries [in the sixteenth century, rather than nature itself] remain the primary source of information on natural history' (see Hoeniger 1985, pp. 145–6). Thus, despite the development of naturalistic art in the Renaissance, many illustrations were still copies of other illustrations rather than drawn from actual specimens; this sometimes paralleled the written text, which too was often a copy of an earlier one.

This point is well documented in two articles by William Ashworth, Jr, 'Marcus Gheeraerts and the Aesopic Connection in Seventeenth-Century Scientific Illustration' (1984) and 'The Persistent Beast: Recurring Images in Early Zoological Illustration' (1985), which also provide a variation on the theme of conventionality in art. In the first article (1984), Ashworth shows that a woodcut of a blacksmith in William Gilbert's *De magnete* (1600) is taken from an illustration in a late sixteenth-century book of Aesop's fables; thus, a scientific illustration is borrowed from a non-scientific source. In the second article, Ashworth (1985, p. 46; cf. Cole 1953) reveals that the famous rhinoceros drawn by Dürer 'was the prototype for virtually all succeeding rhinoceros illustrations for the next two hundred years.' Moreover, this example is 'not the exception but the rule'; in the rest of the article, he shows the 'persistent beast' motif for the shark, Egyptian mongoose, sloth, and 'man-ape' (recall Gesner, again).

In the milieu of Renaissance naturalism, why did such copying take place at all? Ashworth's answer (which is not entirely convincing) is that the artists believed that the original illustrations were drawn from life (which was true, in many cases), and hence, being firsthand observations of the specimen in question, they could be copied as such (Ashworth 1985, pp. 65–6). This may indeed have been the attitude of the artists (although Ashworth presents it as a hunch, not with documented evidence). But perhaps more importantly it is a manifestation of something else that was happening among artists in general at the time. With the development of naturalism in the fifteenth and sixteenth centuries, there followed a proliferation of artist's sketchbooks, patternbooks, and manuals (such as Dürer's, noted above); these provided a storehouse of stock images for novice artists who were learning to draw. They were also used by accomplished artists and formed the basis of what later became the 'Academic' style (Gombrich 1968). Ironically, then, these naturalistic images gave impetus to the practice of copying.

But, one is pleased to report, this was not always true. In the sixteenth century, some illustrations *were* drawn from life (witness the 'real' salamander in Gesner). Yet, as Paul Hulton (1985) has observed, conventionality still was not entirely eliminated; for example, in sixteenth-century depictions of non-Europeans (such as North American aboriginals) drawn from life for ethnological texts, the artists employed conventional poses from classical art (fig. 7.4). This reveals a source of theory-ladenness from art *itself* (that is, from the various conventions of

7.4 Indian in body paint (John White, ca. 1585).

depiction) and not from science (that is, the written text). This situation is analogous to the picturesque mode of depictions of non-European scenes in topographical illustration, noted before. (Of course, this simply underscores the fact that all image-making contains elements of conventionality – a point that many artists seem to have acknowledged [cf. Constable's views on this in the discussion below]). Still, these artists of ethnology were at least attempting to draw what they saw – the scene or native before them – despite their employment of Western motifs. Indeed, such attempts signal an important shift in illustration towards the study of nature itself – supposedly what the work of the naturalist and ethnologist is all about.

It should not be forgotten, moreover, that naturalism in the depiction of some subjects actually began *before* the Renaissance. G. Evelyn Hutchinson's study of the depiction of birds and other animals in some medical manuscripts at the end of thirteenth and beginning of the fourteenth centuries focuses on the emergence of a new naturalism. In 'Attitudes toward Nature in Medieval England: The Alphonso and Bird Psalters' (1974), Hutchinson attempts to identify specific species of birds (and some other animals) within these manuscripts (cf. White 1947). This naturalism began with the representation of plants in sculpture and spread to illustrations in herbals (Pächt 1950); as well, increasing naturalism in animal representation replaced the stylized forms as seen primarily in the bestiaries. Of course, the context of many of the images was still symbolic. Hutchinson, rather, is concerned with how accurately the species are depicted; he specifies when certain depictions are inaccurate and when, on the contrary, the artist's work 'shows a real taxonomist's power of observation' such that the art was probably 'made from actual specimens' (1974, p. 23). Hutchinson saw his work as a form of natural history, or a naturalist's foray into the history of science (1974, p. 6) – in contrast to the work of the art historian, who catalogues such depictions and specifies their symbolic significance (e.g., Klingender 1971; Toynbee 1973).[4]

5. REMARKS ON THE MATTER OF DEMARCATION

A question has been looming behind much of this discussion so far: by what criteria can the demarcation between a work of art and a scientific illustration be determined? This matter is raised indirectly by Knight, when discussing a change in natural history since the eighteenth century. As biologists replace naturalists in the mainstream of science, less emphasis is placed upon the overall external features of organisms: 'illustration has become less crucial than it was – fine discrimination may depend on the bones of the ear of a bird (which can be illustrated, but may not generate great art) or chemical investigation of egg-white proteins, which does not lead to pictures at all' (1985, p. 124). With a change in science itself, Knight goes on, an older illustration may lose its value as a work of science, but 'if it is the work of a great artist, it may pass time's test and live on, passing into "art"' (1985, p. 124). Ironically enough, Knight seems to miss the point that his argument only holds for 'science' as defined by contemporary standards, for surely historians of science, like himself, still view past illustrations as 'science.' Furthermore, why should it take a change in science for the illustration to pass into art; cannot the picture be so viewed at any time?

Another variation of this issue arises in Kemp when he discusses the relationship between the English artist George Stubbs, renowned for his paintings of animals (particularly horses), and the physician William Hunter. In September 1770, Hunter commissioned Stubbs to paint a moose (fig. 7.5), recently imported from Quebec, and to make 'an exact resemblance of the young animal itself, and a pair of horns of the full grown animal' (quoted in Kemp 1990, p. 133). The painting now hangs as a work of art in the Hunterian Art Gallery in Glasgow. But Hunter's original reason for commissioning Stubbs grew out of his interest in comparative anatomy and the pre-evolutionary debate over the possibility of extinct species (indeed, he commissioned other such works); he wished to 'preserve' the image of the moose as the basis of comparison with fossil remains of the Irish elk. After telling us of this scientific reading of Stubbs's painting, Kemp goes on to discuss another of Stubbs's animal pictures – the famed *Horse Being Attacked by a Lion* (ca. 1762) – and speaks of it as 'a more "purely artistic" aspect' of his work (1985, p. 134). That Kemp places 'purely artistic' in quotation marks reveals his apparent ambivalence regarding the demarcation between art and science. Both pictures focus on anatomical accuracy and hence have a 'scientific' component; yet, the *Horse* is found in art books and hence seems to have a more 'artistic' intent. But what of this demarcation problem: is there a clear distinction here between art and science?

A.R. Hall (1960, p. 30) seems indeed to postulate such a distinction when he remarks in passing that 'the romantic nature-lover may not be a good biologist.' Unfortunately, Hall does not develop this idea. The distinction is also implied by Philip C. Ritterbush (1985) when he

7.5 *The Moose* (George Stubbs 1770).

asserts that mere depiction of form is not science, since science requires analysis. Ritterbush attributes this view to De Santillana, although I cannot corroborate his reading (De Santillana 1969, passim). But, without a doubt, Ritterbush (1985, p. 162) believes in the distinction, for he declares: 'Representations of objects become scientific not when they become recognizable but when they serve to convey knowledge of nature.' Regrettably, he does not elaborate upon the phrase 'convey knowledge.' I would assert, on the contrary, that recognizable images do convey visual knowledge.

Furthermore, the argument appears to be misguided: it implies a simplistic ahistorical view of science, such that natural history is not 'real' science until biology (as we know it today) arose. But, as queried before: is it really viable to assert that biologists, because they dissect and analyse, are scientists, and naturalists, because they don't, are not? Of course, this issue is a subset of a much larger philosophical matter – namely, the demarcation between science and non-science (or pseudo-science).

The complexity of this issue, and how it is related to the blurring of the distinction between not only science and pseudo-science but between art and science themselves, may be seen in the following case of the eighteenth-century Dutch artist and anatomist Petrus Camper. His work on physical anthropology (published posthumously in 1791) later formed the basis of craniometry – that is, the theory of human typology by measurement of the head – and, in particular, the racist theory of the so-called facial angle (basically the slope of the profile from the forehead through the upper jaw; fig. 7.6). Scientific racists exploited Camper's ranking of human types, in which (not surprisingly) white Europeans were deemed closer to the ideal beauty of ancient Greek sculpture and Africans were closer to the apes. Although I do not wish to vindicate Camper in any way, it is nevertheless important to point out that later racist applications of his treatise fundamentally distorted his original work. His ranking was not about intellect, but beauty; in particular, he wanted artists to have a schematic means for drawing Africans correctly (not as black Europeans, as was usually done), especially the black Magus in Nativity scenes. It is ironic, therefore, that Camper 'became a villain of science, when he tried to establish criteria for art' (Gould 1991, p. 240). So, again, an artistic artifact (Camper's treatise, as originally intended) is interpreted scientifically (or, more correctly, in this case, pseudo-scientifically) by later interpreters. Most surely, historians should cringe at what racists have done. Nevertheless, we are compelled to admit that from an epistemological viewpoint the racists' handling of the 'text' is no different from Hutchinson's. In short, it seems, that a scientific or artistic 'reading' of a picture is fundamentally a matter of context, not content (cf. Topper 1990a).

6. FROM GEOLOGY TO LANDSCAPE ART

Thus far, the discussion of scientific illustrations has been confined mainly to the subject of natural history, the life sciences, and some allied fields. This is reasonable since specimens of flora, fauna, and the human body are most amenable to visual illustration. Closely related is

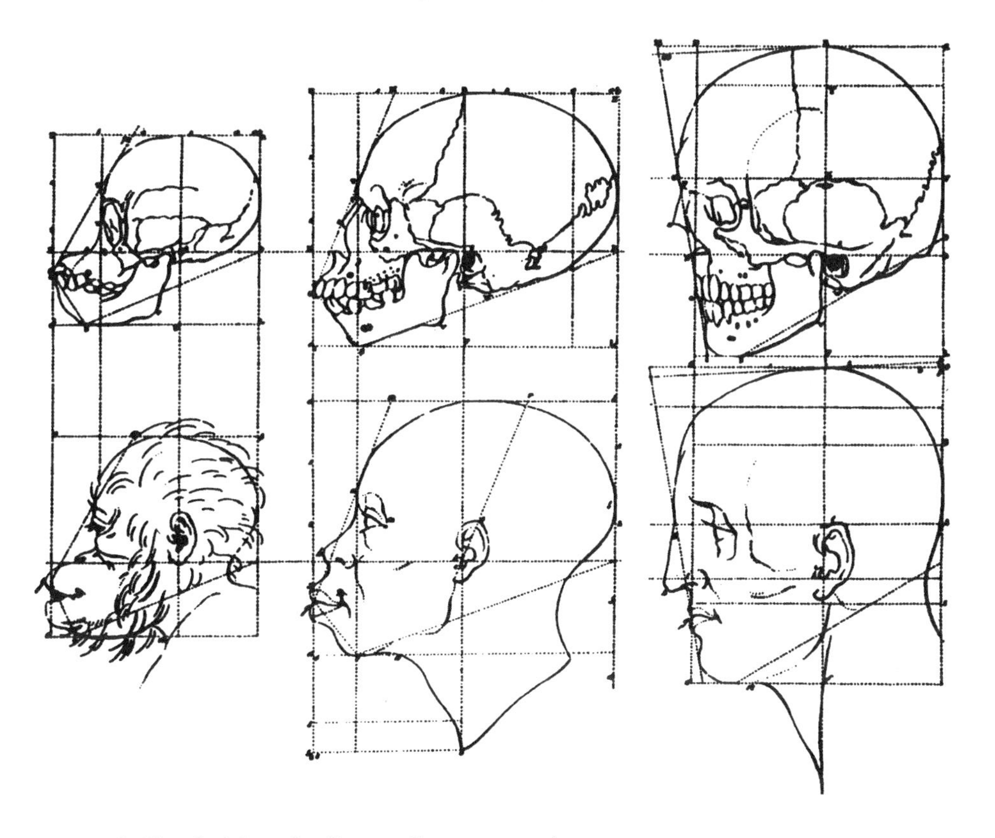

7.6 The facial angle (Petrus Camper 1791).

geology, which, until the Scientific Revolution, was classified as 'natural history'; minerals were one of the three kingdoms (animals and vegetables being the other two) in the organic world-view inherited from the ancients. Rudwick's article on geology, 'The Emergence of a Visual Language for Geological Science' (1976), shows how a visual language arose in the field. This extends the range of subject matter considered so far. Not only are specimens (rocks, minerals, fossils, along with flora and fauna) depicted, but also maps (comprising horizontal and vertical sections and aerial views) and landscapes – in general, 'configurations that could not be adequately conveyed by words or mathematical symbols alone' (1976, p. 151).

Not surprisingly, geological images entail theory-ladenness: 'even the most innocuously "documentary" landscape inevitably embodied some kind of theoretical content' (1976, p. 175). As an example, listen to Rudwick's analysis of an engraving of the eruption of Mount Vesuvius in 1767 (after a painting by Pietro Fabris, from William Hamilton's *Campi phlegraei* [1776]):

> Fabris's dramatic contrast between the violence of an erupting Vesuvius and a calm neo-classical foreground could not help conveying an implicit message about the power of 'the operations of nature' in relation to the life of man. (Rudwick 1976, p. 175)

This brief quotation shows further why it is necessary to consider matters of media and style, along with content, when considering the epistemological nature of scientific illustration.

Rudwick also reveals (with numerous examples) an interesting parallel between geological depictions and those of flora and fauna. As he shows, there was a historical progression from a detailed, specific, or concrete depiction to a more abstract, formalized, or ideal one that utilized various symbols, keys, and colour codes. Much of this, he argues, was borrowed from the drawings of maps and the depiction of geological sections by mineral surveyors and mining engineers (Rudwick 1976, pp. 159, 183n8). This development culminated in the 1830s, when geology attained a 'standardized visual language' (1976, p. 181). This archetypical way of depicting everything from rocks to bugs has continued among naturalist-artists to the present day – as a glance at any 'field guide' will show. This example also reveals that despite the invention of photography about a century-and-a-half ago – and its displacement of most earlier forms of graphic art – the artist still has a role to play in illustration, for the camera captures an individual specimen (the particular) whereas an artist can depict the archetype. This is one reason why naturalist-artists, after the ascent of photography, were still employed for illustrating natural history and anatomy (such as *Grey's Anatomy*). Needless to say, artists are also required for depicting things not photographable, such as diagrams of the solar system and other astronomical depictions, or extinct creatures (such as dinosaurs) – in fact, anything requiring imagination rather than just a keen eye.

Rudwick's discussion of the topographical landscape is of interest, especially in light of the previous mention of the demarcation problem. Popular topographical motifs included volcanoes, basalt and other rock formations, coastal cliffs, and other geological subjects. As he notes, the stylistic elements in this genre of pictures grew out of the picturesque tradition of the eighteenth century (recall the illustrations in Stafford's book) and the later influence of neoclassicism and Romanticism. Within the Royal Academy's hierarchy of artistic subjects (with historical and biblical themes at the top), landscape ranked near the bottom, and topography was below that (Twyman 1970b, p. 169). Accordingly, topographical artists, along with many naturalist-artists, worked outside the mainstream of art and hence had little impact on it. As well, more often than not, the styles of such illustrations were from earlier eras.

This relative 'social' or 'institutional' insularity between art and illustration does not, however, entail a necessary *conceptual* distinction between them, particularly regarding the epistemological problem. Or, put as a question: why can't some landscape painting be classified as scientific illustration of the topographical variety?

The case of John Constable provides an affirmative answer (see Topper 1990b, pp. 302–7). Constable's depictions of skies, particularly his cloud studies, were done with meteorological precision. Yet among contemporary scholars they have become a source of some controversy. Central to the debate is the question of whether Constable's art was based upon direct experience with atmospheric phenomena or on prior knowledge of cloud types, such as that of Luke Howard.[5] This debate clearly involves the problem of the theory-ladenness of depiction. Interestingly enough, Constable himself was aware of this problem (or, at least, a variation thereof), for he realized that artists cannot entirely avoid conventional forms (such as those derived from workbooks or patternbooks). He called these 'mannerisms'; and, since (according to Constable) the aim of great art is the imitation of nature, the less one relied on these conventions or mannerisms, the greater the artist. From this point of view, he wrote, 'painting is a science' and landscape art may 'be considered as a branch of natural philosophy' (quoted in Topper 1990b, p. 304). Modify Constable's thesis in the terms of this paper and it reads: landscape art should be considered a branch of topographical illustration, and thus may be classified as scientific illustration. So again the demarcation is blurred.

Now this may come as a surprise, but the same may be said of Vincent Van Gogh. Although art historians categorize him as a 'post-impressionist' or a 'pre-expressionist' – since they prefer to emphasize how Van Gogh departed from an accurate depiction of nature in his drawings and paintings – recent research from the opposite point of view has revealed the observational acuity of his landscapes, especially his skies. Gedzelman's (1990a) study of the meteorological features of Van Gogh's art shows that he portrayed unique features of cloud formations with 'climatological accuracy' (p. 111). In short, his art reveals a 'sensitivity to the state of the atmosphere' (1990a, p. 114; cf. Gedzelman 1989). Others (Boime 1984 and Whitney 1986) have studied his night scenes and have identified possible star constellations.

Of course, a similar reading of the landscapes of Leonardo would come as no surprise, since we've come to expect such accuracy in his visual description of nature. For example, in a recent collaborative study of some of Leonardo's botanical drawings, an art historian and a botanist have reported on Leonardo's 'remarkable sense of proportion and visual acuity' while nonetheless noting 'his willingness, on occasion, to adjust a plant's natural appearance in acquiescence to his art' (Meyer and Glover 1989, p. 76). On the other hand, a series of late drawings, commonly referred to as depictions of the biblical *Deluge* or *Visions of the End of the World*, have been consistently interpreted as visionary pictures of the world destroyed by torrents of water. But Gedzelman (1990b, p. 650) discloses that in these drawings Leonardo 'accurately and deliberately represented thunderstorm downbursts that he had discovered and diagnosed through a combination of careful observation, incisive experiments, and cogent reasoning.' Thus, contrary to common opinion, even these 'visionary' pictures are visual depictions of reality. This empirical approach to the art of landscape, skyscape, or seascape represents a fertile area for further study.[6]

7. EXTENDING THE RANGE OF ARTIFACTS

Thus far, the category 'scientific illustration' has been confined mainly to printed or drawn artifacts in their final form, although mention has been made of notebooks and workbooks. I should now like to expand the discussion, arguing that notebooks, workbooks, sketchbooks, and other such artifacts – along with completed paintings – are viable candidates as scientific illustrations. In fact, I would go so far as to assert that any visual scribble of a scientist or artist is a bona fide artifact – from field-drawings in natural history, through geometrical diagrams in laboratory notebooks, to any conceptual schema jotted down in visual form. A few specific examples that immediately come to mind are Galileo's diagrams, notes, and calculations of parabolic projectile motion (fig. 7.7) in his working papers on motion (Drake 1989, p. 57); Leonardo's sketch for a set of ball-bearings (fig. 7.8) in *Codex*

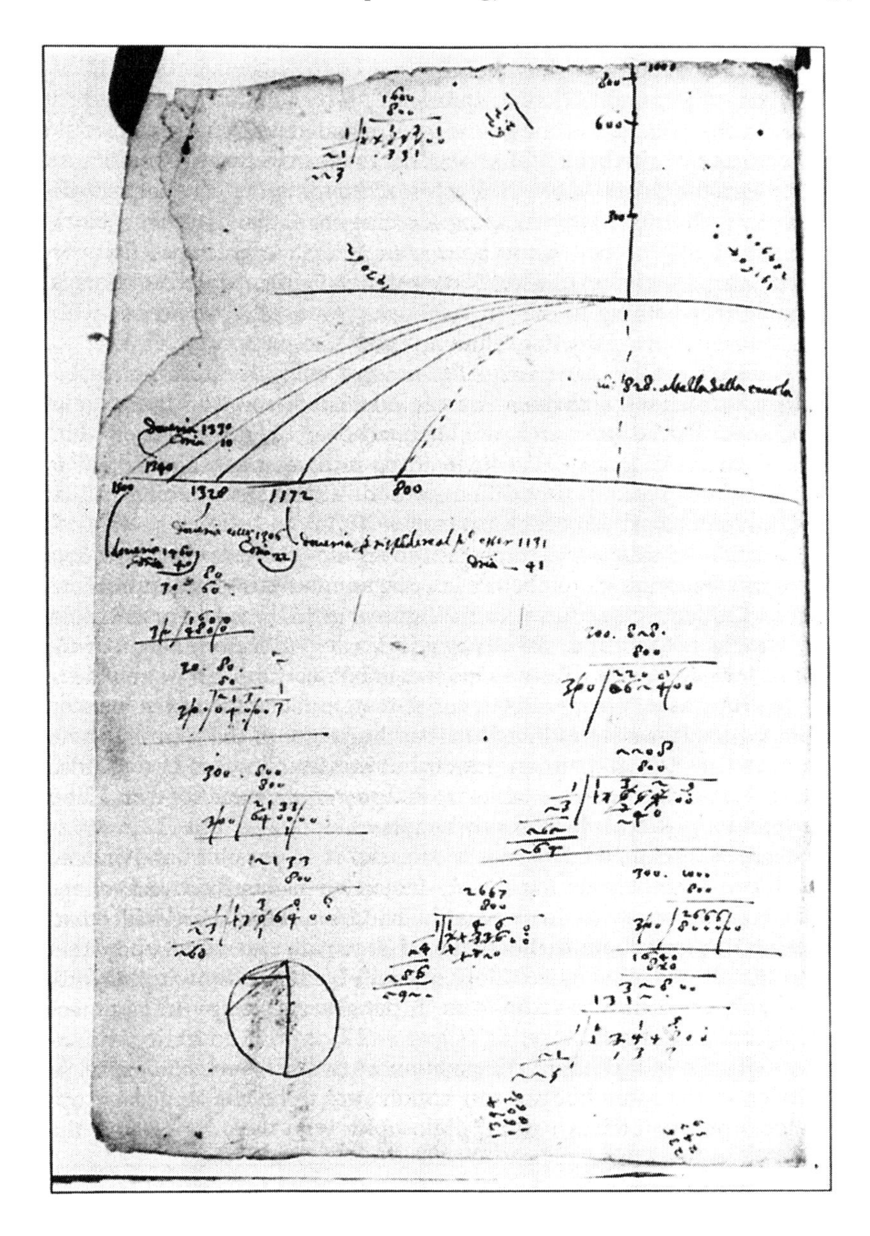

7.7 Galileo's first notes on the parabolic path of projectiles.

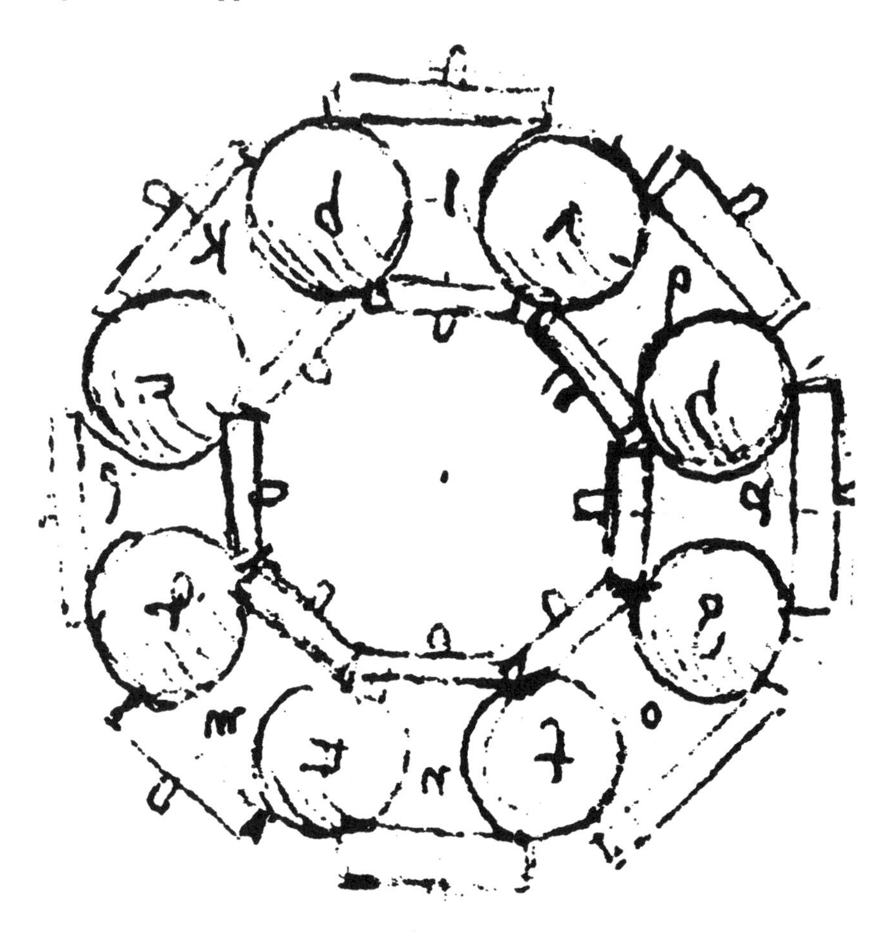

7.8 Ball-bearings. Sketch by Leonardo da Vinci.

Madrid I (Reti 1980, p. 181); Darwin's diagrams of the 'tree of nature' (fig. 7.9) in his first notebook (Gruber 1981, pp. 128–9); and Percival Lowell's drawings of the canals of Mars (fig. 7.10) from his logbook (Sheehan 1988, p. 267). As well, recent scholarship in the epistemology of experimentation (e.g., see Gooding, Pinch, and Schaffer 1989) has helped to highlight the role of visual thinking in laboratory diagrams and notebooks. In sum, such material constitutes an important class of artifacts deserving serious study as scientific illustrations.[7]

Related to this is the subject of 'thought experiments,' the role of which has been studied and often stressed by philosophers and his-

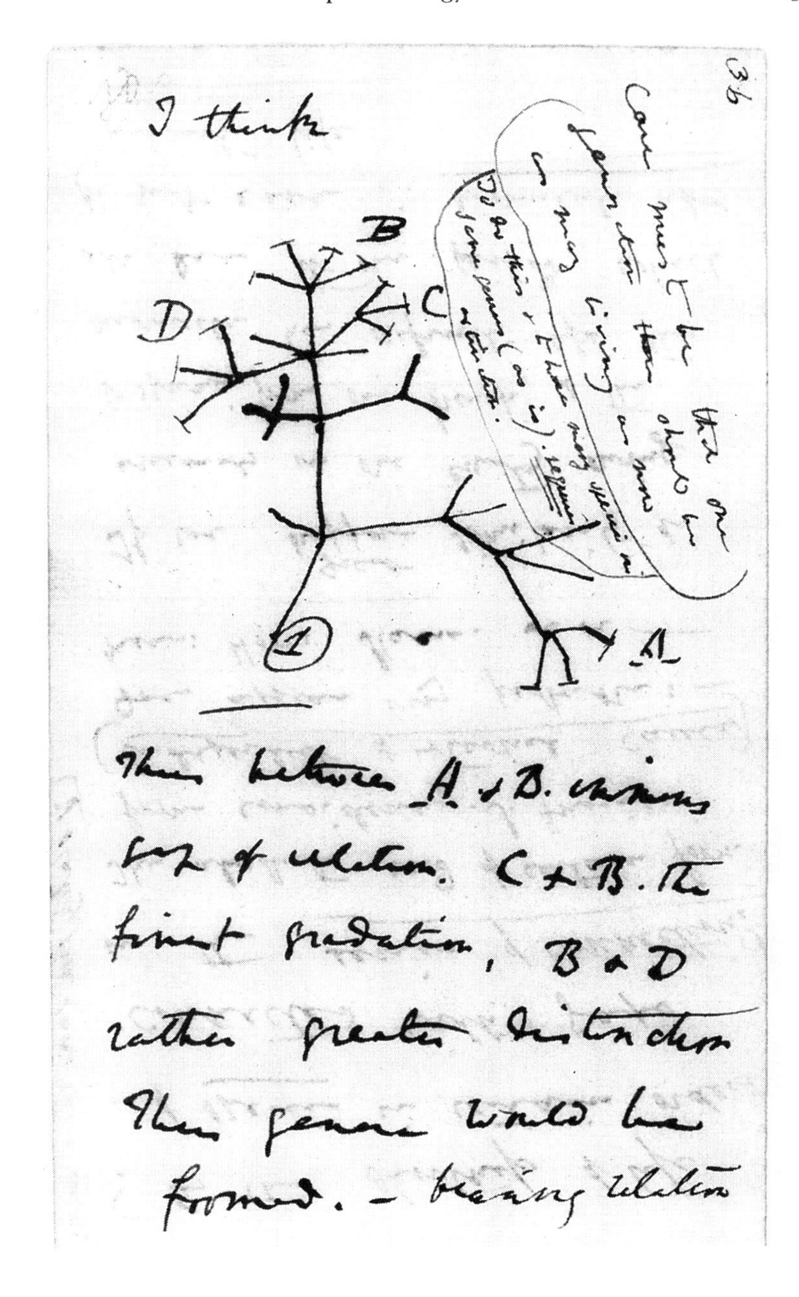

7.9 Darwin's third tree of nature diagram (Charles Darwin ca. 1827–38).

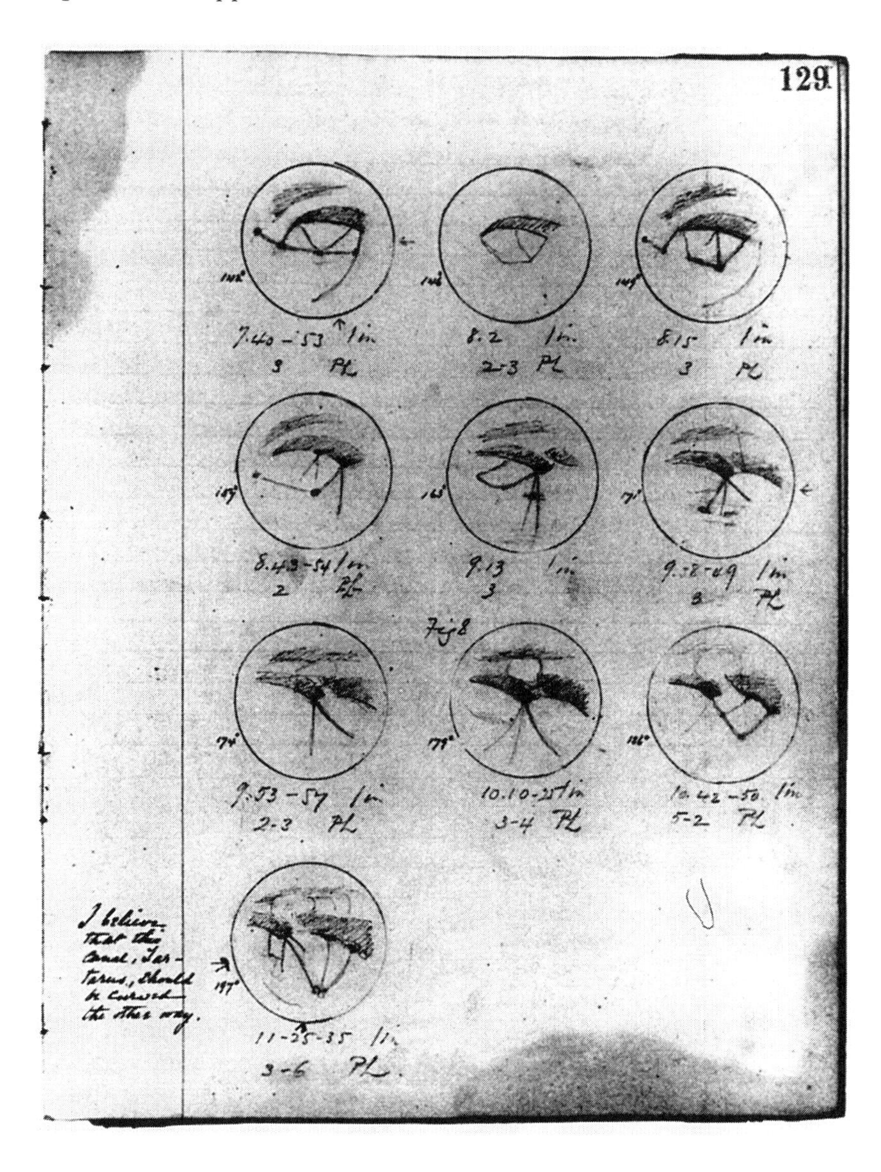

7.10 Impressions of Mars (Percival Lowell 1894).

torians of science. These 'experiments' are often accompanied with diagrams and notes in a scientist's workbook. Such diagrams – the scribbles from the hands (and minds!) of scientists – are manifestations of visual thinking; that is, they are thinking diagrams accompanying thought experiments. The case of Einstein, rightfully, comes to mind, since he claimed that his thinking process involved imagery. Einstein (1949a, p. 7) spoke of thinking in terms of 'memory-pictures' emerging, and submitted that 'certain signs and more or less clear images' were 'psychical entities' in his 'mechanism of thought' (Einstein 1949b, p. 142).[8]

8. TECHNOLOGY AND ANCILLARY SUBJECTS

The dissolving of conventional boundaries has been a theme of this paper. Borders have been breached – between science and pseudoscience, between art and illustration, and between science and art. What then of technology – a topic already broached with the mention of Leonardo's notebooks?

Illustration in technology, as noted at the start of this paper, is the subject of Eugene S. Ferguson's article 'The Mind's Eye: Nonverbal Thought in Technology' (1977). Ferguson begins by saying that the belief that all technological knowledge is derived from science (i.e., that technology is applied science) is just 'folklore.' This then allows him to consider other non-scientific factors in technology; for example, in the field of design, when problems of form arise, the designer may use what Ferguson (1977, p. 827) calls 'nonscientific modes of thought.' His main point is that there are 'nonverbal' (namely, visual) processes involved in designing, such that inventors 'manipulate in their minds devices that as yet do not exist' and which 'cannot be reduced to unambiguous verbal descriptions' (1977, p. 827). Here is his presentation of what may be called a conceptual example:

> The designer of a diesel engine is a technologist who must continually use his intuitive sense of rightness and fitness. What will be the shape of the combustion chamber? Can I use square corners to gain volume, or must I use a fillet to gain strength? Where shall I place the valves? Should I design a long or short piston? Such questions have a range of answers that are supplied by experience, by physical requirements, by limitation of available space, and not the least by a sense of form. (Ferguson 1977, pp. 827–8)

To buttress his case, Ferguson supplies a range of historical examples from the Renaissance to the present. As expected, there is considerable discussion of the 'picture book' tradition, from Biringuccio's *Pirotechnia* (1540; containing about 85 wood engravings) and Agricola's *De re metallica* (1556; containing over 250 wood engravings) through Diderot's *Encyclopedia* (with about 3,000 full-page plates) and beyond. And, importantly, he also refers to engineering notebooks, pamphlets, and manuals, which often contain exploratory, novel, or imaginary devices. This point is eloquently made by Bert S. Hall (1979b, esp. pp. 52–4) when he speaks of technicians 'inventing "on paper"' or Leonardo thinking 'over his creations on paper through the use of visual imagination.' In short, Ferguson and Hall have uncovered the role of visual thinking in technology.

In a related subject, Panofsky and others (as noted above) acknowledged the role of linear perspective in the development of naturalism. But Ferguson (1977) and Edgerton (1980 and 1985) develop this further by pointing to other projective techniques created in the Renaissance: isometric projection (where parallel lines remain parallel), orthographic projection (depicting three views of an object), and the 'exploded' view (showing how a device is put together, and invented by Leonardo) – all of which remain in common usage. Edgerton also emphasizes this work of the artist-engineer, adding to the list of projective methods the 'rotated' view and the 'transparent' or 'cutaway' view. Part of his thesis is that these and other such methods form a network of depictive techniques that were utilized in a range of fields, such as engineering (mechanical drawing, mining, and metallurgy), anatomy, and architecture (cf. Booker 1963).

These various projective systems would seem to be clear-cut cases of techniques categorized as 'science' or 'technology' – except, perhaps, when (as noted before) the technique is sketched by an artist such as Leonardo. But is this an exception or the rule? Artifacts – most assuredly, as has been seen – are not so rigidly categorized. A recent trend, for example, of exhibiting the working drawings of architects in a gallery setting transforms these artifacts into art objects. Another type of transformation – in the same direction, so to speak, that is from science to art – is the use of scientific illustrations as stock motifs for art works. A case in point is some of the works of the twentieth-century artist Max Ernst. As a means of breaking away from traditional subject matter (nudes, landscapes, and such), Ernst used illustrations from science magazines for motifs upon which to base his paintings. Images, such as cutaway views of objects, fluid flow around obstacles, or magnetic lines of force, became the basis of surrealist and abstract paintings – a transformation in context (Stokes 1980), at once from illustration to art, and science and technology to art.

Now surely a recurring query in this paper involves the range of categories for scientific illustration. To be sure, something else that Ferguson mentions extends this range – namely, the role of models employed in many areas of science and technology (such as wind tunnels), as well as in architecture and related fields. Are they not logically related to illustrations – say, as sculptures are to other (artistic) pictures? Should they not, therefore, be classified as scientific illustrations? Why, in brief, is the category usually confined to two-dimensional illustrations?

Clearly three-dimensional models have played key roles in the history of science, the earliest examples being those models representing the heavens and the earth. Orreries and various mechanical gizmos that demonstrate the motions of celestial bodies (either based upon the geocentric or later the heliocentric system) should rightly be classed as a subset of what we mean by scientific illustration (even though they usually 'represented' how angels controlled the heavenly motion!). Of course, two-dimensional 'models' are by their nature more reproducible and mobile, and hence most 'scientific illustration' is on paper. Thus although the spherical earth has been, and still is, represented as a three-dimensional model or globe, most 'maps' are two-dimensional projections.

Cartography is both historically and conceptually related to the Renaissance development of perspective (Edgerton 1975), and thus the history of cartography is also a subset of the history of scientific illustration. In one sense, this is easily accepted: maps (as already noted) employ captions, codes, and symbols; and especially earlier maps, which were extensively embellished with 'art' motifs of various sorts – mythological creatures, border designs, emblems, and the like – so that cartography, a branch of science, is historically associated with the history of art, as is natural history. But, as has been seen, natural history illustration is also 'found' in art (recall Stubbs's *Horse*). A counterpart for cartography appears in the paintings of Vermeer, for the many maps on the walls of his interior scenes contain details of specific places in Holland; they have been found to be real maps (Welu 1975; cf. Woodward 1987). This example from Vermeer thus shows again that art-artifacts may be viewed as science-artifacts.

Perhaps the most ambitious (but, unfortunately, problematic) articles on the subject of scientific illustration are the two (noted above, in passing) by Edgerton: 'The Renaissance Artist as Quantifier' (1980) and 'The Renaissance Development of the Scientific Illustration' (1985). In these articles, Edgerton, as others, points to the rise of naturalism in the Renaissance as providing the requisite pictorial framework for a truly informative scientific illustration. And, as almost all the authors surveyed, Edgerton too focuses on the theory-ladenness (although he does not use the term) of these conventions. However, rather than lament this fact – recall the previous discussion of the double-edged nature of perception and imagery – Edgerton stresses the potential information contained in such illustration, once the 'language' is learned. Contrasting this with pre-Renaissance and non-European art (which, he submits, cannot convey such information), he then leaps to the conclusion that the new space portrayed in Renaissance art was a causal factor in the Scientific Revolution of the seventeenth century. At least that is how I read his pronouncement that 'Galileo could not have done what he did' without Renaissance art: 'He [Galileo] needed precisely the kind of visual education, the familiarity with Renaissance-style pictures in contemporaneous textbooks, only available in the schoolrooms of sixteenth century western Europe' (1985, p. 194). Renaissance art, in other words, provided a formal cause (in Aristotle's sense) for post-Renaissance science. This, it seems to me, is Panofsky's thesis revisited – and, of course, elaborated.

Now it may be reasonable for Edgerton to argue (as Ferguson also did) that Renaissance art provided a means for Renaissance artist-engineers to depict machines, surgeons to depict anatomy, naturalists to depict flora and fauna, and so forth. Nonetheless, Edgerton's more specific thesis is seriously challenged by the historian of science Michael S. Mahoney in his article 'Diagrams and Dynamics: Mathematical Perspectives on Edgerton's Thesis' (1985). Mahoney first undercuts the engineering part of Edgerton's argument by asserting that the machines themselves, as depicted in Renaissance picture books, were not intrinsically different from earlier machines – although they were drawn differently. Of course, the drawings could (and surely did) convey visual information about the machines to the viewer but, in light of Ferguson's argument for non-verbal technological thought, Mahoney's critique of Edgerton is that Renaissance illustration did not convey the thinking process that went into the development of machines. If Mahoney is right, then the discussion should be directed to the workbooks and notebooks of the engineers (mentioned before by Ferguson), where perhaps the new art may have played a role in the non-verbal thinking process.

Mahoney's next important observation also tempers Edgerton's argument. Mahoney admits that Renaissance art improved the depiction of machines, so that they 'looked increasingly like their three-dimensional models as seen in action, even as the artist exploded them, bored through to their internal parts, and twisted and turned their components.' Nevertheless, he offers this qualification:

> ... to show what machines do or how they are assembled is one thing; to show how they work is quite another. However accurately and fully a complex mechanism may be portrayed, an understanding of its operation as a whole rests ultimately on familiarity with the operations of its basic components. Treatises of the genre under discussion took the familiarity for granted. Their authors could not do otherwise, given the nature of their medium. A picture of a windlass, or a system of pulleys, cannot in and of itself set forth the laws that define the device's mechanical advantage. (Mahoney 1985, p. 201)

This is another variation on the theme of the theory-ladenness of illustration, pointing to the limitations of pictorial communication.[9]

Mahoney's last point is probably lethal to Edgerton's thesis. Mahoney argues that Renaissance mechanics was based upon the concept of a machine as 'an abstract, general system of quantitative parameters linked by mathematic relations' which was far 'removed from the physical space the artists had become so adept at depicting. Those terms could not be drawn; at best, they could be diagrammed' (1985, p. 200). Such diagrams (recall fig. 7.7, Galileo's notes on projectile motion) consist of lines, triangles, and curves corresponding to parameters of motion (e.g., distance, velocity, and time). 'Whatever the mathematician's eye is seeing' in these diagrams, says Mahoney (1985, p. 209), 'it has little to do with new pictorial techniques for the accurate representation of physical objects in three-dimensional space. It is the mind's eye that is looking here ...' Edgerton, it seems, expected too much of pictures. Nevertheless, in his most recent work (1991, pp. 15–16; cf. Topper 1988), Edgerton (although acknowledging in passing Mahoney's critique) reiterates his thesis.

9. REMARKS ON EMPIRICISM

Related to the limits of communication of information in pictures is another topic that has cropped up throughout this paper – what may be called the 'empiricism of imagery.' This certainly was central to Hutchinson's natural history approach to illuminated manuscripts when trying to identify species of birds. Constable's and Van Gogh's landscapes and skyscapes have also been noted, as well as have Leonardo's cloudbursts and Vermeer's maps. One could also cite Bruegel's paintings of the mini ice age of his time in his winter landscapes, or Michelangelo's anatomical depictions (Elkins 1984), and much more (cf. Topper 1990b). Surely this is another facet of the topic deserving more study. Here I wish merely to stress the fact that artifacts within the category of 'art' may exhibit this empirical component, and thus scientific illustrations do not have a monopoly on empiricism.

After all, scientific illustrations may, on the contrary, be deficient in their empirical content. On this point, I should like to quote a few of my favourite lines from the writings of Kepler. Here he is assailing Robert Fludd, a contemporary scientist of the more hermetic variety; in particular, Kepler is comparing the illustrations in his books with those in Fludd's:

> I have compared my diagrams with your pictures. I must confess that my book is not as adorned as yours, nor will it appeal to the taste of any future reader. I have an excuse for this defect by my profession, for I am a mathematician. (quoted in Yates 1969, p. 443; translated from the Latin by my colleague Robert Gold)

Kepler, of course, was himself tinged with the mystic science; but even knowing this, we may still appreciate the sarcasm in Kepler's distinction between an apparently meaningless 'picture' and a scientific 'diagram.'

Ultimately, then, the empirical component is fundamental to a complete understanding of the epistemology of scientific illustration – the point, interestingly enough, with which Panofsky long ago began this discussion. Moreover, what has emerged from this analysis – and is important here – is that the empirical matter is not confined only to the genre of scientific illustration, but is germane to all visual imagery – be the medium a picture, a model, or even a film or video – and is independent of its classification as 'art,' 'science,' or even 'pseudo-science.'

10. SUMMARY

Scientific illustration is customarily viewed as a form of art. Only recently, and in a few disparate sources, has scientific illustration been studied as a branch of science – as a means of conveying information. Previously, the epistemological nature of scientific illustration either was taken for granted as being unproblematic or was ignored because of the belief in the supremacy of the written word for recording and conveying information. However, recent studies of scientific illustration as a form of what is called 'visual thinking' have raised a number of critical issues.

In the above overview of the literature on scientific illustration, there has emerged a fairly clear specification of some features of such illustrations, including the following: images invariably accompany a written text such that this bipolar relationship entails theory-ladenness; their history is bound up with the history of media (especially after the invention of print-making); formalistically the illustrations almost always employ styles, motifs, or conventions from the 'fine arts' (sometimes these are contemporary, but they may also be from an earlier period); the images tend historically to shift from the depiction of individuals to types (but there are exceptions; see Kemp 1993); illustrations as artifacts may include an extremely wide range of possibilities – paintings, drawings, prints, models, workbooks of various kinds (sketchbooks, patternbooks, fieldbooks, logbooks), and personal notebooks (thinking diagrams, scribbles of ideas, doodles). Any further study of scientific illustration must confront these and other features of the genre.

As has been seen, broaching scientific illustration from these viewpoints opens up questions regarding not only the nature of scientific illustration itself – its theory-ladenness and empirical content – but also regarding the nature of all types of illustration and imagery, and the demarcation of science, pseudo-science, and art – in sum, the very nature of what we mean by 'science' and by 'art.'

NOTES

1 Another key theorist, Sir Ernst Gombrich, is well known for his seminal work, *Art and Illusion* (1968). The lectures upon which his book is based were titled 'The Visible World and the Language of Art.' One of the essential themes of this book is the role of various stylistic conventions in pictures. Gombrich's article 'The Visual Image' (1972) is a short, clear, and (I think) brilliant essay on the process of 'reading' pictures.

More recently, Edward R. Tufte has published two engaging books on the visual communication of information: *The Visual Display of Quantitative Information* (1983) and *Envisioning Information* (1990). Some of the artifacts that Tufte considers are charts, diagrams, tables, graphs, and instructions.

2 In his examples, Knight has added artifacts from the pseudo-sciences, which, I think, is legitimate in light of the changing definitions of science over the centuries. Of course, this raises a further issue as to the distinction between a scientific illustration and other genres of illustration; this, in turn, is related to the more general problem of the demarcation of science – confronted later in the paper.

3 This leads Kemp to a consideration of Gesner's concern with symbolism, in particular his devotion to the study of nature as revealing 'the creator's divine plan to man' (p. 131) – in short, Gesner's version of the Design Argument. This belief in the revelation of the workings of God in all natural things was held by most scientists at the time – as well as for the next two centuries or so, at least from Kepler through Newton to (but not including) the mature Darwin. Yet Kemp's misgivings, I submit, are misguided. Though any depiction of any thing – or even the thing itself – may be viewed symbolically (you and I, in the context of life, stand for numerous categories: teachers and students, drivers and pedestrians, readers and writers), this fact is irrelevant to whether a depiction of the 'thing' is naturalistic or not. Actually, Kemp knows this, for elsewhere he recognizes the fact that images may continue to be 'read' symbolically, even though they are depicted naturalistically; thus, for example, a naturalistic drawing of a lily by Leonardo may ultimately have been destined as a symbol of the Virgin in a painting. Or consider a favourite example of mine: Rembrandt's *Anatomy Lesson of Dr Tulp* (1632) is at once a depiction of an anatomy lesson (a 'scientific illustration'), a visual expression of the Design Argument (the function of Tulp's hand is compared with that of the corpse), and a *memento mori* (a pictorial reminder of death) – not to forget, of course, that it also is a 'work of art.' These different levels of meaning (Schupbach 1982) point to the fact that naturalism and symbolism are not mutually exclusive categories.

4 This contrast in viewpoints between historians is real. Nevertheless it is important to realize that the distinction resides in the investigators, not the investigated. The naturalist's search for species and the symbolist's search for meaning are not mutually exclusive; or, said another way, both investigators may find what they are looking for whether they observe images in books and manuscripts or search among flora and fauna in a wood. Consequently, there seems to be nothing inherently unique about a scientific illustration; a wide range of images can be viewed from a scientific point of view – as the example of Hutchinson's work shows.

5 For citations regarding this debate, see Topper 1990b, p. 310.
6 See, for example, Gedzelman's overviews of atmospheric phenomena in art in 1991a and 1991b.
7 This seems to have been the one class of artifacts forgotten by Knight in the paragraph quoted at length near the start of this paper. Also, a useful way of thinking about all these artifacts, I believe, is by recalling the philosophical distinction between the context of discovery and the context of justification; imagery in notebooks and such may be considered part of the former, and graphics in books and articles classified within the latter.
8 This, in turn, may lead to the topic of mental imagery – a highly controversial matter, particularly among psychologists. Within the limits of this paper, it is sufficient to note that this has been a topic involving case-studies of some scientists, such as J.J. Thomson (Topper 1980) and Einstein (Holton 1986), and among quantum physicists such as Bohr, Pauli, Heisenberg, and Schrödinger (Miller 1978 and 1984). Surely this is a fertile area calling for further work.
9 Not unrelated to this issue is the well-known aphorism that 'a picture is worth a thousand words.' Now, to be sure, there is – as the history of scientific illustration makes clear – much truth to this, especially when the information being communicated involves physical artifacts. As Leonhart Fuchs wrote in the preface to his illustrated treatise on plants (1542): 'It is the case with many plants that no words can describe them so that they can be recognized. If, however, they are held before the eyes in a picture, then they are understood immediately at first glance' (quoted in Reeds 1976, p. 529n41). But this statement is not true when the information involves logical propositions. For example, a picture alone cannot communicate a negation because visual language is non-propositional; prohibitive images (such as no-smoking signs) must be supplemented with captions or conventional codes. In these instances, one negative word ('No') is, so to speak, worth a thousand pictures. Few pictures, of course, are ever 'read' in isolation; the context usually provides a framework for the correct reading. But still the reader requires a knowledge of that context: if that knowledge is partial or erroneous, a misunderstanding may result.

8. Illustration and Inference

JAMES ROBERT BROWN

1. PROOFS AND PICTURES

Diagrams play an underappreciated role in the sciences. In mathematics their role is especially curious. The following diagram (see fig. 8.1)

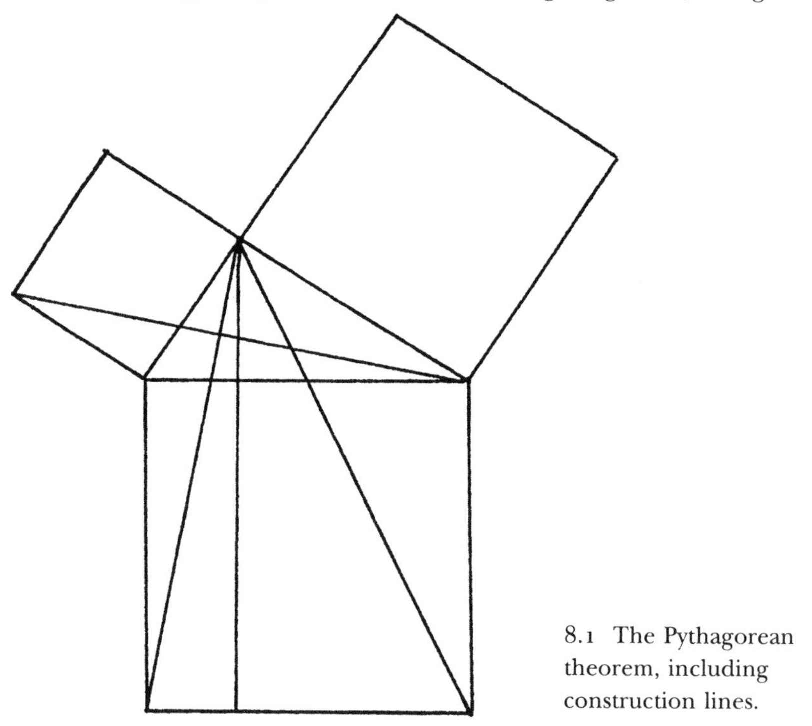

8.1 The Pythagorean theorem, including construction lines.

accompanies the proof of the Pythagorean theorem in Euclid's *Elements.* We can stare at it for days and still not see that the theorem is true; we need a proof – a traditional proof.

The common view of diagrams in mathematics is this: they provide a heuristic aid, a help to the imagination when following a proof. But they are commonly thought of as no more than this. In particular, diagrams cannot justify; they are not to be confused with real proofs, which are formulated in words and symbols. At most, illustrations play a psychological role, and should never be used for making inferences.

For numerous examples such as the Pythagorean theorem, the standard account of diagrams seems right. But there are a few rare and remarkable examples where something quite different is going on. The following theorem is from number theory; it says the sum of the odd numbers, $1 + 3 + 5 + ...$, up to $2n - 1$ is equal to the number n^2. It has a standard proof (by mathematical induction) which uses no diagrams at all; but it can actually be proven with a diagram. (Take a moment to study the proof, to see how it works.)

Theorem: $\sum_{i-1}^{n} (2i-1) = n^2$

Proof:

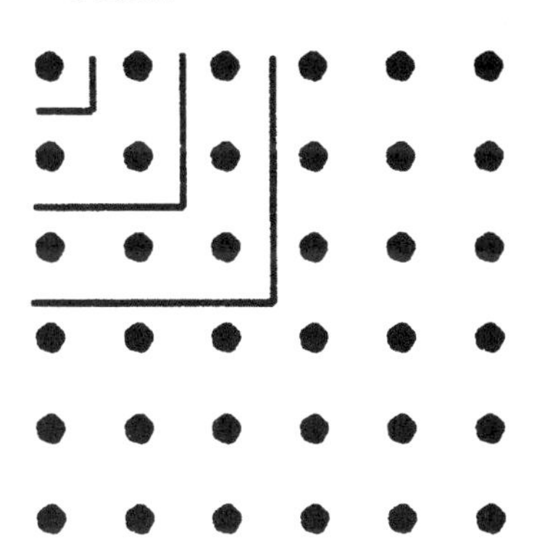

8.2 Picture which serves as a proof of a theorem from number theory.

Of course, there is lots of interpreting going on to make this a proof. For example, we must consider the individual unit dots as numbers, and we must bring some background information from geometry to the effect that a square with sides of length n has area n^2. But these sorts of interpretive assumptions are no less innocuous than those made in a typical verbal/symbolic proof.

Here's another example, this time a result about infinite series.

Theorem: $\frac{1}{2} + \frac{1}{2^2} + \frac{1}{2^3} + \ldots = 1$

Proof:

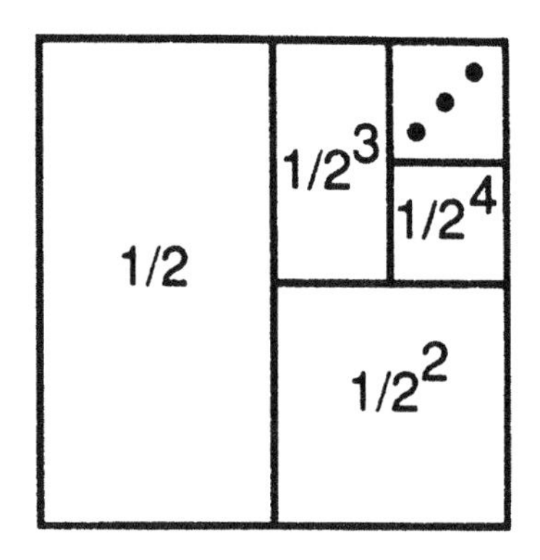

8.3 Picture which serves as a proof of a theorem about infinite series.

The usual proof of this theorem uses standard ε–δ techniques. But nothing like that is used here. Instead, we can simply see a pattern; we can see that the inner boxes are getting smaller, that they will eventually exhaust the unit square (without remainder), and so we see that their sum is equal to 1.

For good measure, here is one more showing the relation between the arithmetic mean (i.e., $a + b/2$) and the geometric mean (i.e., $\sqrt{ab}$) of two real numbers, *a* and *b*. (In understanding the diagram below, the arithmetic mean is straightforward; to interpret the geometric mean, one needs to know the Pythagorean theorem and that the dotted lines form a right-angled triangle on the diameter.)

Theorem: $\frac{a+b}{2} \geq \sqrt{ab}$

Proof:

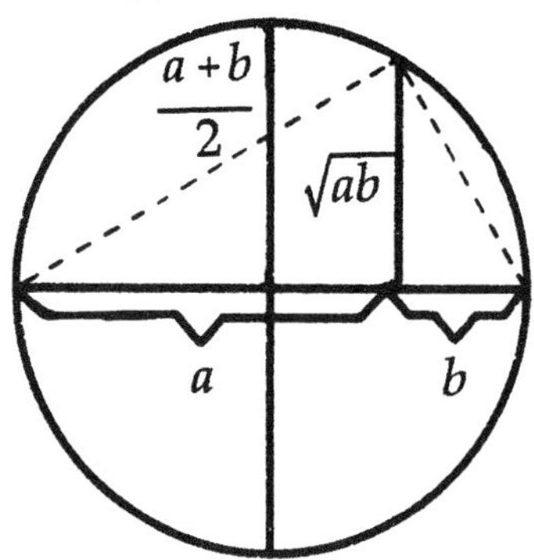

8.4 Picture proving the relation between the arithmetic mean and geometric mean.

The moral I think we should draw from examples like these is simple but profound: *we can prove things with pictures.* In spite of the fact that the number theory diagram seems to be a special case ($n = 5$), still we can see all generality in it. And the proof does not work merely by suggesting the 'real proof,' since in the diagram there is nothing which clearly corresponds to the passage from n to $n + 1$, which is the key step in any proof by mathematical induction.

Those who hesitate to accept the picture as a proof might think that the picture merely indicates the existence of a 'real proof,' a standard proof by mathematical induction. Perhaps they even wish to appeal to the well-known distinction between discovery and justification: the picture is part of the discovery process, while true justification comes only with the verbal/symbolic proof. But consider: would a picture of an equilateral triangle make us think there is a proof that all triangles are equilateral? No. Yet the above picture makes us believe – rationally believe – that there is a verbal/symbolic proof of the theorem. The picture is (at the very least) evidence for the existence of a 'real proof' (if we like to talk that way), and the 'real proof' is evidence for the theorem. But we have transitivity here; so the picture is evidence for the theorem, after all.

Let me put this connection between theorems and pictures in the background for now and turn to my main concern in this paper, phenomena.

2. DATA VS PHENOMENA

My point of departure is a notable recent analysis of phenomena by James Bogen and James Woodward, who make a 'distinction between phenomena and data' (1988, p. 305). The former are constructed[1] out of the latter:

> Data, which play the role of evidence for the existence of phenomena, for the most part can be straightforwardly observed. However, data typically cannot be predicted or systematically explained by theory. By contrast, well-developed scientific theories do predict and explain facts about phenomena. Phenomena are detected through the use of data, but in most cases are not observable in any interesting sense of that term. (1988, p. 305)

> Data are ... idiosyncratic to particular experimental contexts, and typically cannot occur outside of those contexts ... Phenomena, by contrast, are not idiosyncratic to specific experimental contexts. We expect phenomena to have stable,

repeatable characteristics which will be detectable by means of a variety of different procedures, which may yield quite different kinds of data. (1988, p. 317)

There are several important features and consequences of this view. Among the more important are these: explanation is not a relation between theories and observable facts; nor is prediction; theories are not tested by comparing them with experience; and observation – whether theory-laden or not – is 'much less central to understanding science than many have supposed' (1988, p. 305).

Typical of Bogen-Woodward phenomena are weak neutral currents. The associated data are bubble chamber photographs. The relevant theory which is supported by all of this is the Weinberg-Salam theory of weak interactions. It is supported, not by the data, but by the phenomena of weak neutral currents. The existence of the phenomena is in turn supported by the data, the photographs. According to Bogen and Woodward, the data are far too messy and idiosyncratic to serve as evidence for any theory; the phenomena play a crucial and irreducible intermediate role in the process of scientific inference.[2]

A cursory glance at the social sciences suggests the phenomena/data distinction is important here, too, perhaps even more so. Social scientists seem to do (at least) two quite distinct things. One is to establish phenomena; examples might include: that there is widespread child abuse; that *x* per cent of the population are homosexual; that suicide rates in some cultures are higher than in others; that *y* per cent of women are physically abused by their male companions; that there is a *z* per cent unemployment rate, etc. These are often extremely difficult to ascertain, as we might imagine, especially when questions of sexuality are involved. Data collecting is typically involved in establishing such phenomena.

The second job is to *explain* these phenomena. And it is indeed phenomena that social theory attempts to explain. Economists try to tell us why we have high unemployment (a downturn in the economy), not why Joe Blow in particular is out of work (perhaps he was an incompetent worker); and Durkheim told us why Protestant societies have higher suicide rates than Catholic ones (they are socially less cohesive); he does not tell us why Joe Blow killed himself (perhaps he was depressed after losing his job).

A number of examples from the physical sciences, even if only briefly described, should help to explain and reinforce the data/phenomena distinction.

3. EXAMPLES

High energy physics abounds with illustrations of the distinction. We are all quite used to having information from this field presented to us twice over, first as data in the form of a photograph, then as phenomena in the form of an artist's drawing. Figure 8.5 is a typical example. What high energy physics does is explain and predict the drawing on the right, not the photo on the left. The chicken scratches on the left are far too variable, idiosyncratic, and downright messy for any theory to deal with. Theories in high energy physics only try to cope with the phenomena as represented in the artist's drawing.

The so-called mechanical equivalent of heat was established by James Joule in the middle of the nineteenth century. That is, he established 'that the quantity of heat produced by friction of bodies, whether solid or liquid, is always proportional to the quantity of force expended.' In a large number of repeated experiments involving a paddle-wheel contraption (fig. 8.6) that heated a quantity of water when the paddles were driven by falling weights, Joule established

> that the quantity of heat capable of increasing the temperature of a pound of water (weighed in vacuo, and taken at between 55° and 60°) by 1°F, requires for its evolution the expenditure of a mechanical force represented by the fall of 772 lbs through the space of one foot. (Joule 1850, p. 82)

This was not the result of any simple observation, but the culmination and processing of an enormous amount of data.

The method of experimenting was simply as follows:

> The temperature of the frictional apparatus having been ascertained and the weights wound up ... the roller was refixed to the axis. The precise height of the weights above the ground having then been determined by means of the graduated slips of wood ... the roller was set at liberty and allowed to revolve until the weights reached the flagged floor of the laboratory, after accomplishing a fall of about 63 inches. The roller was then removed to the stand, the weights wound up again, and the friction renewed. After this had been repeated twenty times, the experiment was concluded with another observation of the temperature of the apparatus. The mean temperature of the laboratory was determined by observations made at the commencement, middle and termination of each experiment. (Joule 1850, p. 66)

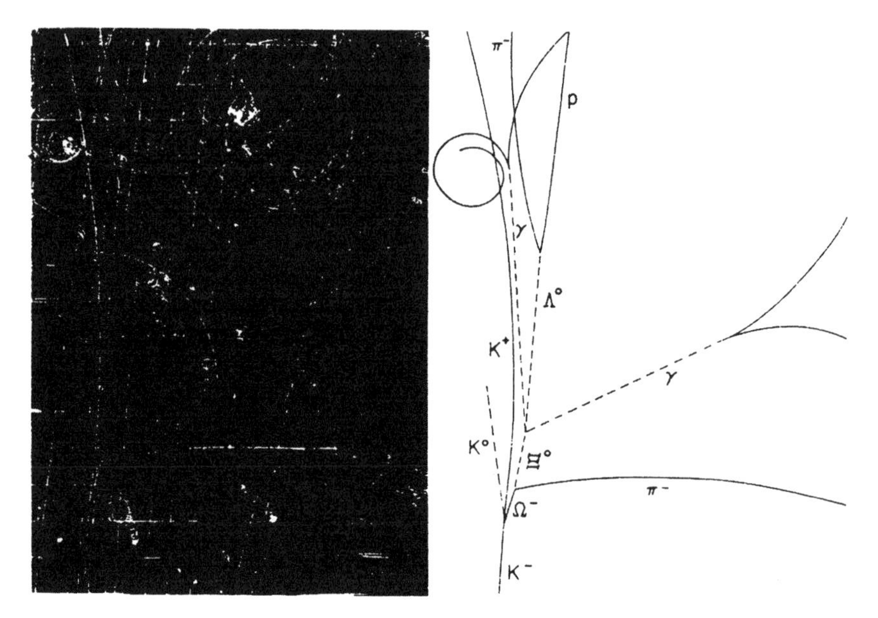

8.5 The first omega-minus event (Brookhaven National Laboratory).

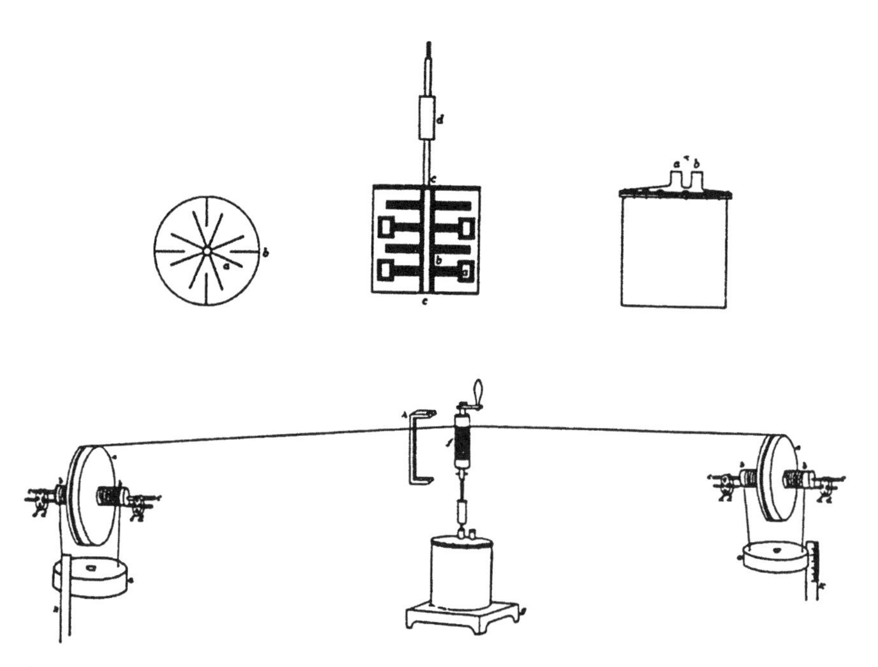

8.6 Joule's apparatus (J. Joule 1850).

It was out of the enormous amount of data that he had that the phenomenon of the mechanical equivalent of heat was brilliantly and painstakingly constructed by Joule.

The Periodic Table of the chemical elements provides yet another illustration. The Periodic Table is a classification scheme of the elements in accordance with their properties. It is a paradigm of the construction of phenomena out of data. The phenomena are the entries in the table – the chemical elements and their properties: atomic weights, atomic numbers, chemical similarities, etc.

There is no algorithm for making phenomena out of data – it is a fallible process. Dmitri Mendeléeff ordered the elements according to their increasing atomic weights. But he noticed that atoms with similar chemical properties recurred periodically at fairly regular intervals. By lumping together those which are chemically similar he created a classification of the elements which was the first Periodic Table.

Though brilliantly conceived, Mendeléeff's taxonomy was somewhat problematic. In the case of a few elements, ordering them by increasing weight was at odds with ordering them in accord with their chemical properties. And the discovery of isotopes (which have different weights but are chemically identical) made matters even worse. This was the background for Henry Mosely's work, begun in 1913.

The characteristic frequencies associated with each of the elements is due, according to Bohr's theory of the atom, to electrons in orbit around the nucleus falling to lower orbits. When they fall from one energy level, or shell, to a lower one, they emit a photon of the appropriate energy, or frequency. Mosely fired cathode rays at several of the heavier elements and recorded the X-ray frequencies produced. He focused on a particular series known as the K_α-lines in a large number of elements. What he discovered is that as the atomic number increases by 1, (i.e., as $Z \rightarrow Z + 1$), the quantity $(4/3 \times \nu(Z) \times R)^{1/2}$ also increases by 1. This led to the following formula for the frequencies of the K_α- series for the element with atomic number Z: $\nu(Z) = (Z - 1)^2 \times (1/1^2 - 1/2^2) \times R$ (where R is the Rydberg constant, known independently, and $1/1^2 - 1/2^2$ is associated with the first and second energy levels).

Mosely's classification and Mendeléeff's coincide except in a few cases. For example, potassium preceded argon in Mendeléeff's table, but Mosely reversed them. This resulted in Mosely's Periodic Table being in full agreement with both the recurring regularities of the chemical properties and with the increasing atomic numbers. There is no internal tension as there was in Mendeléeff's taxonomy.

The history of the Periodic Table illustrates all sorts of interesting things about phenomena. In it we see, of course, the construction of precise elements out of the hodge-podge of data. And we also see that the Table is not explanatory; it is just a taxonomy. But there is one more feature of the Table which strongly supports the data/phenomena distinction. When the Table was being constructed there were 'gaps' in it; that is, nothing had ever been observed which corresponded to certain places in the table (e.g., germanium, Ge). Any theory (such as Bohr's) that attempted to explain the features of the Table would be required to account for every place in the Table, including the gaps. (Or explain why the gaps had to exist, as quantum mechanics does in the case of the very heavy elements – they are unstable.)

I've been saying 'gaps in the Table.' Strictly, this is wrong: the Table is complete; the gaps are in the data. This means that the entries in the Periodic Table cannot be identified with what is actually observed, with data – since there is none (or was none at the time) – but must instead be thought of as phenomena.

4. PHENOMENA AND NATURAL KINDS

The world is full of data, but there are relatively few phenomena. My suggestion is rather simple: *phenomena are abstract entities*[3] *that correspond to visualizable natural kinds.* When scientists construct the phenomena out of a great mass of data, what they are doing is singling out what they take to be genuine natural kinds. In Plato's gruesome metaphor, they are trying to cut nature at its joints. To this I would only add: at nature's visualizable joints. (I should add that an equally strong case can be made for non-visualizable phenomena, too. In this essay, however, my concern is with the picturable only.)

The shift in the ordering structure of the Periodic Table, from atomic weights to atomic numbers, shows the complexity and ingenuity that are sometimes involved in constructing phenomena out of data. But it also shows the importance of natural kinds and their essential properties in scientific thinking. Mosely expressed it well when he summed up his experimental work:

> We have here a proof that there is in the atom a fundamental quantity, which increases by regular steps as we pass from one element to the next. This quantity can only be the charge on the central nucleus, of the existence of which we already have definite proof. (quoted in Trigg 1975, pp. 32f)

Notice that Mosely is not claiming to have discovered that the nucleus has an electronic charge, any more than he is denying that the elements have an atomic weight. His claim is about which of these existing properties is 'fundamental,' or essential (chemically) and which is not.

5. THOUGHT EXPERIMENTS

Thought experiments deal with phenomena.[4] Obviously, they do not deal with actual experimental data – this much is true by definition. But the fact that they involve picturable processes suggests that we need to keep something observation-like centrally involved.

In Einstein's elevator (fig. 8.7), to cite one important example, the observer inside cannot tell whether she is in a gravitational field or accelerating (Einstein and Infeld 1938, pp. 214ff). A beam of light passing through would bend downward if the elevator were accelerating, so that, by the principle of equivalence, it would also bend downward in a gravitational field. The conditions required to make such an observation are so extreme that any actual observer would be a puddle on the floor of the elevator. The observation in this thought experiment is of phenomena, not data.

Newton's (1934, pp. 6ff) bucket thought experiment provides an instructive example in a different way. The thought experiment asks us to imagine the different stages of a bucket partly filled with water as it is released and allowed to 'rotate' (fig. 8.8). The water and bucket would be initially at rest with respect to one another, and the water surface is flat. Next they would be in relative motion. In the third stage, they would again be at rest with respect to one another, but this time the surface of the water would be concave.

Why the difference between stages one and three? Newton's explanation is simply this: in the first stage, the water and bucket are at rest with respect to absolute space, and in the third, they are rotating with respect to absolute space.

After Leibniz, Newton's most forceful critics were Berkeley (1710) and Mach (1883). Did they deny that absolute space was the best explanation for the observed difference? Not really; instead, they denied the alleged observable difference in the condition of the water. They denied that in a universe without distant masses (the fixed stars) the water would climb the walls of the bucket. Clearly, Berkeley's and Mach's fight with Newton is not a dispute over empirical data; it's not even a fight over rival explanations of what is given in the thought experiment – it is a fight over the phenomenon itself.

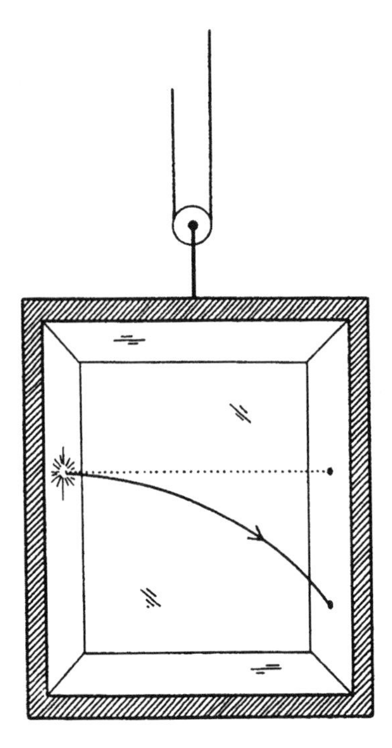

8.7 Einstein's elevator thought experiment.

An example of a rare type of thought experiment is Leibniz's (1686) account of *vis viva.* Rare because it both destroys an old theory and erects a new one at a single blow. When classical mechanics was being constructed in the seventeenth century, it was commonly agreed that something was conserved during certain types of processes. Descartes and Leibniz agreed that *motive force* is conserved, but just what is it? Descartes took it to be *quantity of motion* (roughly, momentum as we would now call it),[5] while Leibniz took it to be *vis viva* (roughly, twice the kinetic energy). In one simple elegant example, Leibniz destroyed the Cartesian view and established his own.

Leibniz starts by making assumptions to which any Cartesian would agree (see fig. 8.9.): first, that the quantity of force (whatever force is) acquired by a body in falling through some distance is equal to the force needed to raise it back to the height it started from; and second, the force needed to raise a one kilogram body (A) four metres (C-D) is the same as the force needed to raise a four kilogram body (B) one metre (E-F). From these two assumptions, it follows that the force

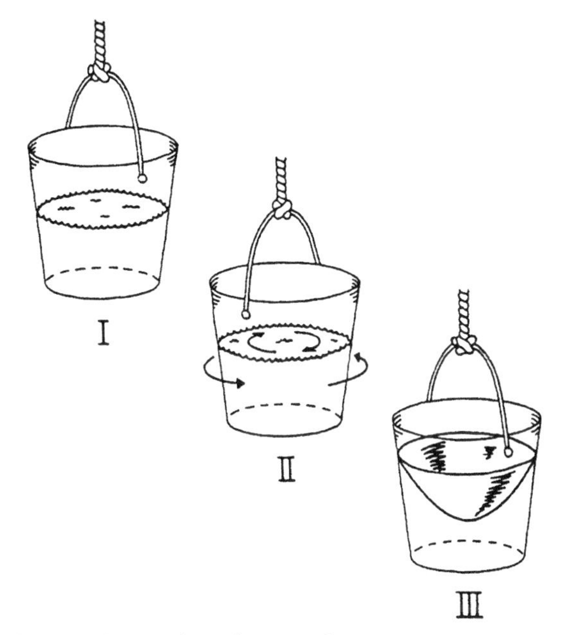

8.8 Newton's bucket theory thought experiment.

acquired by A in falling four metres is the same as the force acquired by B in one metre. Having set the stage, we can now compute the quantities of motion of each using a relation established by Galileo. The velocity of A after falling four metres will be two metres per second. Multiply this by its weight, one kilogram, and we get a quantity of motion for A of $2 \times 1 = 2$. The velocity of B after falling one metre is one metre per second, so that its quantity of motion is $1 \times 4 = 4$.

This simple example refutes the Cartesian claim that force is quantity of motion. But this is merely the first step; Leibniz goes on to give us the right answer. It is elicited from the fact that the distance any body has fallen is proportional to the square of its velocity. So Leibniz's answer to the question What is this motive force which is conserved? is *vis viva*, i.e., mv^2. It is easily verified in this or any other similar example.

What we can see from these thought experiments is that phenomena must be playing a role in scientific inference, a role which is distinct from data. Though phenomena are picturable, they exist at a high level of abstraction.

In passing, a word about the theory-ladenness of observation. No one nowadays believes in raw data; observations are always conceptualized.

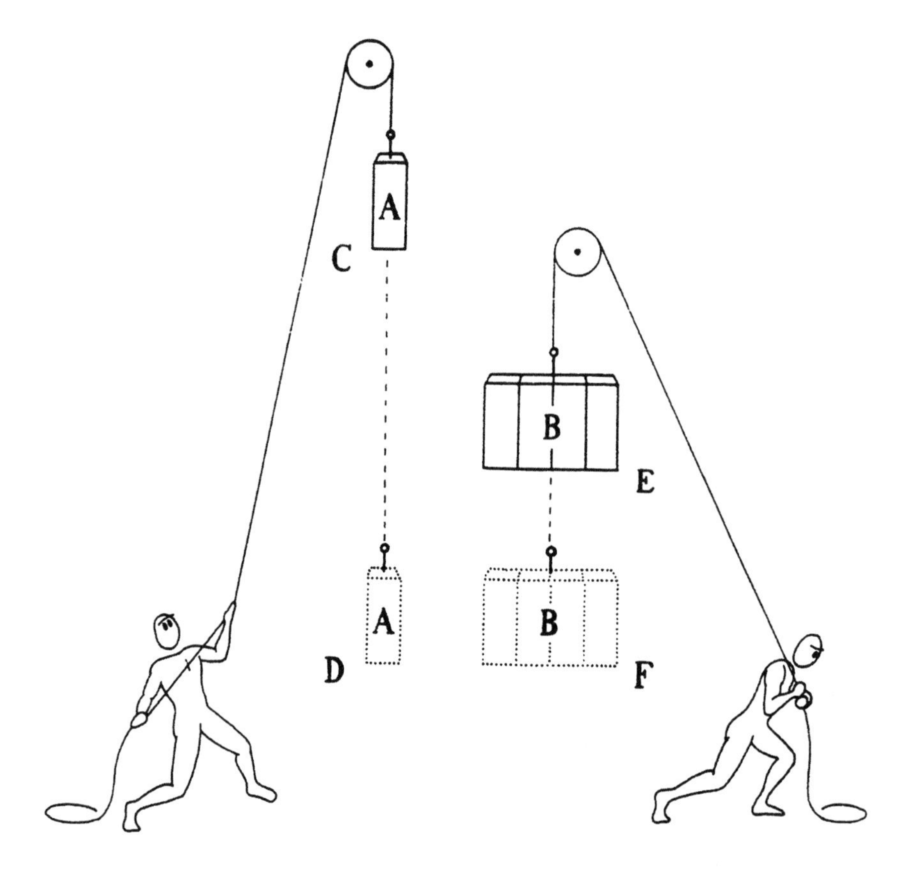

8.9 Leibniz's thought experiment.

(This is undoubtedly one of the great results of modern philosophy of science, due to Hanson, Kuhn, Feyerabend, Sellars, and many others.) Is 'phenomena' not just another name for theory-laden data?[6] In many cases, the distinction between phenomena and conceptualized or theory-laden data will seem artificial. (High energy physicists like to say they can just see the sub-nuclear process right in the bubble chamber photo.) But there are clear cases which cannot be treated as theory-laden observations. The elements of the Periodic Table are phenomena, and no doubt some of them – for example, Fe (iron) – might successfully be treated as observable in some theory-laden way. But there are others – for example, Ge (germanium) – which (at the time of Mendeléeff) were simply not seen at all. Similarly, the phenomena of a thought experiment, such as the light bending in Einstein's elevator, are not actually seen at all either. So these examples of phenomena cannot be reduced to actual observable data, theory-laden or not.

6. PHENOMENA AND INFERENCE

How is it possible that a great and grand theory can seem to be justified by only a tiny bit of sketchy visualized information? Recall the mathematical examples from the beginning of this paper. A lesson about inference can be learned from that rare form of mathematical proof. The claim made there was that a diagram can be a perfectly good proof. One can see complete generality in the number theory picture, for example, even though it only illustrates the theorem for $n = 5$. And one can see the sum of the infinite series, even though one sees the representation of only a finite number of the terms of that series. The diagrams do not implicitly 'suggest' a 'rigorous' verbal or symbolic proof. The regular verbal/symbolic proofs of these theorems are by mathematical induction, by ε–δ limit processes, and by algebraic manipulations, respectively. But the diagrams do not correspond to or 'suggest' these types of proof at all.

One of the morals to be drawn from these examples is of great importance for the philosophy of mathematics, especially concerning the nature of proof. But the moral I want to draw here is just this: we can in special cases correctly infer theories from pictures, that is, from visualizable situations.

The great inductive leap is really from data to phenomena; once we have the phenomena, the further inference to theory often can be actually rather small. This is because of the following feature of natural kinds:

Any natural kind has an essential property (or set of properties) that makes it the kind that it is. If any member of a kind has essential property ϕ, then every member of the kind has ϕ.

For instance, if any sample of water has chemical composition H_2O, then all samples do. But notice our reluctance to make a similar inference about, say, the colour of ravens. We balk at: if any raven is black, then they all are. We hesitate because of our belief that colour is not an essential property of ravens. We do believe that all ravens are black, of course, but this belief is based on the observation of an enormous number of ravens. By contrast, our beliefs about, say, the mass of intermediate vector bosons are based on only a small handful of scratchy bubble chamber photographs. So, either our physics colleagues have absurdly low standards when compared to bird watchers or something remarkably different is going on in each case. Clearly, it is the latter. There is a profound difference between the two cases and it has to do, I suggest, with phenomena being natural kinds. In particular, if any vector boson has mass m, then they all have mass m. Natural kind inference is quite different than enumerative induction, the principle used in inferring the colour of all ravens.[7]

While I have invoked natural kinds and their essential properties to account for some of the aspects of phenomena, my commitment to natural kinds is not too deep. Perhaps *patterns* would be a better notion.[8] Instead of seeing phenomena as constructed out of data, I should take patterns to be so constructed. First, patterns would avoid the controversial metaphysics of essences; second, patterns are obviously abstract and hence clearly different from observable data; and third, inferences from patterns are quite unproblematic, thus lending themselves to quick conclusions of the sort we see in the mathematical and thought experiment examples given above. For now, at any rate, I prefer to remain agnostic and leave this an open question; but for the sake of consistency I will stick with talk of natural kinds.

Of course, the question arises whether we really have a natural kind (or an essential property of a natural kind) on our hands or not. Is mass really an essential property, and colour not? It seems like an *a priori* assumption, and to some extent I dare say that it is. But the view that the colour of ravens is not an essential property while the micro-structure of water is, is at least in part based on very broad experience and the past success of various classes of theories that we hold. Theories based on micro-structure have been enormously successful, while those based on colours have not. So the construction of phenomena out of data is based on more than the immediate data themselves. It is theory-laden, but it need not be laden with the theory that it will subsequently be used to test.

This raises an interesting and important point that I can only mention here – the construction of pseudo-phenomena. Many scientific works are replete with drawings. E.O. Wilson's *Sociobiology*, for example, has almost no photographs but has several beautiful drawings of animals in various activities. One of these shows two dinosaurs fighting. Needless to say, this was seen by no palaeontologist. It is not a datum, but a phenomenon. But is it a real phenomenon? I will leave to others the assessment of sociobiology. However, I will point out that the theory – like any other – is in the business of explaining phenomena, not data. Perhaps it even does this brilliantly, which is why many find sociobiology persuasive. But there is a lot of room to ask pointed questions about the construction of such phenomena (is it a pseudo-phenomenon?) and the role values may have played.

7. FEYNMAN DIAGRAMS

When Richard Feynman was working on quantum electrodynamics in the late 1940s, he created a set of diagrams to keep track of the monster calculations that were required. Though they were intended for his personal use only, 'Feynman diagrams' have become an enormously powerful and popular tool in all areas of high energy physics (for a popular account, see Feynman 1985). Feynman is thought to be one of the most 'visual' of modern physicists (see Schweber 1985), and his diagrams would seem to be a paradigm example of visualization in physics. In a sense this is certainly true. But in another important sense it's quite misleading.

The transition from an initial quantum state to a final state could happen in any of a number of different ways. Each of these different ways can be represented by a diagram, and there are mathematical expressions associated with each. To calculate the final probability for the transition from one state to another, one would just calculate the expression associated with each of the diagrams. (As a practical matter, only the first few will be calculated to get a reasonably accurate answer.) Figure 8.10, for example shows the first few diagrams depicting the perturbation series containing the different possible sub-processes in electron-positron scattering.

Feynman diagrams look something like cloud chamber pictures, and they are often called space-time diagrams. This leads to the confusion. In fact, the diagrams do not picture physical processes at all. Instead, they represent probabilities (actually, probability amplitudes). The argument for this is very simple. In quantum mechanics (as normally understood), the Heisenberg uncertainty relations imply that no particle could have a position and a momentum simultaneously, which means there are no such things as trajectories, paths, through space-

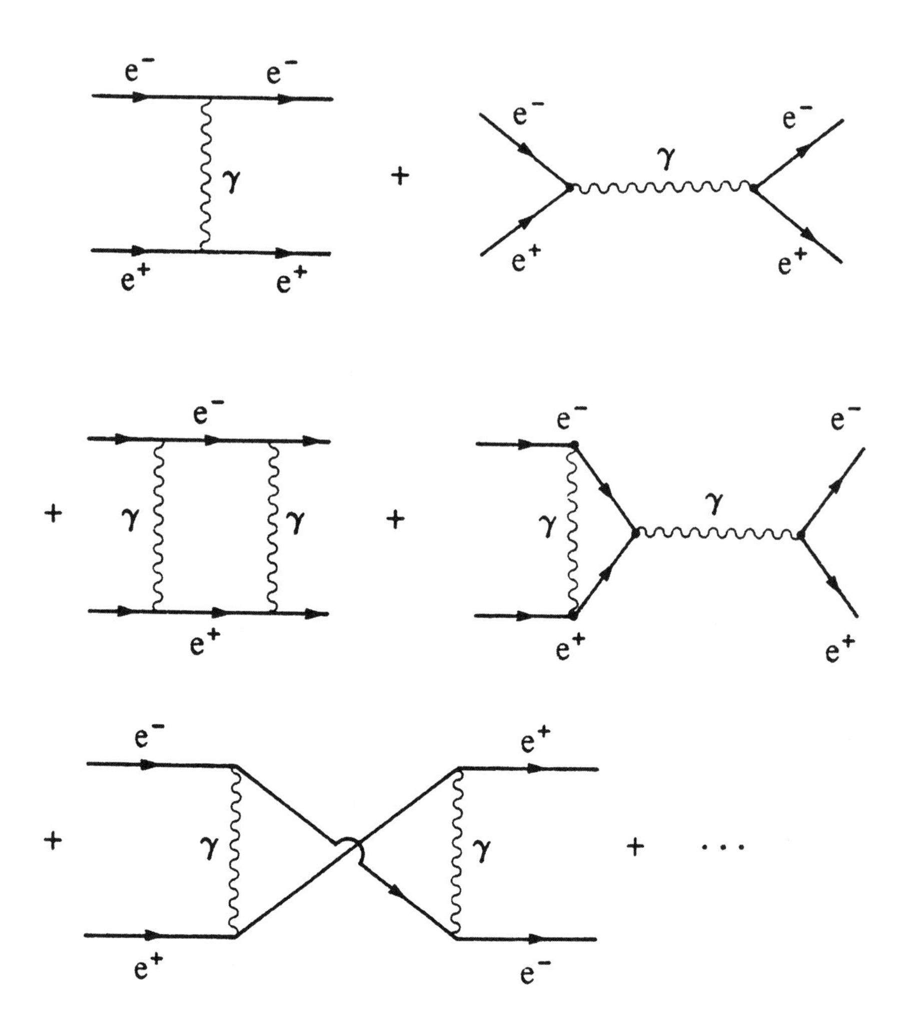

8.10 Feynman diagrams.

time. So the lines in a Feynman diagram cannot be representations of particles and their actual paths through space-time.

So what, then, is being visualized? I think the answer is simply this: Feynman diagrams are geometric representations of probability functions. As such they are quite different from other types of pictures, diagrams, and illustrations I have been discussing above. They are not pictures of phenomena. Of course, visual reasoning plays a role in their use, but this is not connected to natural kind reasoning as I hold the other types to be. Instead, a Feynman diagram is more like a Venn diagram. We depict, for example, that all *A*s are *B*s by representing the set of *A*s and the set of *B*s as circles, with the *A*-circle wholly contained within the *B*-circle. Clearly we can visualize the relation between the *A*s and the *B*s using the Venn diagram, but such visualization is different – though no less important – than the type of visualization involved in the construction of phenomena. In a thought experiment, for instance, we perceive the abstract natural kind; but in a Venn diagram we see some circles. Similarly, the Feynman diagram geometrically represents (often brilliantly) a mathematical function which is linked to a physical process. We see the lines in the diagram; we do not visualize the physical process itself, nor any sort of abstract version of it.

By contrast, phenomena are to be distinguished from data, the stuff of observation and experience. And even though they are relatively abstract, they have a strongly visual character. They are constructed out of data, but not just any construction will do. Phenomena are natural kinds that we can picture. As such they resemble data in a visual way that Feynman diagrams and Venn diagrams do not. In other words, there are different types of visual reasoning in science, and what I am here calling phenomena by no means exhaust visual thinking. But what are the relations between these different types? That seems a suitable challenge on which to close.[9]

NOTES

1 This term is unfortunately loaded. Sociologists of science often use 'construction' to mean 'social construction,' the very opposite of an independently existing object or fact. For instance, Pickering (1984) does this when he speaks of 'constructing quarks.' However, I'm using the term in a more innocuous sense perfectly compatible with describing the objective truth. For example, when a mathematician 'constructs' a function, she is not creating it anew, but merely (though perhaps very

cleverly) characterizing it in terms of other already given mathematical objects.

2 For further discussions, rich with examples, see Woodward (1989), Harper (1990), and Kaiser (1991).

3 By 'abstract' I mean not in space and time; numbers, properties, and propositions are typical abstract entities. Nominalists take such entities to be mere words; realists take them to be genuine objects existing independently of humans. My view of phenomena is, of course, a realist account.

4 For more on thought experiments, see my *The Laboratory of the Mind: Thought Experiments in the Natural Sciences* (1991).

5 But not exactly momentum, since Descartes eschewed mass. For him 'quantity of motion' would be more like 'size of matter times speed.' Leibniz did not have a clear notion of mass either, though unlike Descartes, he was not wedded to a purely kinematic physics.

6 This challenge came from Simon Blackburn.

7 For more on natural kind reasoning, see Harper (1989). Generally, this is unexplored territory and deserves a great deal more attention.

8 Thanks to Mary Tiles for this point. I'm grateful to her for helpful discussions on a number of other topics as well.

9 The themes of this paper are treated in my book *Smoke and Mirrors: How Science Reflects Reality* (1994). I am also very glad to acknowledge the financial help of the Social Sciences and Humanities Research Council of Canada, and finally I wish to thank David Kotchan for drawing several of the diagrams.

9. Visual Models and Scientific Judgment

RONALD N. GIERE

1. INTRODUCTION

When reading scientific papers or watching presentations by scientists, nothing is more obvious than the use of *visual* modes of presentation for both theory and data. This not a new phenomenon, although it has been emphasized recently by the development of computer graphics. One finds a widespread use of various visual devices going back to the Scientific Revolution. Newton's *Principia*, for example, is full of diagrams used in his geometrical demonstrations. But why should anyone be particularly interested in the use of pictures and diagrams in science? Specifically, why should a *philosopher of science* be interested in this particular aspect of the practice of science?

It is my view that studying visual modes of representation in science provides an entrée into fundamental debates within the philosophy of science, as well as in related fields such as the history, psychology, and sociology of science. I will begin by indicating the nature of these debates and pointing out the relevance to these broader issues of the role played in science by visual modes of presentation. In the latter part of the paper, I will use some diagrams that played a central role in the twentieth-century revolution in geology in order to illuminate these general themes.[1]

2. GENERAL ISSUES

Within the English-speaking world, the logical empiricist image of science, and the projects it generated, dominated philosophical thought about science for a generation following the Second World War. Two fundamental aspects of this image are relevant here. First, scientific knowledge consists primarily of what is encapsulated in scientific theories, and theories are ideally to be thought of as interpreted axiomatic systems. It follows that the primary mode of representation in science is *linguistic* representation. Second, the reasoning which legitimates the claims of a particular theory as genuine knowledge has the general character of a logic. That is, there are rules which operate on linguistic entities yielding a 'conclusion' or some other linguistic entity such as a probability assignment.

In the framework of logical empiricism, then, there can be no fundamental role in science for non-linguistic entities like pictures or diagrams. Such things might, of course, play some part in how scientists actually learn or think about particular theories, but unless their content is reduced to linguistic form, they cannot appear in a *philosophical* analysis of the content or legitimacy of any scientific claims to knowledge.

Like so many other aspects of post–Second World War Western culture, the logical empiricists' picture of science began to blur in the decade of the 1960s. A major stimulus for change, and focus for opposing views, was Thomas Kuhn's *Structure of Scientific Revolutions* (1962). The initial rejection of Kuhn's views by philosophers of science was to be expected because he rejected the major assumptions of logical empiricism. According to Kuhn, for example, general statements organized into axiomatic systems play little role in the actual practice of science. There is thus little to be learned about science by reconstructing theories in a logical empiricist mould. Moreover, the relative evaluation of rival paradigms is not something that can be reduced to any sort of logic. It is fundamentally a matter of choice by scientists acting as individuals within a scientific community. For Kuhn, science is primarily a puzzle-solving activity. Scientific revolutions are the result of many individual scientists making the judgment that a particular type of puzzle, or way of approaching puzzles, is no longer fruitful, and that another approach provides a more promising basis for further puzzle-solving activities.

Kuhn himself did not highlight the role of visual or other non-propositional modes of representation in science. Indeed, he avoided talk about representation. I surmise that was largely because he, like most everyone else, thought of representation in propositional terms, and that leads immediately to the concept of truth. His picture of science as a puzzle-solving activity was meant to be an alternative to the view of science as producing truths. Moreover, his emphasis on the incommensurability of terms in the languages of rival paradigms shows his tendency to think of science in linguistic categories.[2]

Nevertheless, Kuhn's approach to understanding science at least opened the door to consideration of non-linguistic representational devices in the practice of science. This was not just because his account was historical, but because it was *naturalistic.* He was trying to explain how science works in terms of naturalistic categories like the psychological make-up of individual scientists and the social interactions among scientists in communities. Thus, whether non-propositional devices like diagrams and graphs play a significant role in science is something to be determined empirically by examining actual cases of science in action.

Philosophers were initially quick to charge Kuhn with having fallen into epistemological relativism, a charge he personally has struggled to avoid.[3] But beginning in the mid-1970s, several groups of European *sociologists* of science have pushed the relativistic aspects of Kuhn's views to their logical conclusion. The slogan of these schools is that science is a *social construct.* The import of the slogan is most quickly grasped by reflecting on the extent to which *society* is a social construct. There is, for example, nothing in the non-human universe that requires representative democracy, an independent judiciary, separation of church and state, or any other of the fundamental structures of American society. These are historically conditioned social constructs. Science, it is claimed, is no different. It follows that the world-view of those we call 'primitive' is in no objective way inferior to ours. It is just different. The only thing special about our scientific world-view is that it is ours.[4]

Significantly, relativist sociologists of science were among the first to investigate the role of pictures, diagrams, and other non-propositional forms of representation in science. Their aim has been to show how images are created and deployed in the social construction of scientific knowledge. The initially plausible view that these various images somehow picture reality is thereby 'deconstructed.'

There are more radical and less radical strains within the constructivist camp. A less radical view is to admit that scientists intend their theories to represent the world and often believe that they have succeeded. It is just that close sociological and anthropological analysis reveals that the intentions are not fulfilled and the beliefs mistaken. A more radical view is that science is not really a representational activity after all. In the twentieth century, painting clearly moved from being essentially representational to allowing forms that are not representational at all. We now have pictures that are not pictures of anything, and were never intended to be. So, it might be claimed, science is now (and maybe always has been) non-representational. Our theories don't picture anything.[5]

My view is that what is needed is a middle way between philosophical positivism and sociological relativism, both of which, in very different ways, deny any genuine representational role for visual images in science. Examining visual modes of theorizing and evaluating data is part of a strategy for developing the desired middle way. Since images could not literally be true or false, this strategy avoids raising questions about the nature of truth. It thus makes possible the pursuit of a naturalistic theory of science which goes beyond puzzle solving to explore ways in which visual models might genuinely represent the real world, and be correctly judged to do so.

As just indicated, there are several major parts to the overall program of developing a naturalistic middle way. A major task, for example, is simply to understand the various ways images and other non-propositional devices can be used to represent the world. Here I will approach this task only to the extent of pointing out how a model-based understanding of scientific theories makes it possible to treat things like diagrams and scale models on a par with the more abstract theoretical models that, on this account, form the core of any scientific theory. The focus of this paper will be on explaining how pictorial presentations of data can be used in judging the relative representational adequacy of visually presented models of the world. Or, to put it in more traditional terms, I want to present (part of) a theory of scientific reasoning in which visual presentations of both data and theory can play a significant role. The 1960s revolution in geology provides a particularly rich context for just such a presentation.

3. MODELS AND THEORIES

For a generation now, a number of philosophers of science have been developing an alternative to the logical empiricist account of scientific theories. This account has several names. It is sometimes called the 'semantic view of theories,' by way of contrast with the supposed 'syntactic' character of theories on the received view. It is also called the 'non-statement' view, the 'predicate' view, or (as I now prefer) the 'model-based' view of theories.[6]

A common way of describing the model-based view is to say that theories include two different sorts of linguistic entities. Some are predicates, which may have a quite elaborate internal structure, as, for example, the predicates 'pendulum' or 'two-body Newtonian gravitational system.' Others are statements of the form 'X is P' where X refers to a real-world system and P is one of the predicates, as in the statement 'The earth-moon system is a two-body Newtonian gravitational system.' The predicates, as such, have no truth values, but the associated statements do.

This way of characterizing the statement and model-based views of scientific theories makes the differences between them seem relatively formal, even trivial. Expressions that function as empirical laws on the statement view, for example, reappear in the definitions of predicates on the model-based view. And there seem to be few if any significant empirical claims that could be formulated in one framework and not in the other. In the present context, however, the main difficulty with this way of formulating the difference is that it overemphasizes the *linguistic* aspects of the model-based approach. A way to redress this deficiency is to shift one's focus away from the predicates to the objects they encompass.

On my understanding of a model-based approach to scientific theories, the predicate 'pendulum,' as it appears in classical mechanics, does not apply directly to real-world objects like the swinging weight in the grandfather clock that stands in my living-room. It applies, rather, to a family of idealized models, the central example of which is the so-called 'simple pendulum.' A simple pendulum is a mass swinging from a massless string attached to a frictionless pivot, subject to a uniform gravitational force, and in an environment with no resistance. This is clearly an ideal object. No real pendulum exactly satisfies any of these conditions. So no real pendulum is a simple pendulum as characterized in classical mechanics. And the same is true for more complex types of classical pendulums: damped pendulums, driven pendulums, and so on. Figure 9.1 shows a family of models of pendulums radiating out from the model of a simple pendulum. So what is the relationship between the idealized model pendulums of classical mechanics and real swinging weights? It is, I suggest, like the relationship between a *prototype* and things judged sufficiently similar to the prototype to be classified as of

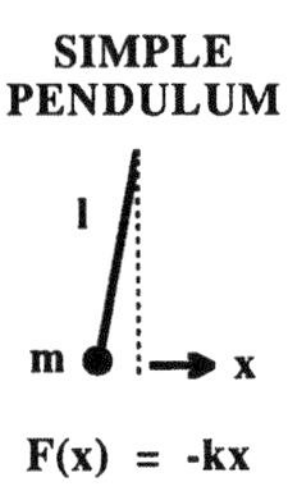

9.1 A family of models of pendulums radiating out from the model of a simple pendulum.

that type. And how are such judgments made? After all, any two objects (idealized or not) are similar to each other in infinitely many ways. Which features count for judgments of similarity to the prototype, and why do some features count more than others? Here there are no simple answers.

To some extent the models themselves provide guidelines for the relevant similarity judgments. The main dynamical variable in any model of a pendulum is the period of oscillation. It is a characteristic of the models that the mass of the bob is irrelevant to the period. Only the length of the suspension, plus the gravitational force, matters. So the mass of the bob should be relatively unimportant in classifying some real swinging weight as a simple pendulum. So should its shape. Yet if one is building an ordinary grandfather clock, a one-pound pie-shaped bob swinging in air and a one-ton spherical bob swinging in water will not be regarded as equally appropriate approximations to a simple pendulum, even though the deviation in period from the ideal might be similar. Other, highly practical, considerations are overriding.

Figure 9.2 is an attempt to picture the relationships among representational devices such as language, models, and objects in the real world. Important and interesting though these relationships may be, I cannot further pursue these general issues here.[7] The main point for present concerns is that, on this view of scientific theories, the primary representational relationship is not the truth of a statement relative to the facts, or even the applicability of a predicate to an object, but the similarity of a prototype to putative instances. This is not a relationship between a linguistic and a non-linguistic entity, but between two non-linguistic entities. Once this step has been taken, the way is clear to invoke other, less abstract, non-linguistic entities to play a similar role.

Consider the sketch of a simple pendulum shown at the top of figure 9.1. I would regard this diagram as a particular embodiment of the abstract model of a simple pendulum. It too can serve as a prototype in judging the similarity of a particular real pendulum to the classical model of a simple pendulum. What holds for this simple diagram should in principle apply to a host of other non-linguistic representational devices. The difficulties in getting from 'in principle' to 'in practice' should not be underestimated, but they are not my main concern here.

4. CRUCIAL DECISIONS

The idea of a 'crucial experiment,' as expounded, for example, by Francis Bacon, was a major cultural achievement of the Scientific Revolution. It deserves to be ranked along with such other achievements as the calculus, the telescope, and the air pump. How it came to have that status is a difficult historical question. Of course, with the hindsight of three centuries, we know that the role of crucial experiments has often been exaggerated, and that the designation of an experiment as 'crucial' often comes long after the fact. But this only shows that the idea of a crucial experiment can play a rhetorical as well

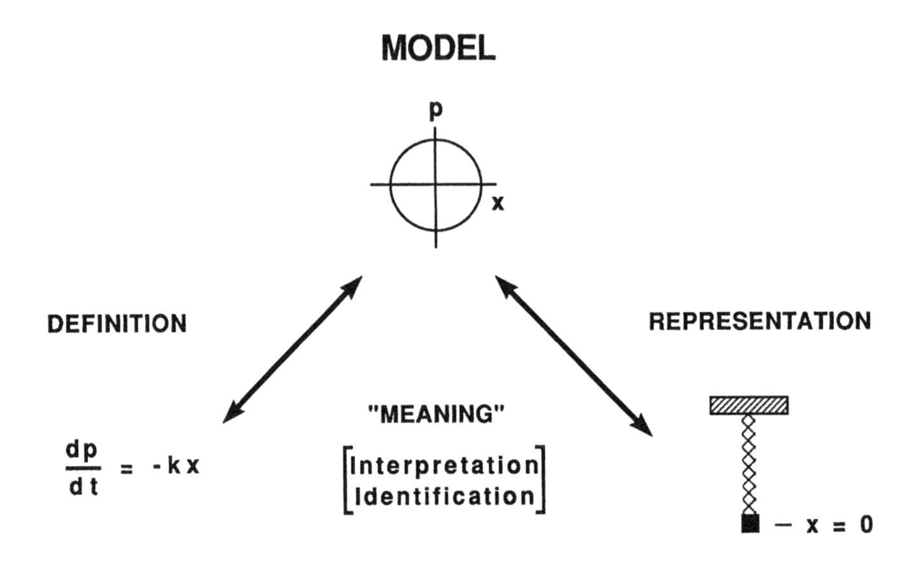

9.2 Relationships among language, models, and objects in the real world.

as an operational role in science. It does not show that, properly understood, it plays no operational role.[8]

But what is the 'proper' operational understanding of crucial experiments? Here there are as many answers as there are approaches to the general problem of theory evaluation in science. For example, on a purely *deductive* account of scientific reasoning, the 'logic' of crucial experiments is just the logic of disjunctive syllogism plus modus tollens. Suppose T_1 and T_2 form an exclusive and exhaustive disjunction. And suppose that T_1 implies O while T_2 implies *Not-O*. The 'crucial experiment' yields the 'observation' O. By modus tollens, T_2 is falsified, thus justifying T_1 by disjunctive syllogism.

Again, on a *probabilistic* account of theory evaluation, we suppose that T_1 and T_2 have comparable initial probabilities. A 'crucial experiment' would be one for which the final probability of T_1 given the observed outcome is much greater than the final probability of T_2. Recently it has been argued that theory evaluation is primarily a matter of the relative *explanatory coherence* of the rival theories with given observations. Here the explanatory coherence of a theory is a function of coherence relationships among statements.[9]

There is no need here to rehearse the many reasons that might be given for rejecting these approaches to scientific reasoning.[10] I wish only to point out that they all assume a *propositional* account of scientific theories. It is statements that are falsified, assigned low probabilities, or cohere. There is therefore no way that visual or other non-propositional forms of information can play a role in the reasoning without first being reduced to propositional form. I will now outline an account of crucial experiments that allows reasoning based directly on visual images. It assumes a model-based understanding of scientific theories along the lines outlined above.

The label 'crucial decisions' already indicates that my account of crucial experiments will be formulated within a general account of human judgment. In developing an account of human judgment, one faces a number of alternatives. One is between an account of judgments by individuals or by groups. I shall focus on individuals.[11] Another alternative is whether the individuals in question are to be regarded as 'rational agents' or simply scientists. I shall focus on scientists who are idealized only in the sense that the objects of any theory (e.g., classical mechanics) are idealized, not in the sense of providing normative standards.[12]

On anyone's account, a crucial experiment is designed to decide between two well-defined alternatives. The alternatives may be highly specific hypotheses or more broadly conceived 'approaches' to the same subject matter. The restriction to two alternatives is not as severe as it might seem. Although, in principle, there are always infinitely many logically possible alternatives, in practice scientists rarely face more than a few. And if there happen to be more than two, they can be dealt with in sequence, two at a time.[13]

So the model of crucial decisions to be employed here is a model of an individual scientist trying to decide between two alternative models, which for the moment we will designate simply as M_1 and M_2. This yields the standard two-by-two decision matrix shown in figure 9.3. We need only be a little careful how we understand the alternatives. The label 'M_1' stands for what decision theorists call 'a possible state of the world' and should be understood as referring to the possibility that the world is more or less like the idealized model referred to as 'M_1,' or at least that the world is more like M_1 than M_2. And conversely for M_2. The label 'Choose M_1' means that the agent chooses to regard M_1 as

	M_1	M_2
CHOOSE M_1		
CHOOSE M_2		

9.3 Decision matrix for a choice between alternative models of the same real system.

providing a satisfactory representation of the world, or at least a better representation than that provided by M_2. And conversely for M_2.

For present purposes we can take an experiment to be a physical process that yields a reading within a specified one-dimensional range of possible readings, as shown in figure 9.4. What makes an experiment *crucial* are the following conditions:

(i) If the actual world is like model M_1, then the experiment is very likely to yield a reading in the range R_1 and very unlikely to yield a reading in the range R_2.
(ii) If the actual world is like the model M_2, then the experiment is very likely to yield a reading in the range R_2 and very unlikely to yield a reading in the range R_1.

The connection between these conditions and the decision matrix is made by the following obvious 'decision rule':

(a) If the experiment yields a reading in the range R_1, choose M_1.
(b) If the experiment yields a reading in the range R_2, choose M_2.
(c) If the experiment yields some other reading, reconsider the whole problem.

That this is the appropriate decision rule can be seen simply by running through the possibilities. If the world really is captured by M_1, then, by condition (i), the experiment will most likely yield a result in range R_1, and, following the decision rule (a), one will choose M_1 as providing the better representation of the world. This is clearly the

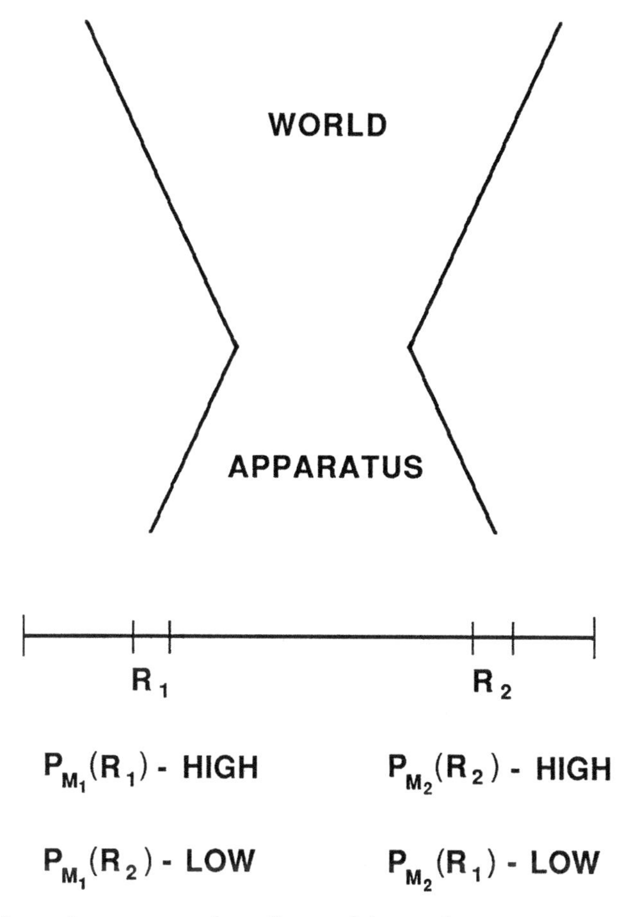

9.4 A schematic representation of a crucial experiment.

appropriate choice. Similarly, if the world really is captured by M_2, then, by condition (ii), the experiment will most likely yield a result in range R_2, and, following the decision rule (b), one will choose M_2 as providing the better representation of the world. This is again clearly the appropriate choice. Either way, one is very likely to make the 'right' choice. Of course, if the reading is something else, there are lots of possibilities, including that neither M_1 nor M_2 is a very good representation of the world, that the experiment was badly done, etc.

There are many things remaining to be said about this understanding of crucial experiments, and many things said that might be disputed.[14]

For present purposes only one aspect requires further clarification. The expressions 'very likely' and 'very unlikely' in the two conditions stated above must refer to *physical probabilities* (propensities) in the world. They cannot refer to degrees of belief or epistemic judgments. That would lead to a completely different account of crucial experiments.

In a single spin of a fair roulette wheel, for example, it is very unlikely that the result will be double zero. It is not just that people attach a low degree of belief to this outcome. Rather, given the physical construction and operation of a roulette wheel, a double zero is physically unlikely. The reason most people give little credence to a belief in this outcome being realized is that they know it is physically unlikely. This is not to say that access to knowledge of physical probabilities is mysteriously direct. On the contrary. These judgments, like all other judgments about the world, are based on more or less definite models of the world. So judgments about physical probabilities, like all judgments about the physical world, are *model-based.*

We are now, finally, ready to proceed to the main objective of this paper, which is to show how visual presentations of both models and data can be used in crucial decisions about which models best represent the real world.

5. IMAGES AND ARGUMENTS

At this point I will narrow the discussion to the example of twentieth-century geology.[15] Here the alternative scientific theories are better thought of as broadly conceived *approaches* to geophysics. One approach, commonly labelled *stabilism,* is that the major geological features of the earth, particularly oceans and continents, originally formed in roughly their current configuration and have remained stable in those positions throughout geological time. The overall mechanism was taken to be cooling, contraction, and solidification of an originally molten sphere. The alternative approach, *mobilism,* is that the relative positions of the continents and oceans have altered in major ways in geological time, that is, since the original formation of solid land masses. It is a standard part of mobilism, for example, that the Atlantic Ocean is a relatively recent product of a separation of North and South America from Europe and Africa respectively.

During the 1920s, the mobilist cause was championed by Alfred Wegener, a German scientist whose earlier work was in meteorology and atmospheric physics. Wegener provided mobilism with many dramatic visual presentations, most notably a series of three world maps picturing the breakup of Gondwanaland, Wegener's original supercontinent containing most of the world's land mass (fig. 9.5). These pictures, which first appeared in the third (1922) German edition of his book, *Die Entstehung der Kontinente und Ozeane*, show the breakup taking place between the Carboniferous (300 million years ago) and the Early Quaternary (500 thousand years ago). I do not claim that these maps constitute the entire content of Wegener's mobilism. Rather, they are visual models which are part of a diverse family of models which all together constitute Wegener's theoretical resources for presenting a mobilist history of the earth.

Wegener gathered evidence for mobilism from many domains, including geology, geophysics, palaeontology, palaeobotany, and palaeoclimatology. Here I will concentrate on just one piece of evidence, the celebrated 'fit' between the eastern coastlines of North and South America and the western coastlines of Europe and Africa. Figure 9.6 reproduces Wegener's sketch of this fit as it appeared in the first (1915) edition of his book. It is a crude sketch. There exist far better drawings exhibiting a better fit dating from over half a century earlier.[16]

What is notable in this sketch is the explicit attention paid to geological features *other* than the fit of the coastlines. In particular, Wegener has marked areas crossing roughly between England and New England, and between South Africa and southern South America, where mountain ranges appear roughly continuous across the postulated border. These congruences play a major role in his presentation, and apparently did so in his own thinking as well.[17]

Referring to the match in coastlines, Wegener at one point remarks that it reminds him 'of the use of a visiting card torn into two for future recognition' (1924, p. 44). This is a highly visual metaphor. A little later he modifies and expands the metaphor. Referring to both the match in coastlines and the match in features across the boundary, he writes:

> It is just as if we put together the pieces of a torn newspaper by their ragged edges, and then ascertained if the lines of print ran evenly across. If they do, obviously there is no course but to conclude that the pieces were once actually attached in this way. If but a single line rendered a control possible, we should have already shown the great possibility of the correctness of our combination. But if we have *n* rows, then this probability is raised to the *n*th power. (1924, p. 56)

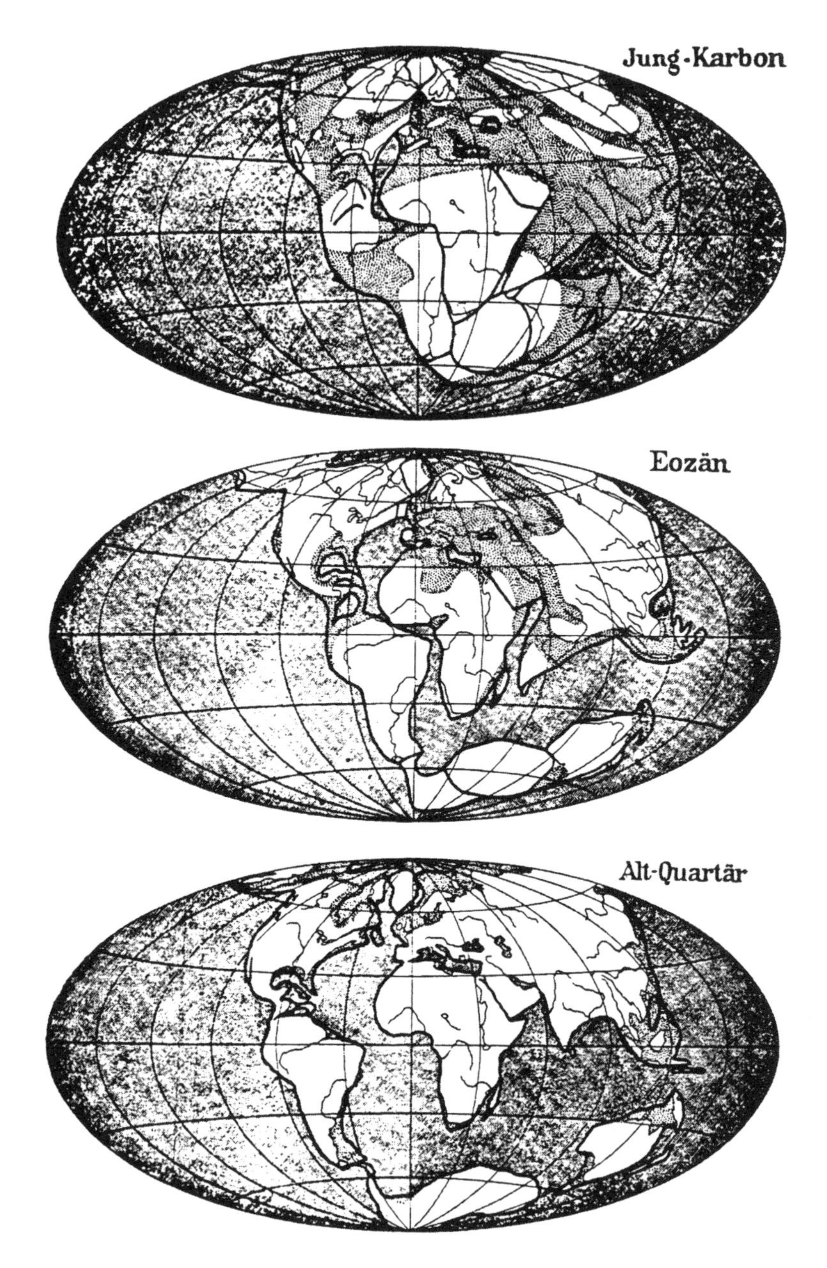

9.5 Wegener's visual representation of the breakup of Gondwanaland (A. Wegener 1922).

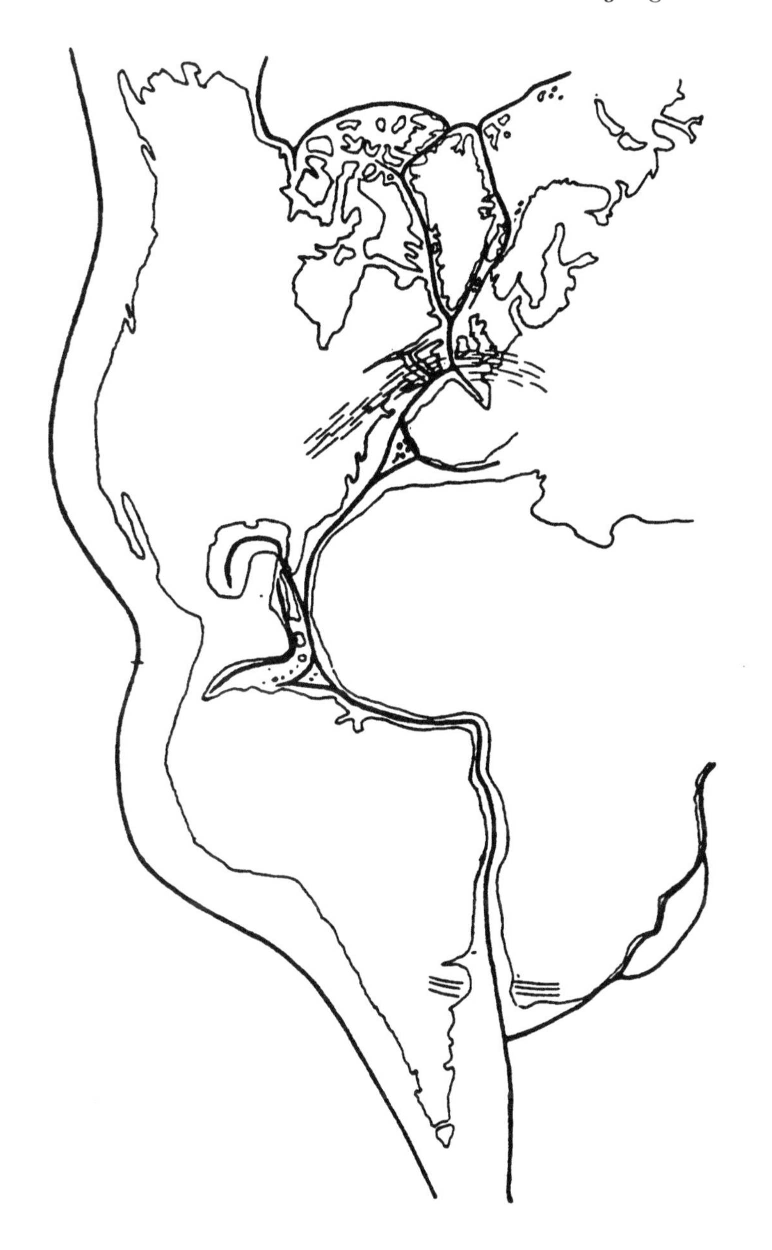

9.6 Wegener's sketch of the Atlantic coastlines indicating continuous mountain ranges across the assumed line of separation (A. Wegener 1915).

Here Wegener appears to go beyond my analysis of crucial decisions, claiming a high probability for his theory itself. But his analysis of the evidence includes the two conditions required for my analysis to be operative. That is, if his mobilist account is correct, then the existence of congruences like those noted is highly probable. Conversely, if stabilism is correct, and the continents formed independently of one another, then such congruences are highly unlikely. Since Wegener clearly believes that the congruences do exist, my analysis could explain why he thinks that mobilism is the obvious choice.

In presenting Wegener's argument, I have employed images he himself utilized. What exactly is the role of the images in the presentation? One cannot argue, I think, that the images are *logically* essential. Any information that can be presented in a two-dimensional image can also be presented in a linear, symbolic form, as the digital encoding of images makes obvious. But we are not here concerned with how logically possible scientists might reason. We are concerned with how actual scientists do reason.[18] Wegener's presentation makes it clear both that the images played a large role in his own thinking and that he expected them to play a role in the thinking of his audience as well. But what is that role?

The images, I suggest, function as partial visual models of the relevant features of the earth. As such they provide grounds for model-based judgments about the physical probabilities that would be operative in the world if it were structured according to the model. Thus, the images provide a basis for the model-based judgments regarding physical probabilities needed in my account of crucial decisions.

Of course not all the information necessary for making the required probability judgments is present in the image itself. In Wegener's case, for example, one must know that mountain ranges are relatively rare. They do not exist all up and down the coasts of Europe, Africa, and the Americas. Moreover, mountain ranges have distinctive characteristics. So finding several mountain ranges that are congruent across the boundary when the coastlines are lined up according to their matching shorelines is indeed physically unlikely if those mountain ranges had been formed independently on continents separated by thousands of miles.

What happens in such cases, I suggest, is that the visual model serves as an organizing template for whatever other potentially relevant information the agent may possess, regardless of how that information is encoded. The visual image guides the agent's recall of stored information by providing a guide to what, within the agent's diverse store of information, is most relevant to the required probability judgments.

For the purposes of the overall argument of this paper, it is not necessary that my suggestions regarding the role of images be scientifically correct. It would be nice, of course, if something like this were indeed the case. And it is in line with some current thinking in the cognitive sciences. But all my argument requires is that some such account be physically (and psychologically) possible. That shows at least that images could play a significant role in scientific reasoning. And that is enough to refute in principle claims that no such role is possible.

Before leaving this example, I would like to illustrate my position with one further image that played a role in the debates over mobilism in the 1920s. Two years after the 1924 publication of the English translation of Wegener's book, *The Origin of Continents and Oceans*, the American Association of Petroleum Geologists sponsored a symposium in New York City on mobilism (van der Gracht 1928). The symposium featured leading scientists from around the world, including Wegener himself. Of the fourteen participants, roughly one-third supported mobilism, one-third were genuinely open-minded, and one-third were strongly opposed to mobilism.

Among the arguments against Wegener in particular was one offered by Yale geologist Chester Longwell based on the map shown in figure 9.7. This map shows a fairly good fit of the coastlines of Australia and New Guinea within that of the Arabian Sea. But no one present, including Wegener, wished to argue that Australia once filled the Arabian sea.

On my analysis of crucial decisions, the real point of this image is to undermine one of the conditions for Wegener's view of the decision in favour of mobilism. Wegener's position requires that it be physically improbable that there be such matching coastlines within a stabilist model of the earth. Longwell's map provides a clear visual presentation of just such a match. One need not invoke much additional information to be led towards the conclusion that, even within a stabilist model, such matches may not be so improbable as Wegener's position requires. As Longwell himself put it:

> This case is worth some study, in connection with the better known case of South America and Africa, in order to convince ourselves that apparent coincidence of widely separated coast lines is probably accidental wherever

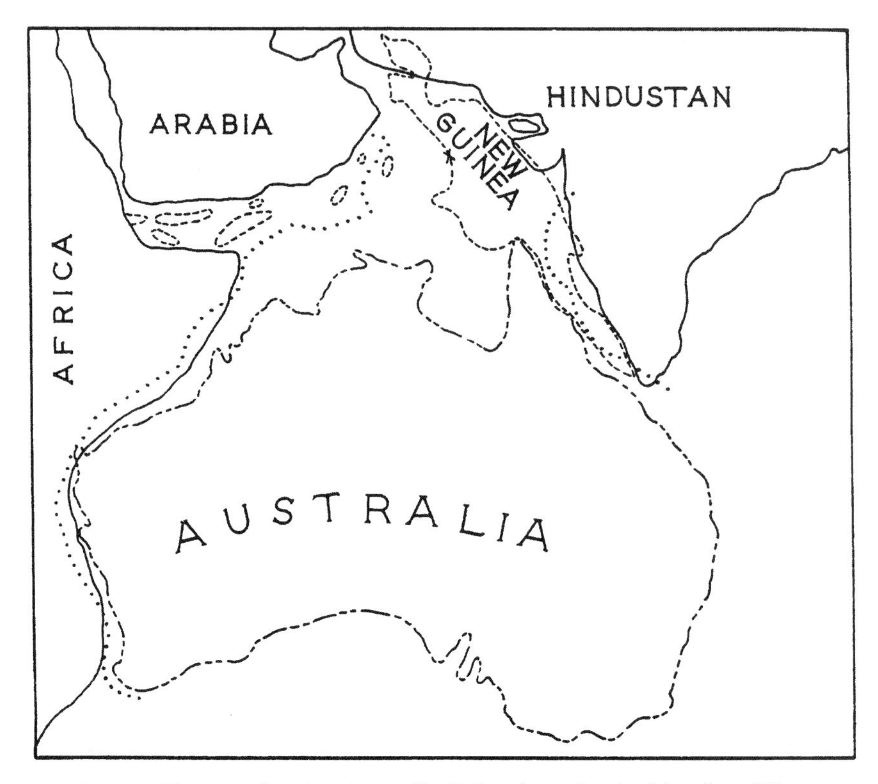

9.7 Longwell's map showing Australia fitting into the Arabian Sea (W. van der Gracht et al. 1928).

found and should not influence anyone unduly in considering the displacement hypothesis. (van der Gracht 1928, p. 153)

In short, there may be visual force on both sides of an argument.

6. THE VISUAL DEVELOPMENT OF THEORETICAL MODELS

Wegener died tragically in 1930 on an expedition to Greenland in search of new evidence for mobilism. About the same time, an English geologist, Arthur Holmes, suggested a new mechanism for mobilism. Inspired by the discovery of natural radioactivity, Holmes reasoned that such radioactivity in the earth might be able to produce sufficient heat to create convection currents of molten minerals just below the earth's crust. These currents could split the crust and move it laterally great distances before turning downward towards the core. Figure 9.8 reproduces Holmes's visual rendition of this model, in which a continental block is ripped in two creating a new ocean where once land had been.[19]

In spite of its visual power, Holmes's model seems to have done little to stave off the general decline of interest in mobilism following Wegener's death. A good part of the explanation for its lack of immediate influence, I would argue, is that this model provides no basis for a crucial decision between mobilism and stabilism. The processes pictured in Holmes's model would be taking place well below the earth's crust, too remote for any then known instruments. And contemporary surface manifestations, if any, would take place too slowly to measure. Holmes himself did not even conjecture a possible crucial experiment.

Holmes's convection model was revived thirty years later by the American geologist Harry Hess. Figure 9.9 reproduces Hess's version of the model (Hess 1962, p. 607). The main difference is that Hess has the convection current rising under the ocean floor rather than a continental block. This was because Hess, unlike Holmes, intended his model to explain the origin of the great ocean ridge systems first explored in the 1950s. The ridges, on Hess's model, are produced directly above the rising convection current, which then spreads out creating a new sea floor. But Hess's model, like Holmes's, provides no basis for a crucial decision between mobilism and stabilism.

The makings of a crucial experiment were provided by a new graduate student in geophysics at Cambridge, Fred Vine, and his recently appointed supervisor, Drummond Matthews. In late 1962, Matthews returned from an expedition to the Indian Ocean, where he had obtained systematic measurements of the total magnetic field at the level of the ocean floor across the Carlsberg Ridge. The task of analysing this magnetic data fell to Vine, while Matthews went off on his honeymoon.

Using then very new computer techniques, Vine determined that the magnetic readings across the ridge showed a small periodic variation as one moved away from the centre of the ridge. Similar periodic variations in magnetic intensity along the ocean floor had earlier been observed in other areas, such as the Pacific Ocean off the coast of North America. Taking these variations in magnetic intensity as a real phenomenon requiring explanation, Vine, early in 1963, set about finding one.

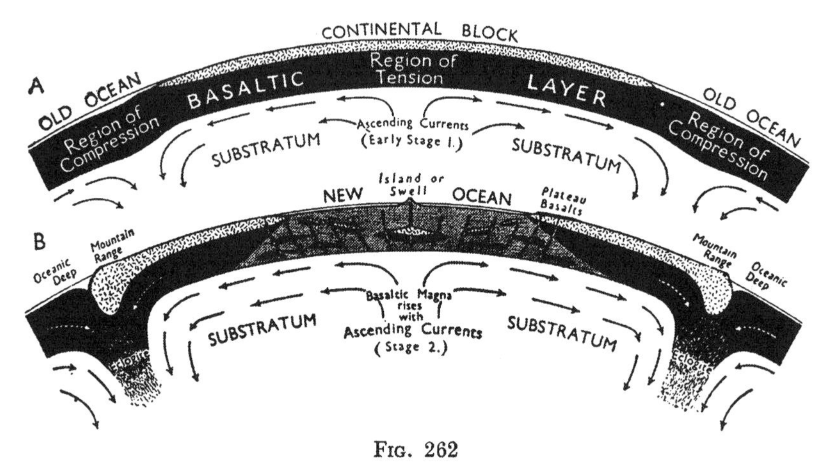

FIG. 262

Diagrams to illustrate a purely hypothetical mechanism for "engineering" continental drift. In A sub-crustal currents are in the early part of the convection cycle (Stage 1 of Fig. 215). In B the currents have become sufficiently vigorous (Stage 2 of Fig. 215) to drag the two halves of the original continent apart, with consequent mountain building in front where the currents are descending, and ocean floor development on the site of the gap, where the currents are ascending

9.8 Holmes's dynamic visual representation of convection currents splitting a continent to produce a new ocean (A. Holmes 1944).

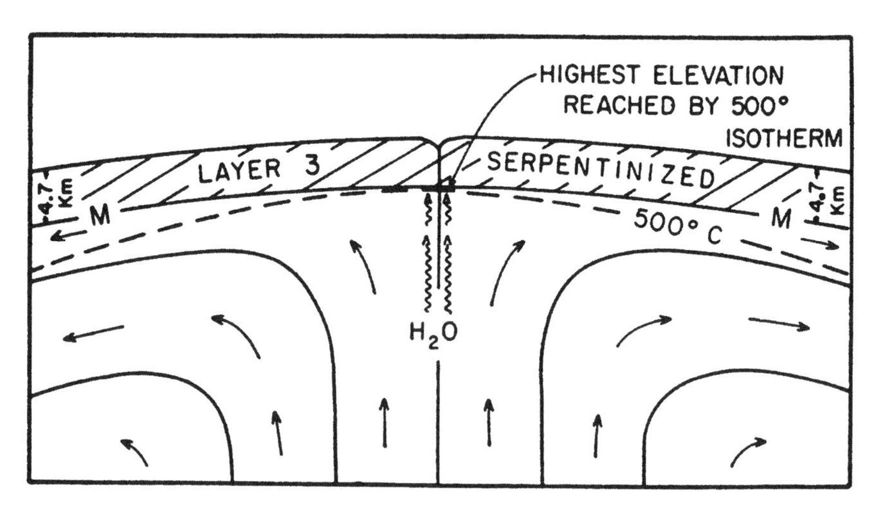

9.9 Hess's dynamic visual model of sea-floor spreading produced by convection currents (H.H. Hess 1962).

Vine was keenly aware of Hess's model of sea-floor spreading, having seen Hess himself present it during a conference at Cambridge in January 1962. A possible link to the observed variations in magnetic intensity was provided by initially unrelated work on palaeomagnetism. Researchers in California, led by Allan Cox, had been examining the direction of remanent magnetism in core samples from lava flows. Such samples provide a measure of the direction of the earth's magnetic field when the examined material was molten since magnetic material in a molten fluid would tend to line up with the existing magnetic field of the earth. What they found, confirming scattered findings dating to half a century earlier, was an apparent change in direction of the earth's magnetic field several times in the past four million years.

Figure 9.10 shows a visual presentation of both the data and the theory in one of the first publications, in mid-1963, of the California group (Cox, Doell, and Dalrymple 1963). The single vertical scale, representing age in millions of years, starts with zero at the top and shows increasing time into the past as one moves down the scale. This arrangement represents the obvious geological fact that in a lava flow the younger materials from recent eruptions are towards the top while the older materials are deeper. Each data point represents a number of rock samples from a given site, with the average age of the samples indicated by the location of the data point relative to the vertical scale. The polarity of the sample, 'normal' or 'reversed,' is indicated by its location in the left or right column. The rival models are simple. They just represent the magnetic field of the earth as having been continuously normal for a time into the past, then being reversed, then being normal, and so on in equal time intervals. One model puts the period of the reversals at a half million years, the other at a million. That both models are consistent with the data can be seen immediately in the graphical presentation.

What is the connection between (i) Vine's data showing regular variation in magnetic field intensity extending out from an ocean ridge, (ii) Hess's model of sea-floor spreading, and (iii) evidence for geomagnetic reversals? The answer cries out for a dynamic, visual model, but none seems to have been published during the crucial years 1963–6. Vine and Matthews's 1963 paper contains, instead, the following verbal description:

> The theory is consistent with, in fact virtually a corollary of, current ideas on ocean floor spreading and periodic reversals in the Earth's magnetic field. If the main crustal layer ... of the oceanic crust is formed over a convective up-

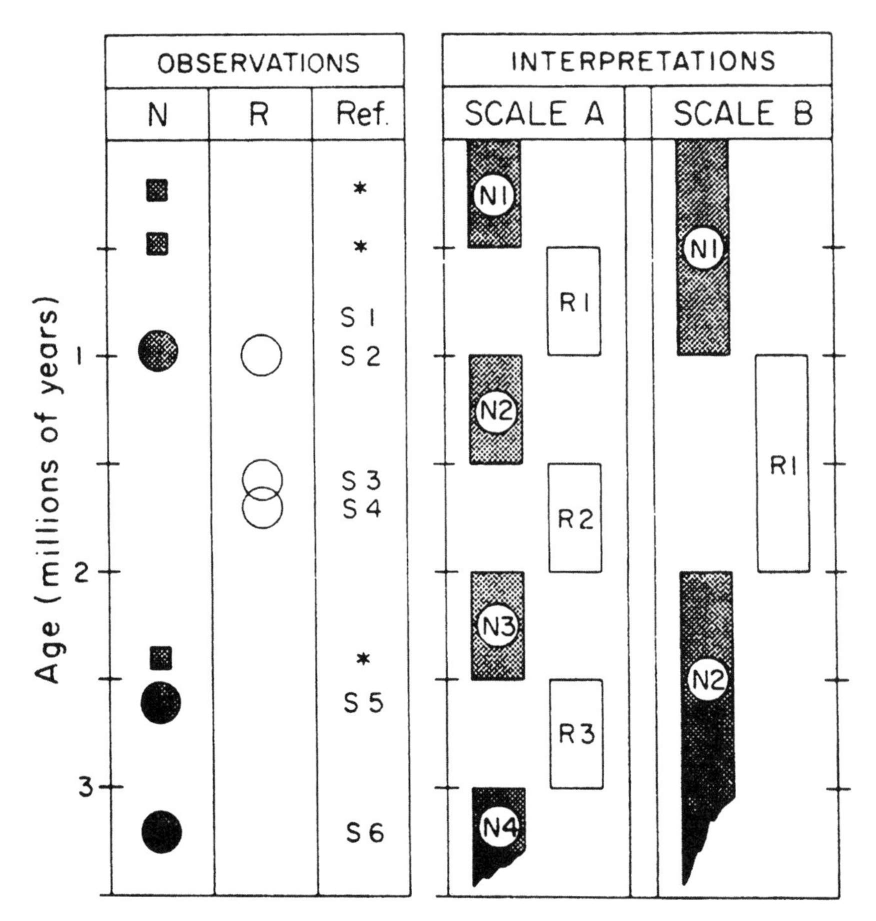

9.10 The first visual presentation of data and models by the California group investigating geomagnetic reversals (A. Cox et al. 1963).

current in the mantle at the centre of an oceanic ridge, it will be magnetized in the current direction of the Earth's field ... Thus, if spreading of the ocean floor occurs, blocks of alternately normal and reversely magnetized material would drift away from the centre of the ridge and parallel to the crest of it. (Vine and Matthews 1963, p. 948)

No one can deny, however, that in reading this description it helps to refer back to the dynamic visual models of Holmes and Hess.[20]

Following the above description is a visual presentation of the magnetic data and a corresponding model of the sea floor across three different ridges (fig. 9.11). This appears to be an adaptation of Cox's model for geomagnetic reversals, except that the blocks of alternately magnetized material are laid out horizontally rather than vertically. This difference, of course, reflects the differing causal processes suggested as having produced the two configurations of differentially magnetized materials.

Publication of Vine and Matthews's paper seems to have convinced almost no one of the reality of sea-floor spreading and the mobilism it implies. Not even they were willing to claim they had proven the case. In the last few lines of their 1963 article, they write:

> It is appreciated that magnetic contrasts within the oceanic crust can be explained without postulating reversals of the Earth's magnetic field; for example, the crust might contain blocks of very strongly magnetized material adjacent to blocks of material weakly magnetized in the same direction. However, the model suggested in this article seems to be more plausible because high susceptibility contrasts between adjacent blocks can be explained without recourse to major inhomogeneities of rock type within the main crustal layer or to unusually strongly magnetized rocks. (Vine and Matthews 1963)

In terms of my model of crucial decisions, they do seem to think that condition (i) is satisfied. The results obtained are fairly probable given a model incorporating sea-floor spreading and geomagnetic reversals. But these results are not wildly improbable if those assumptions are mistaken and stabilism is correct. So there is no adequate basis for making a crucial decision in favour of sea-floor spreading and mobilism.

Of course, the noted possibilities for stabilist explanations of the data are not directly contained in the visual presentation of their model of the sea floor or of their data. But realizing that the simple periodic structure of the model was just read off the similarly simple periodic structure of the data makes it easy visually to assimilate suggested alternative models. So the visual presentation facilitated the judgment that the prospects for a crucial decision were not yet compelling.

During the next two years, the conditions for a crucial decision improved in one respect, but declined in another. In 1964 the California group (Cox, Doell, and Dalrymple 1964), having acquired data from several new sites, published a new scale of geomagnetic reversals (fig. 9.12). Two major differences from the earlier scale are immediately evident. First, they have given up the assumption of equal time periods of normal and reversed polarity. The major normal and reversed 'epochs' are now of irregular duration. Second, they have

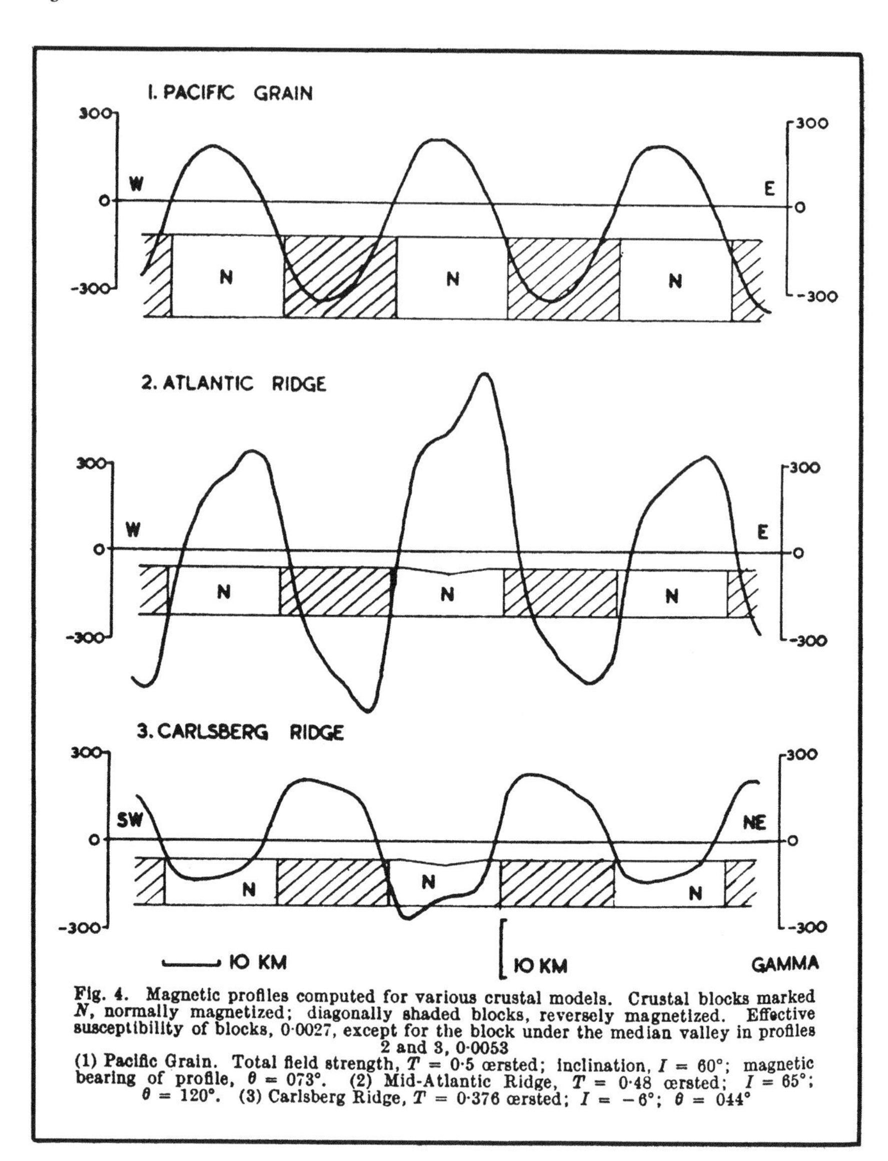

Fig. 4. Magnetic profiles computed for various crustal models. Crustal blocks marked N, normally magnetized; diagonally shaded blocks, reversely magnetized. Effective susceptibility of blocks, 0·0027, except for the block under the median valley in profiles 2 and 3, 0·0053

(1) Pacific Grain. Total field strength, T = 0·5 œrsted; inclination, I = 60°; magnetic bearing of profile, θ = 073°. (2) Mid-Atlantic Ridge, T = 0·48 œrsted; I = 65°; θ = 120°. (3) Carlsberg Ridge, T = 0·376 œrsted; I = −6°; θ = 044°

9.11 Vine and Matthews's models for magnetic profiles near ocean ridges (F.J. Vine and D.H. Matthews 1963).

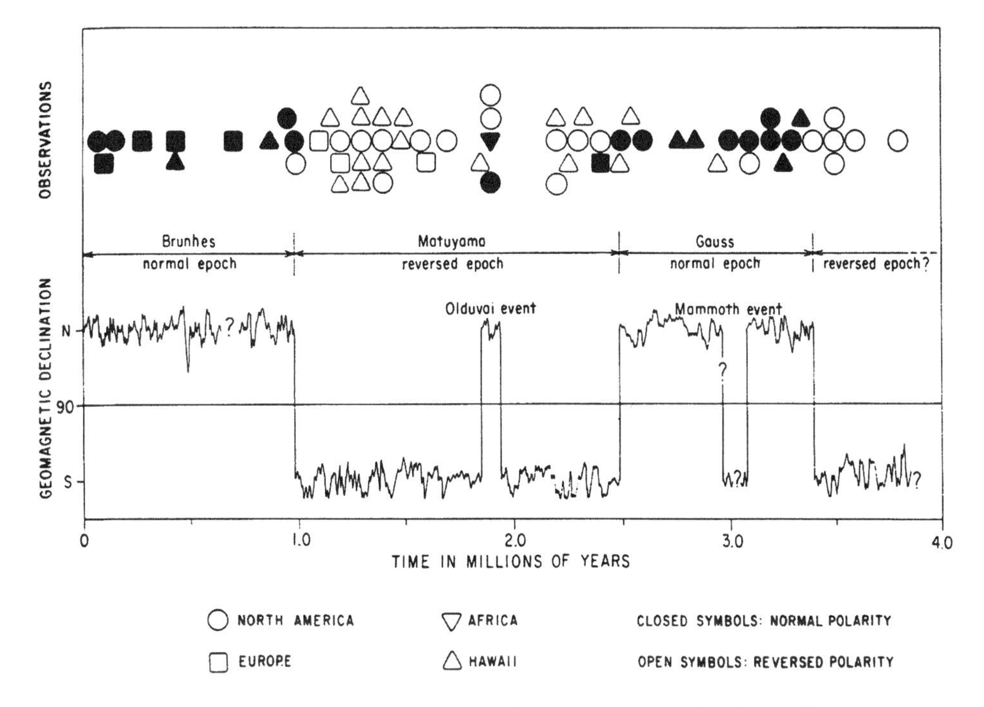

9.12 The second visual presentation of data and models by the California group investigating geomagnetic reversals (A. Cox et al. 1964).

refined the scale to include several brief (100 thousand year) 'events.' The Olduvai event, around 1.9 million years ago, is a brief period of normal magnetism within a long epoch of reversed magnetism. Similarly, the Mammoth event, around 3 million years ago, is a brief period of reversed magnetism within an epoch of normal magnetism.

Meanwhile, Vine and a visiting senior Canadian geologist, Tuzo Wilson, were busy analysing magnetic data from yet another ridge system, this one in the Pacific off the coast of Vancouver. Figure 9.13 reproduces one of their visuals as published in 1965 (Vine and Wilson 1965). Following the California group, their models now exhibit both unequal epochs and several briefer events. In terms of my own model for crucial decisions, the good news was that an aperiodic pattern of reversals with intervening small events is very unlikely to appear in scattered places around a stabilist earth. On any stabilist model, it would take a near miracle for the possible sources of magnetic variation

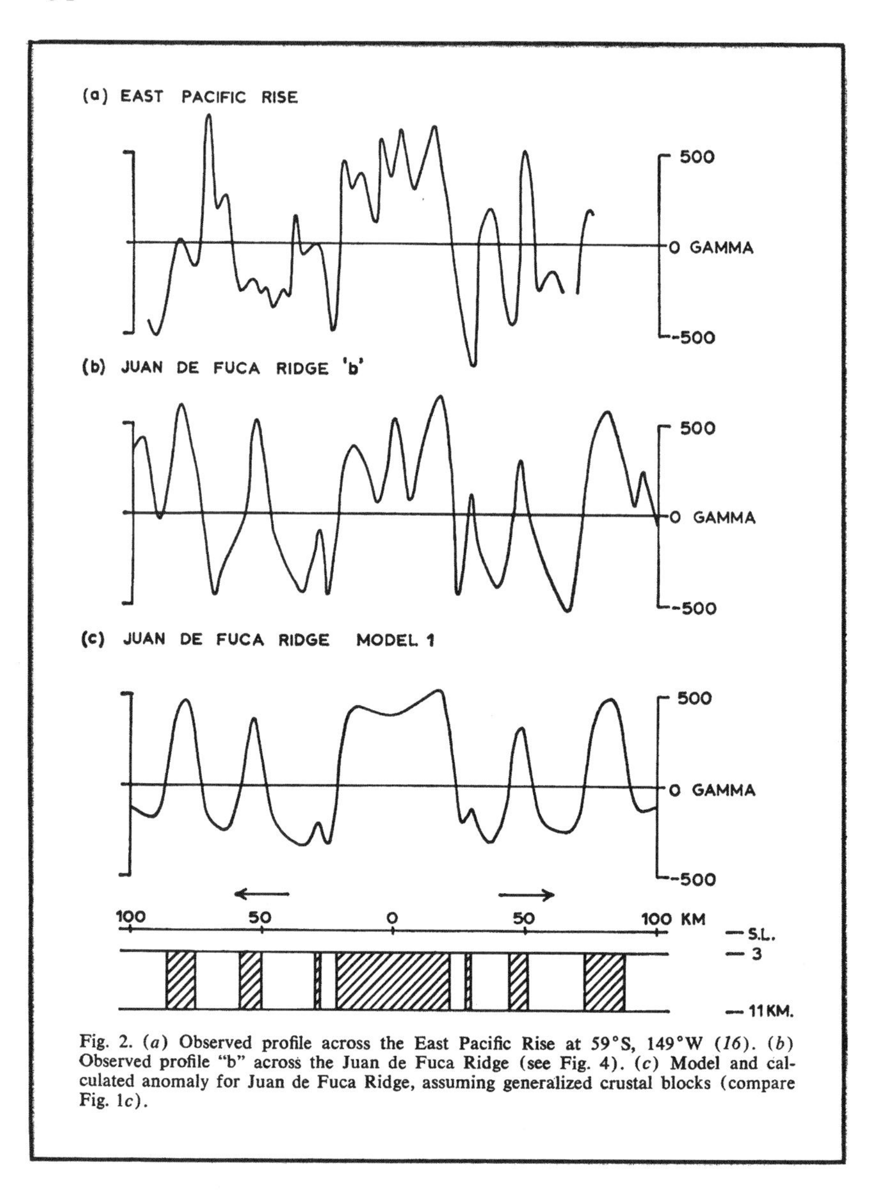

Fig. 2. (*a*) Observed profile across the East Pacific Rise at 59°S, 149°W (*16*). (*b*) Observed profile "b" across the Juan de Fuca Ridge (see Fig. 4). (*c*) Model and calculated anomaly for Juan de Fuca Ridge, assuming generalized crustal blocks (compare Fig. 1*c*).

9.13 Vine and Wilson's visual comparison of magnetic data and crustal model for the Juan de Fuca Ridge (F.J. Vine and J.T. Wilson 1965).

in the sea floor near ridges to produce the same complicated irregular pattern near several different ridges lying in different oceans.

The bad news was that the pattern they were finding in the magnetic sea floor data was not exactly what the Cox scale would lead one to expect if the Vine-Matthews model were correct. If one holds to the constraint that the spreading rate of the sea floor has been roughly constant, there was no way to match up the observed magnetic readings across the ridge with the new scale of reversals published by the California group. The various epochs and events simply did not line up as expected. What was gained in the satisfaction of one condition was lost in failure to satisfy the other.

7. THE PERSUASIVE POWER OF IMAGES

Within a year, the situation had changed dramatically. In late 1965 a research vessel operated by the Lamont Geological Observatory of Columbia University returned from a new geological survey of the Pacific-Antarctic Ridge with the dramatic magnetic profile shown in figure 9.14 (Pitman and Heirtzler 1966, p. 1166). Whereas earlier profiles and geomagnetic time scales had extended out to around four million years ago, this profile extended out a distance corresponding to ten million years, revealing a continuing pattern of reversals never before detected.

The bilateral symmetry of the profile is of particular significance, as is the method used to make it visually obvious. The centre profile shows the magnetic readings moving from west to east at the right of the diagram. The top profile is just the middle profile reversed, with west on the right of the diagram. Merely by scanning visually across the diagram and comparing these two profiles, one can see just how amazingly symmetrical the profile is. That it should be symmetric is an immediate consequence of the Vine-Matthews model, since the sea floor should spread out equally on both sides of a ridge. The lower profile is derived from the model shown at the bottom of the diagram.

About the same time as the data from the Pacific-Antarctic Ridge were being analysed, another group at Lamont was busy analysing the magnetic orientation of sedimentary materials in core samples taken from the ocean floor near the tip of South America. Like cooling lava, sediment traps magnetic materials in their existing spatial orientation as the sediment packs more tightly. If the pattern of geomagnetic reversals is real, it should also be recorded in such sediments. Again the

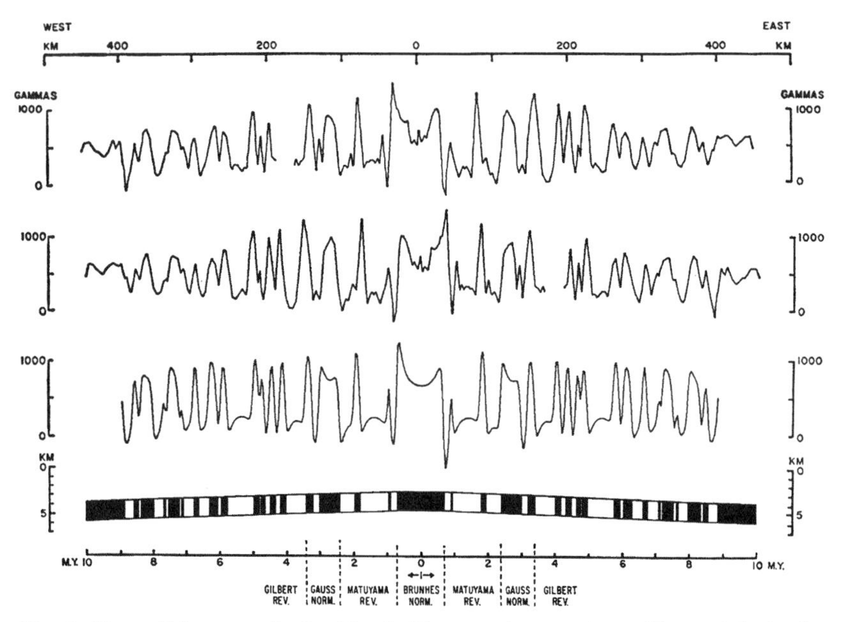

Fig. 3. The middle curve is the *Eltanin*-19 magnetic-anomaly profile; east is to the right. The upper anomaly profile is that of *Eltanin*-19 reversed; west is to the right. On the bottom is the model for the Pacific-Antarctic Ridge. The time scale (millions of years ago) is related to the distance scale by the spreading rate of 4.5 cm/yr. The previously known magnetic epochs since the Gilbert epoch are noted. The shaded areas are normally magnetized material; unshaded areas, reversely magnetized material. Above the model is the computed anomaly profile.

9.14 A visual comparison of magnetic profiles of the Pacific-Antarctic Ridge with a corresponding model. Note especially the symmetry described in the text (W.C. Pitman and J.P. Heirtzler 1966).

evidence, as presented visually in figure 9.15, is dramatic (Opdyke et al. 1966, p. 350). Just by inspecting the diagram, one can see almost immediately that the pattern formed by regions of normal and reversed magnetism within the core samples closely matches that of the magnetic profiles across ocean ridges.

But what of the mismatch between the geomagnetic times scales of the California group and the sea floor profiles which had plagued Vine and Wilson just one year earlier? That too was resolved. Working with rock samples discovered near Jaramillo Creek in New Mexico, several members of the California group discovered that the current period of normal magnetism extended not one million years into the past, but

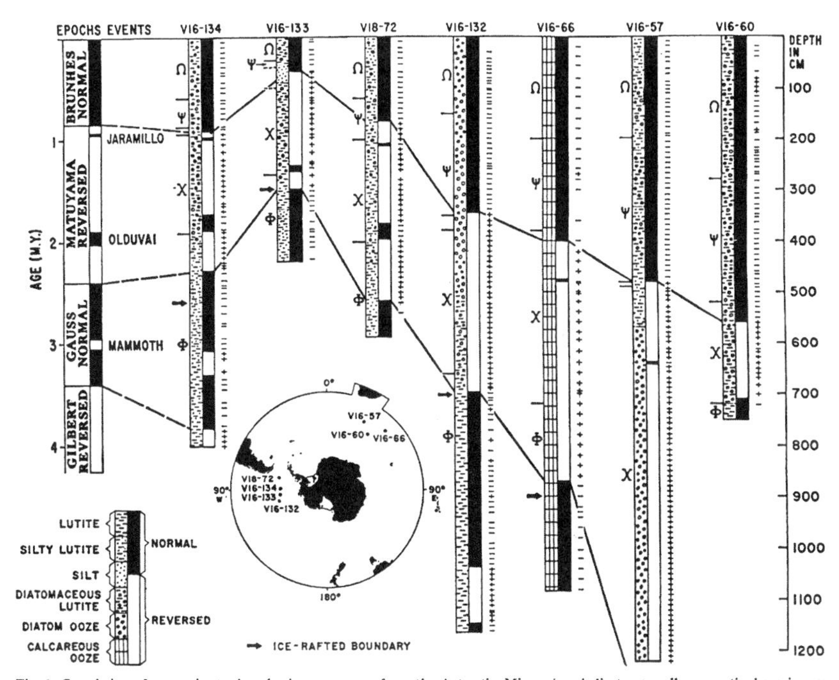

Fig. 1. Correlation of magnetic stratigraphy in seven cores from the Antarctic. Minus signs indicate normally magnetized specimens; plus signs, reversely magnetized. Greek letters denote faunal zones (17). Magnetic stratigraphy in left-hand column is from (1, 2). Inset: sources of cores.

350 SCIENCE, VOL. 154

9.15 A visual comparison of magnetic reversals in deep-sea sediments with the time scale for reversals in terrestrial lava flows (N.D. Opdyke et al. 1966).

only about 0.7 million years. It was followed by a brief period of reversed magnetism and then an event of normal magnetism extending between 0.9 and 1.0 million years ago. Vine and Wilson had gone astray because they had identified the first normal event in their profiles as the nearly two-million-year-old Olduvai event rather than the one-million-year-old Jaramillo event. Figure 9.16 presents Cox's retrospective summary of his group's results during the crucial years 1963–6 (Cox 1969, p. 239).

Versions of these last three images were all presented in a historic session of the April 1966 meeting of the American Geophysical Union. By all accounts, the effect was dramatic. And so it should have been if my account of crucial decisions is correct. That the data obtained were

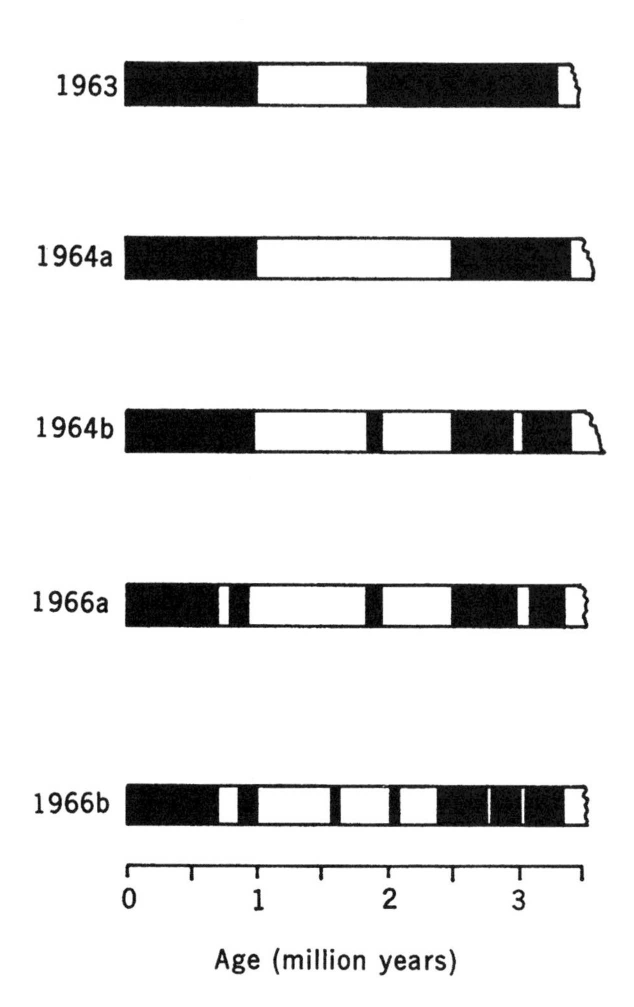

9.16 Cox's summary of refinements in the California group's scale for geomagnetic reversals during the years 1963–6 (A. Cox 1969).

to be expected if the Vine-Matthews model is correct was well known. What made these presentations especially dramatic was that they showed how utterly improbable the data would be on any stabilist model. What stabilist process (short of divine creation) could possibly have produced that visually dramatic and detailed signature pattern simultaneously in widely scattered continental lava flows, deep sea sediments, and the floor of several different oceans? This was visually obvious to all, regardless of their particular research specialties. As Allan Cox, who chaired the April AGU session, later summed it up, '... there was just no question any more that the seafloor-spreading idea was right' (Glen 1982, p. 339).

8. CONCLUSION

This paper connects several recent themes in the philosophy of science, and in science studies more generally: (i) a model-based picture of scientific theories; (ii) a naturalistic account of crucial decisions; and (iii) interest in the use of visual images in scientific thinking. To structure the paper I assumed a model-based account of scientific theories and used the fact that it could accommodate visual information to support my naturalistic account of crucial decisions. But the argument need not have been structured this way. Probably the most appropriate conclusion is that these three themes are mutually reinforcing, and together support a move away from an exclusive reliance on propositional modes of analysis for understanding the workings of modern science.

NOTES

1 The original version of this paper, under the title 'The Visual Presentation of Theory and Data: A Cognitive View,' was presented at a meeting of the Society for Social Studies of Science in November 1987, and for the Committee on History and Philosophy of Science at Johns Hopkins in December. In February 1989 a later version, under its present title, was presented at the Science Studies Units of the Universities of Bath and Edinburgh, and for the Department of Logic, Methodology and Philosophy of Science at the University of London. A still later version, 'Visual Models in Science: Lessons from the Revolution in Geology,' was presented for the Centenary Conference in the History of Science at the University of Oklahoma in September 1990. The current version, once

again extensively rewritten, owes much to colleagues at all these institutions. I am especially thankful to Professor Brian Baigrie for providing the opportunity for me finally to put these ideas into print. The support of the National Science Foundation is also gratefully acknowledged.

2 Here it is worth noting that Kuhn's book contains not a single illustration.

3 For good examples of the initial philosophical reaction to Kuhn's work, see Shapere 1964 and Scheffler 1967. Kuhn's implicit reply is scattered throughout the essays reprinted in *The Essential Tension* (1977).

4 The now classic references to the new sociology of science include Barnes 1974, Bloor 1976, Latour and Woolgar 1979, Knorr-Cetina 1981, Collins and Pinch 1982, and Collins 1985. This early phase is nicely summarized, with many references, in Barnes and Edge 1982 and by Shapin 1982. For more recent developments, see Latour 1987, 1989, Woolgar 1988, and the contributions in Pickering 1992.

5 For these developments within the sociology of science, see Lynch and Woolgar 1990, particularly the editors' introduction and the essays by Latour and Tibbetts. I myself have reviewed these essays in some detail (see Giere 1994a).

6 The roots of the model-based view go back to the work of J.C.C. McKinsey, Evert Beth, and John von Neumann in the 1930s, '40s, and '50s. It came to prominence in the philosophy of science in the 1960s, '70s, and '80s through their followers, Patrick Suppes, Bas van Fraassen, and Frederick Suppe, respectively. In addition to reprints of his own papers, Suppe's recent book (1989) provides a good bibliography and a useful participant's overview of these developments (ch. 1, 'Prologue').

7 I have pursued one line of development in Giere 1994b.

8 For an elaboration of the view that crucial experiments are typically after-the-fact reconstructions, see Brannigan 1981.

9 Paul Thagard (1991) is a vigorous exponent of the view that scientific revolutions are to be understood as reflecting the greater explanatory coherence of the victorious theory.

10 I have criticized probabilistic accounts of human judgment in Giere 1988, chapter 6, and discussed Thagard's coherentist approach in Giere 1989a and Giere 1991.

11 For a justification of the strategy of focusing on individuals, see Giere 1989b.

12 I have argued the virtues of a naturalistic rather than a normative account of human judgment in Giere 1988, especially chapters 1 and 6, and further defended this approach in Giere 1989c.

13 Sequential testing, however, introduces the possibility that which model ends up being chosen is a function of the particular order in which alternatives were considered.

14 In this account I have omitted (i) justification for the lack of prior probabilities in the model; (ii) explicit reference to the overall utilities of the decision maker; and (iii) the need for a supplemental decision strategy, such as satisficing, which justifies the obvious decision strategy in terms of satisficing relative to various expected utilities. These aspects of crucial decisions are discussed in Giere 1983, and 1988, chapter 6.

15 Twentieth-century geology, particularly the 1960s 'revolution' in geology, is fast becoming a standard test case for science studies. Among recent books in which it features in whole or in part are those by Le Grand (1988), Stewart (1990), Thagard (1991), and myself (Giere 1988, ch. 8). The articles are already too numerous to list here. One article (Le Grand 1990), however, deserves mention as it explicitly uses images from the 1960s revolution in geology to argue the case for the importance of visual imagery in science.

16 Marvin (1973, p. 43), for example, reproduces an 1858 engraving by Antonio Snyder showing a much too good fit. Speculation regarding the fit of the two hemispheres seems to have followed shortly upon the production of maps of the New World comparable in detail to those of Europe and Africa. This fact fits Latour's (1986) thesis about the importance of 'centers of calculation' for scientific progress. Direct comparisons of the coastlines become possible for a single observer only after many measurements have been brought together in one place and rendered graphically on a single piece of paper for easy viewing.

17 I suspect that Wegener may have been the first to emphasize such geological congruences, but I am not myself sufficiently familiar with the historical sources to vouch for this suspicion.

18 Herbert Simon (1978) distinguishes 'informational' from 'computational' equivalence for representations. Informational equivalence is a generalization of logical equivalence. Computational equivalence means that the same information can be extracted from the representation with the same computational resources. In terms of this distinction, a linear, digital encoding of an image, as for television transmission, may be informationally equivalent to the reconstructed image on a screen. But these two representations are not computationally equivalent. In particular, an ordinary human would find it physically very difficult, if not physically impossible, to extract particular spatial information from the digital representation. But simply by looking at the pictorial representa-

tion, anyone could easily determine, for example, that the cat is on the mat. I would prefer a slightly different terminology, saying, rather, that the two representations are 'logically' equivalent but not 'cognitively' equivalent. But the fundamental idea is the same.

19 This image is taken from the first (1944, p. 506) edition of Holmes's textbook. A very similar diagram appeared in Holmes 1930.

20 It is an interesting question why a dynamic version of Hess's model incorporating geomagnetic reversals does not appear in the literature until after 1966. My suspicion is that publishing conventions in professional journals like *Nature* at that time favoured diagrams and graphs which presented data, as opposed to those that merely pictured speculative models.

10. Are Pictures Really Necessary? The Case of Sewall Wright's 'Adaptive Landscapes'

MICHAEL RUSE

1. INTRODUCTION

Biologists are remarkably visual people. I have before me a flyer from a major publisher, promoting the new edition of an (apparently) highly successful college text in cell biology, co-authored by (among others) the Nobel laureate David Baltimore (Darnell, Lodish, and Baltimore 1990). The 1,105 pages include no less than 1,050 illustrations; the people asked to flack the book harp on the virtues of the pictures ('I appreciate the use of data and actual micrographs. The artwork, and especially the use of color, is outstanding');[1] and instructors adopting the book as a text get a free set of overhead transparencies, with the opportunity to buy more.

Nor is this love of the pictorial confined to the pedagogical. If you look at the papers that biologists produce, and even more at their books, you find them chock-a-block full of photographs and drawings, of graphs and figures, of maps and of stylized tables. Moreover, thanks to advances in technology – photography, computers, printing – the use of pictures of one sort or another is, if anything, increasing rather than otherwise. Bursting with vibrant coloured photographs, some publications seem to owe as much to Walt Disney as they do to Charles Darwin.

Biological illustration has been around for a long time – plenty of time for the philosophers, whose self-appointed task is the understanding of science, to react to it, delving into its nature and significance. So, let us ask about what they have to say – and the answer, I am afraid, is 'remarkably little.' To the best of my knowledge, the classics of logical empiricism never raise the general question of scientific illustration, and the same seems to be true of non-classics of that era devoted explicitly to problems of biology.[2]

Moreover, one suspects that the silence was, if anything, actively hostile. People did not talk about biological illustration because they did not judge it to be part of 'real science.' This enterprise produces statements or propositions, ideally embedded in a formal system. It may be *about* the real world, but it is not in any sense *of* the real world, in being a copy or mirror image. Like Plato in the *Republic* and (many years later) Pierre Duhem (1954), who contrasted the admirable French mind of pure reason with the grubby English fondness for concrete models, philosophers recognized that regretfully human weakness demanded the visual. But it was judged at best a prop. And in the discussion of physical models – about the closest that the logical empiricists ever did get to the visual – one was warned constantly of the dangers of illicitly identifying aspects of the artifactual with aspects of reality (see, for instance, Braithwaite 1953, Hempel 1965, and Bunge 1967; although see also Achinstein 1968).

I am a philosophical naturalist, thinking that one's philosophy must be informed and in accord with the methodological dictates of science. I believe that one must be true to the real nature of science, not to an idealized preconception. I am, therefore, made most uncomfortable by this tension between the reality and the theory. I say this with even more discomfort because, admittedly in a very minor way, I myself have been responsible for the tension (especially in Ruse 1973). The aim of this discussion, therefore, is to start to make amends. At the very least, so voluminous an item as biological illustration demands philosophical attention, whatever one's ultimate conclusion.

As a philosophical naturalist, my scientific-type inquiry is focused on science itself.[3] As the biologist studies organisms, so I study what the biologist thinks and produces about organisms. Hence, my starting point here has to be with actual examples of biological illustrations or diagrams. I shall, indeed, look at but one example; although, I hope that its great importance in the history of science will justify such selectivity, even to the point of allowing me to draw some general conclusions. From among the many candidates – Richard Owen's vertebrate archetype, Charles Darwin's tree of life (not to mention Ernst Haeckel's), the chromosome maps of T.H. Morgan and Company, the million exemplifications of the double helix – I chose the adaptive landscapes of the great population geneticist Sewall Wright. And the question I ask is: what was/is their status and role within evolutionary biology?

I start by looking first at the landscapes themselves, asking about their nature and history. Then I go on to inquire into their significance. Next come some thoughts about the quality of the science in which they are embedded. My discussion concludes with a few comments of a more general nature. One example cannot justify a whole theory of (scientific) knowledge, but it can set us in a certain direction. Technically speaking, my concern is with the first actual public presentation of an illustration of an adaptive landscape by Sewall Wright. Since he and others repeated the performance many, many times, unless confusion would ensue I shall refer indifferently to the class of such illustrations.

2. ADAPTIVE LANDSCAPES

Sewall Wright's first job after leaving graduate school (Harvard) was with the U.S. Department of Agriculture. In 1926 he was appointed to the faculty at Chicago, and it was about this time that he wrote the major paper in evolutionary theory (Wright 1931) on which his reputation (justly) rests. His biographer, Will Provine (1986), suggests that the motivation might have been Wright's desire to prove himself as a real academic, but, as it happens, the paper was not published until 1931. By then, especially in response to dialogue with R.A. Fisher, there had been some modifications to the text, although one understands that they were not drastic.

Much of the text of this paper is given over to complex mathematics – at least by biological standards, especially by biological standards of the day. Wright concerned himself primarily with the fate of genes in populations, under given conditions of selection, mutation, and so forth, and he was interested in the consequences of population sizes being genuinely finite and thus subject to random factors in breeding (errors of sampling). He was able to show that if population numbers (or rather 'effective' population numbers, taking into account such things as sex ratios) are large enough, and the forces are strong enough, then selection and like factors determine the fates of genes. For instance, a favoured gene or gene combination will establish itself in a population. However, what Wright was able to show also is that if population numbers are small (judged against the other factors), then genes will 'drift' either to total elimination or total fixation – despite counter-forces of selection and the like. Chance becomes a real phenomenon for change.

To illustrate the mathematical points, Wright gave graphs showing possible effects, and these together with the formal conclusions were used to launch Wright's own particular theory of evolutionary change: the 'shifting balance' theory (fig. 10.1). Wright argued that very small populations would suffer from significant drift and rapidly go extinct. However, conversely, large populations under fairly uniform selective pressures would not truly be candidates for any significant change, good or bad – or at least they could incorporate only *very* slow and stately change (see fig. 10.1).

For significant change, within realistic timespans, one needs a more dynamic mechanism. This is provided by the breaking of a species into sub-populations, of a size-order where drift could be effective – but not of a size so small that drift could be too effective! Every now and then, such a sub-population would, by chance, come up with a highly adaptive gene complex, and then this combination could take over the species, either by direct selective elimination of rivals or by interbreeding.

In formulating this theory, we know that Wright drew heavily on his knowledge of animal breeding. This point is not of great importance to us here. What is of importance is the fact that, presumably like his knowledge of animal breeding, Wright's theory transcended his formalisms. It was based on them, but was not identical. It was more inclusive (more falsifiable, in Popper's terminology). There was nothing in the formalisms about species' subdividing, about new adaptive complexes being hit upon, about insufficient time for selection in large groups, and so on. This was added. Significantly, Wright and Fisher agreed on the mathematics, but because Fisher added different non-formal elements, he came up with a very different theory of change. (Most importantly, Fisher [1930] believed that selection in large groups *did* hold the key to evolution. I will be returning to this point at the end of this discussion.)

Wright's paper, a long paper, appeared in the journal *Genetics*. The next year (1932) he had a wonderful opportunity to promote his theory, because he was asked (by E.M. East, his doctoral supervisor) to participate in a forum (with Fisher and with the third great theorist, J.B.S. Haldane) at the Sixth International Congress of Genetics, at Cornell. Normally, Wright was as given to long mathematical demonstrations in lectures as he was in print, but here he was forced to keep his presentation very short – and urged to keep it simple. To do this,

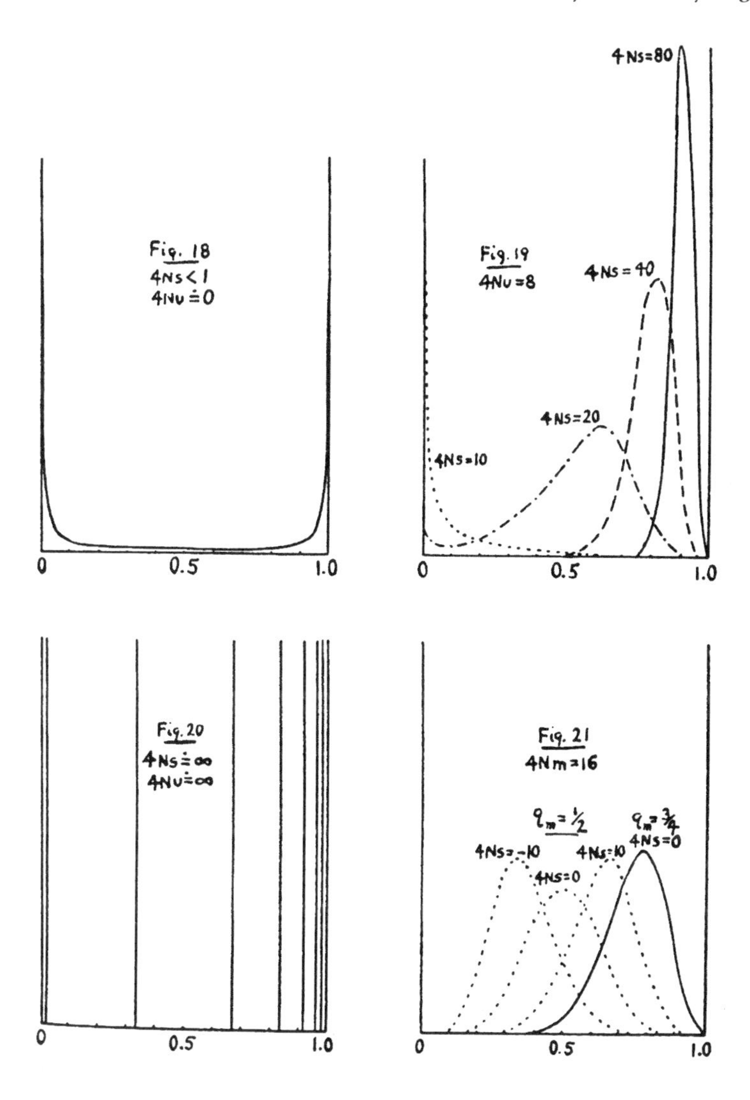

10.1 Four figures illustrating the various fates of a gene in a population (Sewall Wright 1931).

he dropped the mathematics entirely, presented his shifting balance theory in words (as he had done in his long paper) and backed up his thinking with a new metaphor, which he presented pictorially: the *adaptive* landscape.

Wright wrote, and illustrated, as follows:

> If the entire field of possible gene combinations be graded with respect to adaptive value under a particular set of conditions, what would be its nature? Figure 1 [fig. 10.2 in the present essay] shows the combinations in the cases of 2 to 5 paired allelomorphs. In the last case, each of the 32 homozygous combinations is at one remove from 5 others, at two removes from 10, etc. It would require 5 dimensions to represent these relations symmetrically; a sixth dimension is needed to represent level of adaptive value. The 32 combinations here compare with 10(1000) in a species with 1000 loci each represented by 10 allelomorphs, and the 5 dimensions required for adequate representation compare with 9000. The two dimensions of figure 2 [fig. 10.3] are a very inadequate representation of such a field. The contour lines are intended to represent the scale of adaptive value.
>
> One possibility is that a particular combination gives maximum adaptation and that the adaptiveness of the other combinations falls off more or less regularly according to the number of removes. A species whose individuals are clustered about some combination other than the highest would move up the steepest gradient toward the peak, having reached which it would remain unchanged except for the rare occurrence of new favorable mutations.
>
> But even in the two factor case (figure 1) [fig. 10.2] it is possible that there may be two peaks, and the chance that this may be the case greatly increases with each additional locus. With something like 10(1000) possibilities (figure 2) [fig. 10.3] it may be taken as certain that there will be an enormous number of widely separated harmonious combinations. The chance that a random combination is as adaptive as those characteristic of the species may be as low as 10(−100) and still leave room for 10(800) separate peaks, each surrounded by 10(100) more or less similar combinations. In a rugged field of this character, selection will easily carry the species to the nearest peak, but there may be innumerable other peaks which are surrounded by 'valleys.' The problem of evolution as I see it is that of a mechanism by which the species may continually find its way from lower to higher peaks in such a field. In order that this may occur, there must be some trial and error mechanism on a grand scale by which the species may explore the region surrounding the small portion of the field which it occupies. To evolve, the species must not be

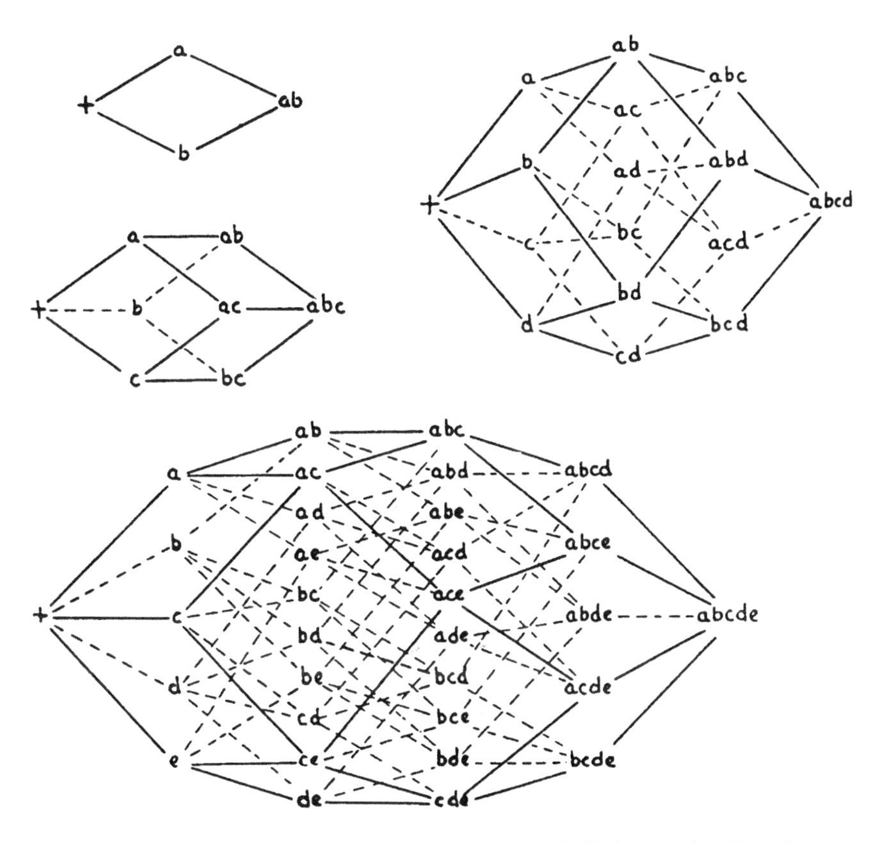

10.2 The combinations of from two to five paired allelomorphs (Sewall Wright 1932).

under strict control of natural selection. Is there such a trial and error mechanism? (Wright 1932, pp. 162–4)

Next, Wright presented (without the mathematical backing) versions of the graphs of gene distribution that had been given in the large paper (fig. 10.4; Wright's figure 3). He showed visually how drift and other phenomena can occur, given the right specified conditions. Then, using the landscape metaphor, Wright showed how the various options might or might not lead to change, and – as before – he opted for a position that involved a break into small groups, drift, and then reasonably rapid adaptive change in one direction (fig. 10.5; Wright's figure 4):

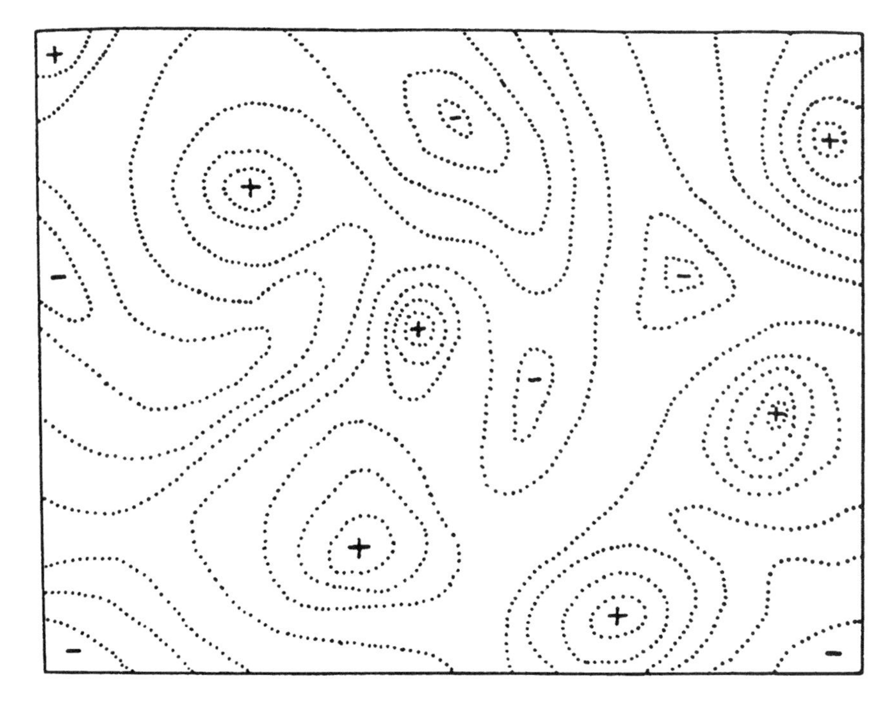

10.3 Diagrammatic representation of the field of gene combinations in two dimensions, instead of many thousands. Dotted lines represent contours with respect to adaptiveness (Sewall Wright 1932).

Finally (figure 4) [fig. 10.5], let us consider the case of a large species which is subdivided into many small local races, each breeding largely within itself but occasionally crossbreeding. The field of gene combinations occupied by each of these local races shifts continually in a nonadaptive fashion (except in so far as there are local differences in the conditions of selection). The rate of movement may be enormously greater than in the preceding case since the condition for such movement is that the reciprocal of the population number be of the order of the proportion of crossbreeding instead of the mutation rate. With many local races, each spreading over a considerable field and moving relatively rapidly in the more general field about the controlling peak, the chances are good that one at least will come under the influence of another peak. If a higher peak this race will expand in numbers and by crossbreeding with the others will pull the whole species toward the new posi-

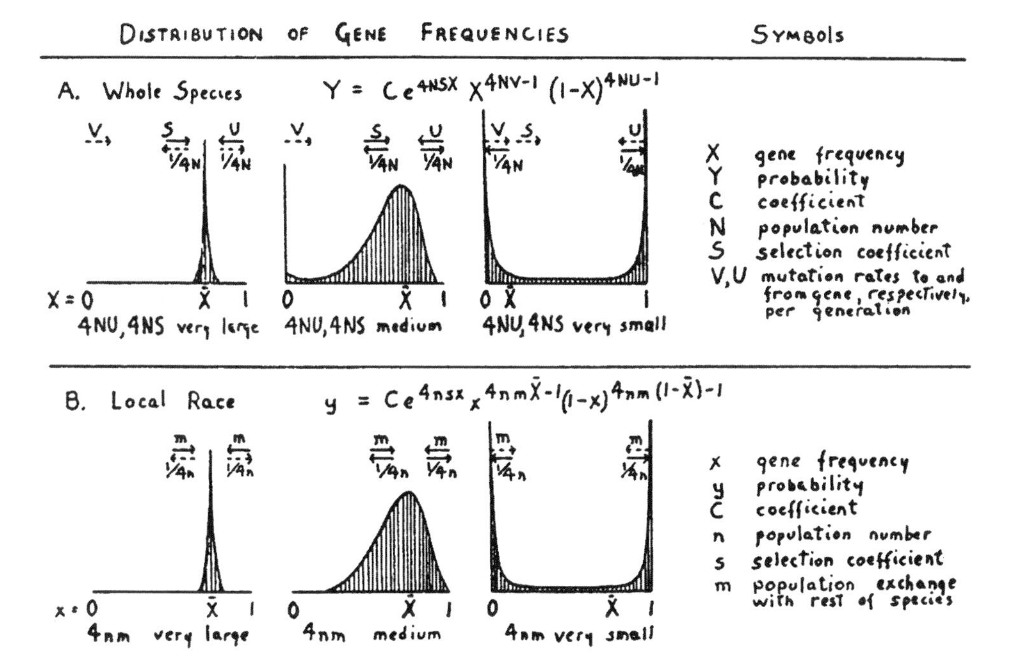

10.4 Random variability of a gene frequency under various specified conditions (Sewall Wright 1932).

tion. The average adaptedness of the species thus advances under intergroup selection, an enormously more effective process than intragroup selection. The conclusion is that subdivision of a species into local races provides the most effective mechanism for trial and error in the field of gene combinations. (Wright 1932, p. 168)

3. HOW IMPORTANT WERE THE ILLUSTRATIONS?

Let us start with the basic historical facts. Wright's talk was a great success.[4] People grasped what he had to say, and they responded warmly to his claims – at least, this seems to have been true of his American audience. Moreover, word seems to have got out, and Wright was flooded with reprint requests. Most important was the fact that among Wright's listeners at Cornell were active and ambitious young

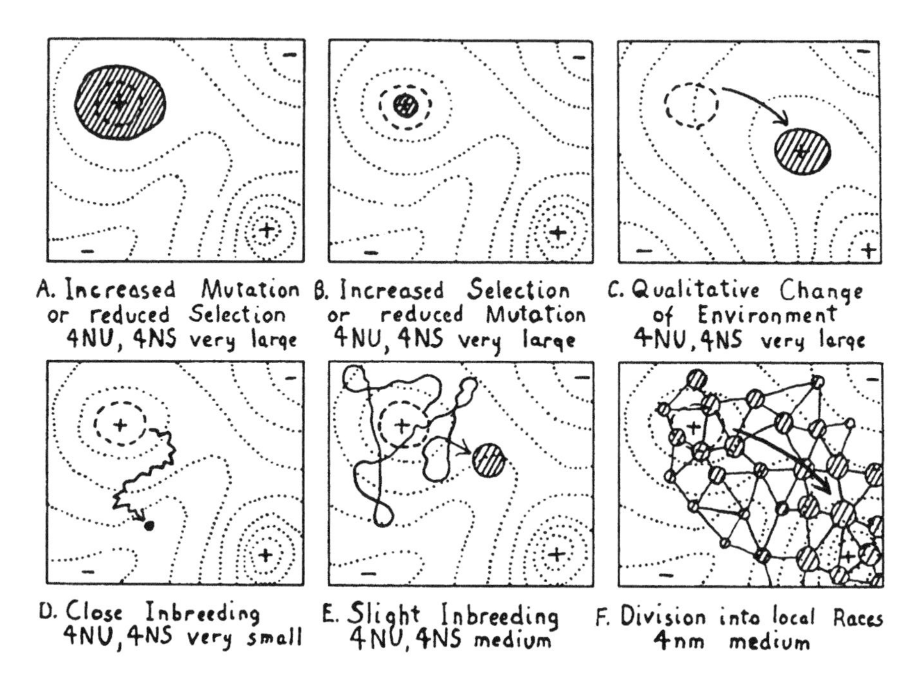

10.5 Field of gene combinations occupied by a population within the general field of possible combinations. Type of history under specified conditions indicated by relation to initial field (heavy broken contour) and arrow (Sewall Wright 1932).

evolutionists, simply desperate for a good theory around which to structure their empirical research.

One of these people was the Russian-born Theodosius Dobzhansky, then working in Morgan's lab at Caltech. In his own words, 'he simply fell in love with Wright,' or at least with the ideas (Provine 1986, p. 328). Thus, when in 1936 Dobzhansky was invited to give the Jessup lectures at Columbia, Wright's shifting balance theory had pride of place, and in the published version next year – *Genetics and the Origin of Species* – Wrightian adaptive landscapes got full treatment. Indeed, it is not too much to say that the metaphor was offered as the crucial key to the understanding of evolution.

Dobzhansky's book had immense influence. It has fair claim to having been the most important work in evolutionary theory since the *Origin.* And with the influence has gone the Wrightian landscape –

reproduced again and again, in work after work (not the least of which were Wright's own writings, which were using the original illustrations right down to the 1980s). In America, all of the major evolutionists used the notion of a landscape. The metaphor also found its way across the Atlantic; although, to be quite candid, people in Britain were not as keen on it, especially inasmuch as it was tied to non-adaptive drift. (More on this point, later.) In America, most people indeed used the actual illustrations, and even with those who did not, the idea can usually be found lurking in the background. In his *Systematics and the Origin of Species*, Ernst Mayr displayed his lifelong churlishness towards genetics. But though the actual illustrations are absent, the idea is there.

Most interestingly, those evolutionists who could not use Wright's landscapes directly adapted them to their own ends. As a palaeontologist, G.G. Simpson (1944) could not work at the genetic level, nor could he think in terms of individual populations of a species. So he hypothesized landscapes of phenetic or morphological difference, and he supposed taxa of higher categories working their ways across the landscapes, down valleys and up peaks. Wright, incidentally, approved of this extension (figs. 10.6 and 10.7).

Actually, by 1951, when Dobzhansky published the third edition of *GOS*, he too had started thinking in terms of multiple species rather than populations with a single species. What is as interesting as this point is the fact that as evolutionists in America – Dobzhansky particularly – became more selectionist in the 1940s (thanks to empirical findings about chromosome polymorphisms), so Wright's picture was retained and reinterpreted. By 1951, in the third (very selectionist) edition of *GOS*, the picture was at its height:

> Every organism may be conceived as possessing a certain combination of organs or traits, and of genes which condition the development of these traits. Different organisms possess some genes in common with others and some genes which are different. The number of conceivable combinations of genes present in different organisms is, of course, immense. The actually existing combinations amount to only an infinitesimal fraction of the potentially possible, or at least conceivable, ones. All these combinations may be thought of as forming a multi-dimensional space within which every existing or possible organism may be said to have its place.
>
> The existing and the possible combinations may now be graded with respect to their fitness to survive in the environments that exist in the world. Some of

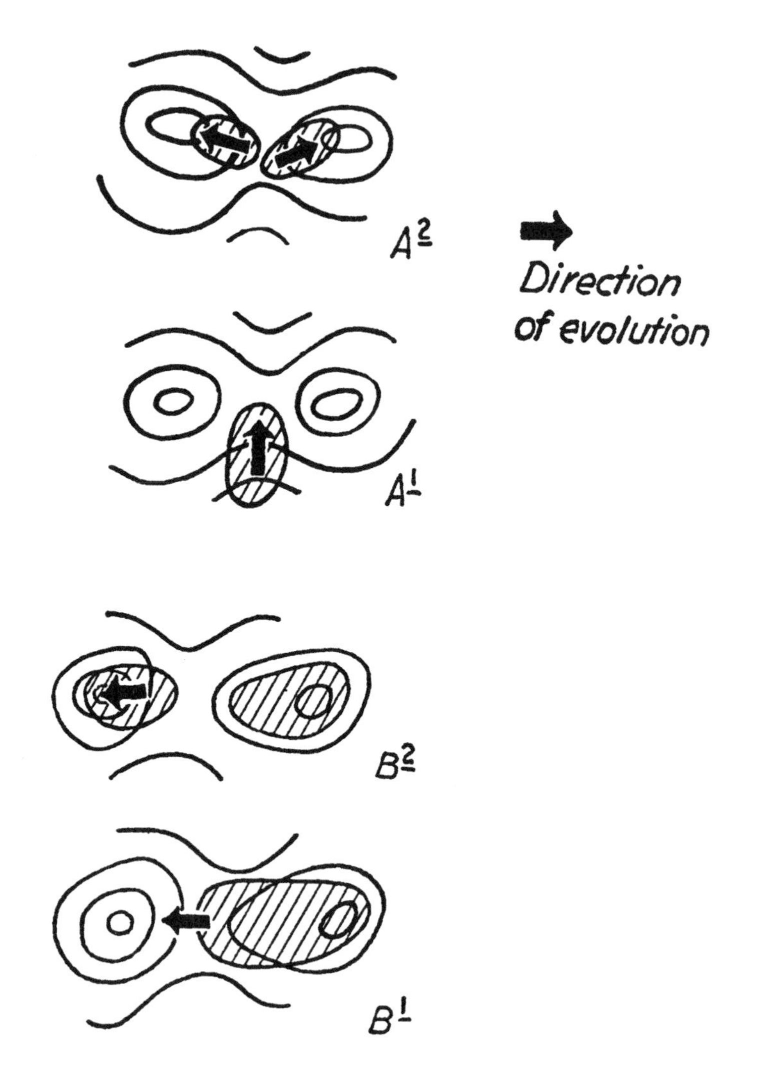

10.6 Two patterns of phyletic dichotomy; shown on selection contours. *A*, dichotomy with population advancing and splitting to occupy two different adaptive peaks, both branches progressive; *B*, dichotomy with marginal, preadaptive variants of ancestral population moving away to occupy adjacent adaptive peak, ancestral group conservative, continuing on same peak, descendant branch progressive (G.G. Simpson 1944).

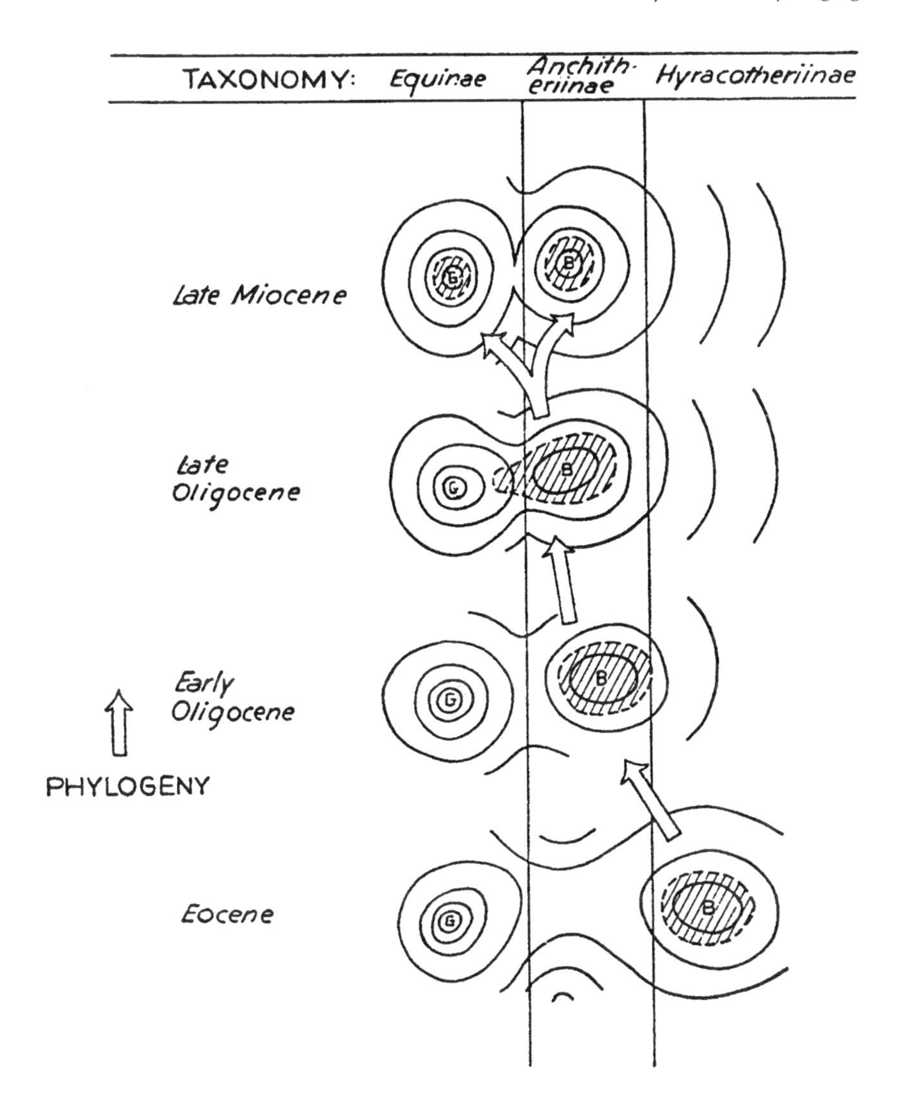

10.7 Major features of equid phylogeny and taxonomy represented as the movement of populations on a dynamic selection landscape (G.G. Simpson 1944).

> the conceivable combinations, indeed a vast majority of them, are discordant and unfit for survival in any environment. Others are suitable for occupation of certain habitats and ecological niches. Related gene combinations are, on the whole, similar in adaptive value. The field of gene combinations may, then, be visualized most simply in a form of a topographical map, in which the 'contours' symbolize the adaptive values of various combinations (Fig. 1) [fig. 10.8]. Groups of related combinations of genes, which make the organisms that possess them able to occupy certain ecological niches, are then represented by the 'adaptive peaks' situated in different parts of the field (plus signs in Fig. 1). The unfavorable combinations of genes which make their carriers unfit to live in any existing environment are represented by the 'adaptive valleys' which lie between the peaks (minus signs in Fig. 1). (Dobzhansky 1951, pp. 8–9)

Diminished now are the drift aspects, and emphasized are the adaptationist aspects.

So much for history. Wright's idea of an adaptive landscape – where by 'idea' I mean at the general level the metaphor, but at a specific level actual pictures, and usually the original pictures of Wright himself – became a commonplace in evolutionary thought. Moreover, note that – identify metaphor and picture if you will – I am not talking about any old adaptive landscape. I am talking about landscapes precisely of the kind as are exemplified by the pictures. Or rather, of *representations* of landscapes as are exemplified by the pictures.

But, speaking now at a philosophical level: were the landscapes *really* part of evolutionary thought? Or, rephrasing the question, since Dobzhansky is generally taken as one of the founders of the 'synthetic' theory of evolution, also known as 'neo-Darwinism': was Wright's metaphor in general, and his pictures in particular, *really* part of the synthetic theory of evolution, of neo-Darwinism?

The answer, of course, depends on what you mean by '*really* part of.' The pictures were around in a big way, so that they are clearly candidates for inclusion in a manner that for instance (to take an object entirely at random) the head of King Charles I was not. The decision for inclusion must therefore depend on how one construes inclusion itself. Let us run through some possible senses.

At the most basic level, the pictures obviously are part of evolutionary thought. Evolutionists thought about them a great deal, and put them in their publications. There is an end to the matter. The pictures were in, and King Charles's head, which went unmentioned, was not. I

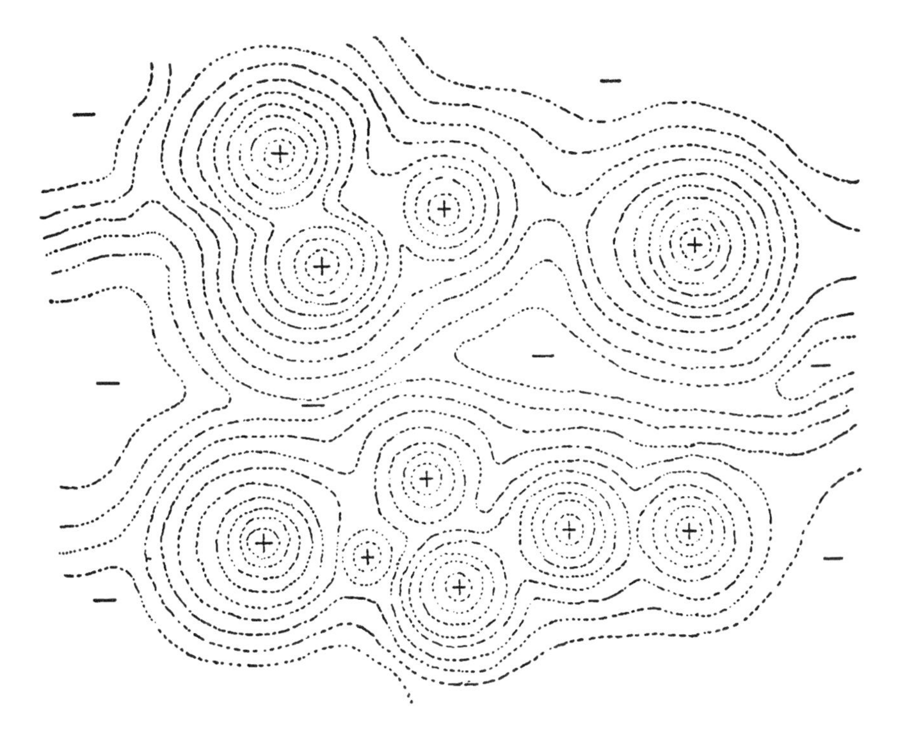

10.8 The 'adaptive peaks' and 'adaptive valleys' in the field of gene combinations. The contour lines symbolize the adaptive value (Darwinian fitness) of the genotypes (T. Dobzhansky 1951).

realize, of course, that many philosophers – all of those of the older cast of mind – will find this answer profoundly unsatisfying. They will claim that the question is not whether people did think about them – we know that they did – but whether they *had* to think about them. Were the pictures an integrally necessary part of the science? Putting matters another way: the pictures were part of evolutionary thought, but were they part of evolutionary *theory*?

Let me say right out that, as a naturalist, I do not find the basic-level answer quite so trivial as all that. While I see a place for philosophy being prescriptive, it should also be descriptive. The illustrations occupied a lot of space – mental space and printed space. An adequate philosophy of science must recognize this fact. But I will accept that this conclusion leaves open the possibility that in some sense the pictures were not absolutely necessary. As established thus far, the science in some fashion could have gone on without them. The process might not have been so fast, but presumably that is the price one pays for conceptual purity – assuming, as I am sure traditional philosophers would assume, that pictures are impure. I add parenthetically that I am not sure how easy, or indeed possible, teaching might have been without the pictures, without the very metaphor. However, for sake of argument, I adopt here the traditional academic stance that teaching does not occupy the first-class mind, anyway.

Return to the question of the status of the pictures. The argument for their necessity can be made a notch stronger. Not only were the pictures part of evolutionary thought, the scientists involved could *not* have done their work without the pictures. I speak now at the empirical level of psychological or intellectual ability. Wright's mathematics was simply too hard for the average evolutionist. It was certainly too hard for that very non-average evolutionist Theodosius Dobzhansky. He admitted again and again that he could not follow Wright's calculations.[5] And he was not alone. G.L. Stebbins, another who heard Wright at Cornell, and later to provide the botanical arm to the synthetic theory, likewise was quite incapable of thinking mathematically.

But, they could understand the pictures! And so, as a matter of empirical fact, this was the level at which these men worked. They seized on the notion of an adaptive landscape and they experimented and theorized around it. Dobzhansky, for instance, studied natural populations of Drosophila, looking for evidence that they have drifted apart in a non-adaptive fashion (Lewontin et al. 1981). At first he did think he had evidence for his hypothesis. Then he found evidence against it. What is important is that, as noted above, in both cases it was at the picture level that he was thinking, because quite frankly he could do no other. In this sense, therefore, history supports the philosophical claim that the pictures were necessary. The science would not have been done without them.

'The science would not have been done without them'? Here the traditionalist philosopher will again enter an objection. The important point surely is whether the science *could* not have been done without the pictures. A philosophical analysis tries to strain out the fallibility of the individual and to aim for the ideal. Remember that Popper (1972) refers to science as 'knowledge without a knower,' meaning, not that science exists independently of individuals – although sometimes his metaphysical speculations about World 3 seem to imply just this – but that the idiosyncrasies, including the intellectual weaknesses, of individual scientists have no place in real science. In this spirit the claim will be that, although the pictures were undoubtedly needed for the real scientists involved, in theory they were dispensable. Moreover, the claim will probably be that the ideal, that which is in some sense preferable, would do away with the pictures. In a perfect world, the pictures could and would go.

I know this kind of claim will be made, because in the past I would have been one to make it. Now, as a naturalist, I find myself very uncomfortable with it. Somehow I feel that even a philosopher should acknowledge the realities of human nature. Of course, there is always the danger of subjectivity or relativism here. No one would (or should) want to argue that the only adequate philosophical analysis is one which embraces everyone who has ever thought scientifically – right down to the most lazy, inadequate undergraduate. But however one makes the cut, in talking about Dobzhansky and Mayr and Simpson and Stebbins, we are talking about the top evolutionists, the men who made the subject. So let me say simply that I find unconvincing the flat *a priori* dictum that the abilities of the scientists involved must necessarily (obviously?) be excluded from any adequate philosophical analysis. To the contrary, my feeling now is that the philosopher should start with the empirical necessity of the pictures and base his/her analysis on that.

However, again for the sake of argument, let us grant the traditionalist the point. Let us be swayed by some such claim as: 'The history of recent evolutionary theory shows that, although the pictures were needed in the earliest days, over time with increased formalism, their use has declined, thus showing that the ideal is a science without pictures.' As a matter of fact, I do not know if this claim is empirically true, but it is certainly the kind of claim that will be/has been made. So let us go along with it.

Still the traditionalist has problems. It must still be conceded that the pictures were important, and may indeed now still be important, if not always in the future. And by 'important' here I do not just mean 'helpful.' We have seen that the formalisms themselves did not express Wright's theory fully. The formalisms alone were shared by Fisher, who had an altogether different theory. The adaptive landscape idea went beyond the formalisms, expressing the notion that drift could generate variation in isolated populations, and that selection could then act to bring about rapid change. Moreover, let me point out that this, more than anything, was the *theory*, so that the traditionalist cannot wriggle out of the claim that the adaptive landscape idea was (and may still be) part of Wright's basic science.

4. WRIGHT'S TWO (1931 AND 1934) PAPERS

The response no doubt will be that although Wright's theory clearly did go beyond the formalisms (because at that stage it was 'immature'?!), the claim for the necessity of the pictures can be jettisoned. After all, in the main 1931 paper there were no pictures or even the metaphor. Everything that needed to be said, could be said and was indeed said, in words, literally.

In reply to this, I will say three things. First, I simply do not know whether or not Wright had the landscape metaphor in mind when he first thought up his theory. We know that it predated publication of the 1931 paper, because it is used in an earlier letter to Fisher (see fig. 10.7). Wright may have had it all along. I do know that the young Wright (and the old Wright, for that matter) was an Henri Bergson enthusiast, and something very much like the adaptive landscape metaphor occurs in *Creative Evolution* (published in 1912). It could well be that Wright was thinking seriously about landscapes even before he began his formalisms. The case for the necessity of the landscapes in the 1932 form of the theory does not depend on this, but I think the critic should tread warily before making sweeping claims about what *must* have been the case, historically. (Towards the end of this paper, I will have more to say about the historical underpinnings of Wright's thought.)

Second, I would challenge the claim that the 1932 version of Wright's theory was simply the 1931 version, without the mathematics. The pictures do indeed add some factual claims – most importantly, that there are going to be some adaptive peaks for organisms to occupy, so long as one drifts far enough. The 1931 version really does not say much about why drift will eventually pay off. I have quoted the relevant passages and they are very vague. Indeed, Wright has already said that one small group drifting will probably go extinct. In the 1932 version, the pictures make it clear that there are all sorts of good opportunities waiting for drifters. Wright could have drawn a peak with a plain all around it, or with lots of (by definition) inhospitable seas or uncrossable rivers or chasms. But he does not, and it is certainly part of the plausibility of his theory that every peak seems to have other relatively accessible peaks in the vicinity.

Third, before it is immediately objected that one could have expressed all of Wright's new (post-1931) claims in words, let me point out that he did not. Moreover, let me point out also that (as people like Mary Hesse [1966] have pointed out generally about metaphorical thinking) there is a heuristic element to adaptive landscapes which escapes a simple list of factual claims that a scientist might make at a particular time (specifically Wright in 1932). Like all metaphors, they are 'open-ended' in a way that the strictly literal is not.

In this context, consider Dobzhansky's own 1951 rendering of the landscape (fig. 10.8). He has peaks clustering together in a way quite absent from Wright. Although, interestingly, he does not acknowledge the fact (that is, he does not write it down in words), he is adding a distinctively new element to the theory – that adaptations are not random and that what works well in one way might have similar (although somewhat different) mechanisms also working well. The point is similar to someone noting the virtues of both gasoline and diesel motors, and noting also what a big gap there is between them and a steam engine or a jet engine.

There is therefore a forward-rolling aspect to Wright's picture. It stimulates you to push ahead with more claims. Just as in real life peaks tend to be clustered (the Alps, the Rockies), so Dobzhansky was stimulated to think of adaptive clustering. In doing this, I suspect that Dobzhansky was following what was already assumed by Simpson (see figs. 10.6 and 10.7, which make significant the spacing of the peaks). Relatedly, as I mentioned earlier, Dobzhansky like Simpson went beyond Wright's thinking about the landscape working *within* a species, to the landscape telling of relationships *between* species. For Wright, it was populations on the road to speciation climbing the peaks. For Dobzhansky, the peaks were occupied by different species. It is in this significant sense, how Dobzhansky pushed beyond Wright's own picture, centring on the heuristic value, that I would deny that Wright's adaptive landscape could, even in theory, be dropped without loss of content.

But what about the final claim of the critic, at least in this line of argument? My original thesis was about the status of pictures in science. However, by my own admission, I have moved freely back and forth between metaphor and illustration, basically counting them as one and the same – or, rather, I have in the specific instance of Wright's adaptive landscapes. Yet there is a difference. The one is a physical picture on a printed page. The other is not. My original claim was about the former, not the latter. Even if we concede the necessity of the latter, it does not follow that the former was necessary. Perhaps the population geneticists did need the metaphor. They did not necessarily need the pictures. Wright could have talked about an adaptive landscape and that would have been enough – for him, for Dobzhansky, and for all the others.

At this point, I give up. 'You win!' Although why anybody should insist on keeping the pictures out, unless their computer could not handle graphics, altogether beats me. However, the victory strikes me as being pretty thin. The case that Wright had an uneliminable (without loss of content) pictorial metaphor at the heart of his (post-1931) theory is unchanged. And that, quite frankly, is good enough for me. Moreover, in line with a point made earlier, I remind you that the identity is not between a picture and an adaptive landscape *per se.* It is between a picture and a particular representation of an adaptive landscape, namely the kind of representation one finds in the picture! I suppose one could describe all of this in words; but somewhere, it seems to me, we would have to have an image at play, even if it were only a mental image.

5. BUT IS IT GOOD SCIENCE?

We cannot yet turn positively to explore the implications of our findings for more general questions about scientific knowledge. There is another line of argument which will tempt the traditional philosopher of science. It will be granted now that at least some science, at some level, incorporates pictures. But the complaint will now be that the *best* science does not. All science, even relatively good science, would be better were there no illustrations. Top quality science is just a formal system.

I confess that my general reaction to this line of inquiry is to query precisely whose criterion of value is being invoked here. Why is the best science non-pictorial? It seems to me that by just about any standard of excellence you might normally raise, the work of Wright and his successors like Dobzhansky rates highly. If anything, it defines the criteria rather than is measured by them. But since I have staked my position so firmly on one single case, perhaps the critic can come back on the basis of this case. Good though Wright's work may have been, there are reasons to think it might have been better without the adaptive landscape idea.

Interestingly – almost paradoxically – Provine (1986) seems to incline this way. He characterizes the general reading of adaptive landscapes as 'unintelligible' (1986, p. 313) and concludes his discussion of the notion on a very negative note: 'I would emphasize in conclusion that Wright's shifting balance theory of evolution in no material way depends upon the usefulness of his fitness surfaces as heuristic devices' (Provine 1986, p. 317). He is very much of the school that as evolutionary science has matured, the need for and value of the surfaces has dropped away. (Since I am about to criticize Provine's position quite strongly, I want to enter more than the conventional disclaimer. Without Provine's brilliant work on the history of population genetics in general and on Sewell Wright in particular, it would be quite impossible for philosophers such as myself to work with any degree of sophistication on the meta-theory of this area of evolutionary biology.)

How might the critic argue the negative point? Most obviously, I suppose, by pointing out that the heuristics of the landscape are all very well, but if they lead one on false trails, their virtues are of dubious status. Take the question of other peaks surrounding any specified peak. Perhaps these represent niches which do truly exist. Perhaps they do not. One has no right to assume, as the metaphor forces on one, that they are always there. In fact, they are probably not.

In response, I would agree that perhaps Wright's picture does suggest false trails. But with respect: 'So what?' No one wants to say that scientific hypotheses – exciting scientific hypotheses – always work or are always true (although sometimes philosophers have a yearning towards this last option). The point is that the theory is fertile and, with respect to something like available niches, can be tested and rejected or revised if necessary. In fact, as comments I have made already clearly imply, one can certainly redraw Wright's landscapes if one finds that niches are not readily available. And if no niches at all are available, then the whole theory must be rejected, not just the pictures. I am not now saying that the empirical evidence is irrelevant to the worth of a theory. I am assuming, what is true, that Wright's work led to a mass of successful empirical research.

I might add in this context that, although treatment of metaphor usually labels implications cleanly as good, bad, or neutral heuristics, in real life (as our example shows) it is often not so easy to decide whether or not implications are such a very good or bad thing. Take the presumed stability of Wright's landscape. Although the possibility of change is certainly mentioned, generally – as with landscapes as opposed to water-beds – the terrain is supposed to be fairly solid. This suggests that organisms will scale ever higher peaks, and that in the long run there will be progress.

However, although many today – like George Williams (1966) and

Stephen Jay Gould (1989) – would consider this the consequence of a negative heuristic, others are not so sure. I am certain that Wright himself endorsed progress. (Look at his figure in the letter to Fisher [fig. 10.9].) Not only is the botanist G.L. Stebbins a progressionist, he has used Wright's ideas to make precisely such a case (Stebbins 1969). And active today, someone like E.O. Wilson (1975) is an organic progressionist and would, no doubt, find any supporting implications of Wright's metaphor most comforting. He does indeed talk of the 'peaks' of social evolution (occupied by the colonial organisms, the social insects, the higher mammals, and humans) and of our own species having 'reversed the downward trend' (where sociality is getting ever looser). We are on the way up to the highest point of all.

The critic might now argue in a slightly different way. Wright himself admits that in his diagrams he is collapsing down a huge amount of information into two dimensions (three if you consider the axis from eye to page). But is this legitimate? One is taking drift from many many dimensions and confining it to two dimensions. One of the things that Wright always prided himself on was his recognition of the fact that genes in combination might well have very different effects from genes taken singly. What right therefore have we to assume that the many drifting genes will combine to behave like one drifting gene (or, rather, a line of such genes)?

There is an important point here – one which shows that although Wright himself may have been sensitive to gene interaction, critics like Ernst Mayr (1959) were not simply revealing their personal prejudices when they accused the population geneticists of undue reductionistic thinking, in treating their subjects as beans in a bag. However, note that if there is a problem here – that the collapse of dimensions is too dramatic – it is one which affects all levels of theory and not just the illustrations. Again, therefore, I suggest that Wright's theory should simply be put to the test, and a check made to see if genes do wander in the way that he suggested.

In fact, as I have intimated, a decade after Wright published, Dobzhansky and others found strong evidence that selection is far more powerful and effective than Wright and others had suspected. (I am not now referring to molecular genes, which by their very nature evolve at levels below the power of selection.) The shifting balance theory required modification. But I am not sure that such modification required/requires rejection of the very notion of an adaptive landscape. One can rework the landscape to show that factors other than drift are

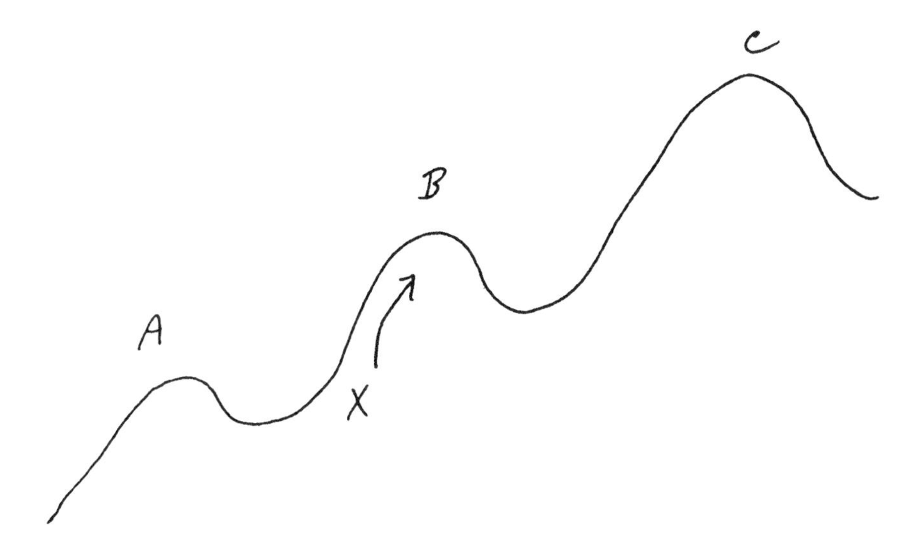

10.9 Two-dimensional fitness surface. From a letter from Wright to Fisher, 3 Feb. 1931 (W. Provine 1986).

significant. This, indeed, was precisely the move of Simpson and Dobzhansky!

None of this is to deny, in line with some of the points made by Provine, that even as it stood, there was some confusion in Wright's thought about selection and adaptation – a confusion reflected in the pictures. Like many around 1930, Wright was torn between adaptationism and non-adaptationism. As one who revered Darwin, he thought that selection was important; but all the (American) naturalists around him were saying that it was not. Hence, in one respect, Wright wanted selection to be important between members of groups, and his pictures rather imply this. In another respect, Wright doubted that there is much adaptive difference between group members – even when the groups are as large as a species or more – and he rather implied that drift proves this also!

However, it seems to me that the correct analysis here is that Wright was trying to have his cake and eat it too. The problems and any weaknesses do not come from the pictures as such. Moreover, as the case for adaptation was strengthened, Wright could and did more firmly opt for his first alternative – and even deny that he ever held the second alternative! The presence or absence of the pictures was irrelevant. (Actually this is not quite true. If selection is completely unimportant, then the adaptive landscape becomes an uninteresting plain. It is clear that Wright always thought that at some level selection is important. He was unsure about the level. What this means is not that the landscape is irrelevant, but more that there is confusion about the status of the groups that hover around the peaks. Are they subgroups or are they full-blown species? As we have seen, people went both ways on this.)

6. INDIVIDUALS VERSUS GROUPS

Let me go at the problem one more time, making the case against Wright's work in a way that I think would be favoured by Provine. There is at least some confusion in Wright's theorizing (paralleling a similar confusion of Fisher's) over whether he is talking about individuals or about groups. Sometimes the theme seems to be that of the fate of a gene (or a string of alleles) in a population. Sometimes the theme seems to be that of the fate of a group, and of the gene ratios varying within that group. In fact, strictly speaking, Wright's early analyses were couched more in terms of the former and later (after the mid-1930s) in terms of the latter – presumably in line with Dobzhansky and others – but Wright tended to slip back and forth. More significantly, sometimes he spoke of his landscapes in terms of the former and sometimes the latter.

Now, in a sense, you might think this is not desperately important. As Provine notes, most biologists simply think the group treatment is the integral of the individual and so (biologically) not much rests on the distinction. But as Provine rightly notes also, in some respects the group perspective does set major questions for the landscape metaphor. What are the coordinates of the (two-dimensional) map? Does one have two sets of gene ratios? If so, what about the (possibly) many hundreds of other ratios? Moreover, how now does one interpret the map? Points are presumably groups. At the least, this is going to require some fairly drastic redrawing.

In fact, as Provine points out, when Wright moved his mathematics to a group level, the theory became highly abstract – calculating the adaptive value of a population (w) 'for more than one locus with two alleles was practically impossible' (1986, pp. 305–6). To be honest, I am not sure whether this point counts against the landscape metaphor, or for it. Does this mean that we can push on only because of our picture – mathematics fails – or that we should not dare to move because the mathematics fails? I cannot see why the first option is necessarily incorrect. (Of course, this is to talk of long ago. Today, with much more mathematical talent in biology, not to mention computers, we are in a much stronger situation.)

Provine stresses that Dobzhansky could never follow Wright's mathematical extensions, but we virtually knew that anyway. He could not follow the mathematics at the individual level. It is true that, even assuming the legitimacy of an extension, the landscape in its original individual-based form remains important – indeed, its mathematical base seems even more crucial. What then of the individual perspective? Provine refers to the diagrams understood at this level as 'unintelligible' – hardly the mark of the best quality science. What are the grounds for this drastic assessment? Let me quote Provine in full:

> The first and most important thing to notice about Wright's first published version of his fitness 'surface' is that his construction does not in fact produce a continuous surface at all. Each axis is simply a gene combination; there are no gradations along the axis. There is no indication of what the units along the axis might be or where along the axis the gene combination should be placed. No intelligible surface can be generated by this procedure. By no stretch of the imagination can Wright's famous diagrams of the 1932 paper be constructed by his method of utilizing gene combinations. The diagrams represent a nicely continuous surface of selective value of individual genic combinations; the method Wright used to generate this surface actually yields an unintelligible result. Thus the famous diagrams of Wright's 1932 paper, certainly the most popular of all graphic representations of evolutionary biology in the twentieth century, are meaningless in any precise sense. (Provine 1986, p. 310)

Is this conclusion well taken? One thing that Provine highlights is the fact that, strictly speaking, we do not have a continuous surface, but a set of discrete points. However, if the points are vanishingly small and jammed in together tightly enough – both conditions that Wright meets – then like the printed version of a photograph, also made of many small discrete points, we have an effectively continuous surface.

A more important point that Provine highlights is that we certainly have no typical linear dimensions along the axes, as one would with a regular map. But even this does not strike me as critically fatal. We never did have a conventional map – although I do concede that probably many read Wright as if we did. We are not thinking quantitatively but more qualitatively (in the third dimension we are quantitative). But maps of this kind are not unknown – those of the world, for instance, which blow up areas with a certain quality and drastically reduce areas without such a quality. One cannot measure regular distances on such a map, but it is still a map in the sense of showing what leads where.

Actually, for all the concessions I am prepared to make, I am not sure that Provine reads the maps altogether accurately. Each axis is *not* a gene combination. Each point is such a gene combination. Therefore, one might perhaps construe the axes as measured in 'unit gene changes' or some such thing. Even though I am not altogether certain what it would mean to say that each gene change was equal, at least one would have properly ordered sequences along the axes. Although our problem here is somewhat compounded by the fact that apparently Wright accepted Provine's criticism – 'When I spoke with Wright about the problem, he thought it over for several days, and suggested that the only way he could see to save something of his original version of the individual fitness surface was to use continuously varying phenotypic characters as the axes of the surface' (1986, p. 311) – it could just be that Wright's response translates back to a rough equivalence to the suggestion I have just made.

All in all, therefore, I conclude that the criticisms of Provine or fellow conservatively minded analysts are not well taken. Wright's work was not perfect, in the sense of being absolutely true or totally without conceptual blemish. But this is a far cry from saying it was not first-rate science. Fortunately, scientific theories are like human beings – they are complex entities, with lives of their own, and the best are the best, not because they never do anything wrong, but because they do so many things right.

7. PUTTING WRIGHT IN CONTEXT

What have I proven? I have certainly not proved that every scientific theory has to have pictures, or that every scientific picture is essential. By my own admission, I have been dealing with a picture of a special kind, namely one which expresses a metaphor. Nor am I claiming here that every scientific theory contains metaphors, although as a matter of fact this is a claim I would be prepared to defend. I am not even claiming that every scientific metaphor gives rise, actually or potentially, to a picture. Indeed, this seems to me to be a false claim. Only in a very limited way do such important biological metaphors as natural selection or the struggle for existence give rise to pictures, and these are usually misleading. (Most struggle is not a literal struggle.)

Nevertheless, some scientific metaphors are pictorial – Wright's landscapes prove this. Especially, the landscapes are crucial when you think (with Dobzhansky) of species and their adaptive relations to each other. And those metaphors/pictures are in an important sense (any sense which is important) essential parts of the science – Wright's landscapes prove this. Moreover, the science containing these pictures can be good science – Wright's landscapes prove this also. These seem to me to be a good set of conclusions with which one could end this somewhat preliminary foray into the philosophical significance of biological illustration. But, on the admirable principle that one should never end a discussion on a safe and reasonable note, let me now push on out into treacherous depths, making a few comments about what my discussion implies for broader questions about scientific knowledge, and indeed knowledge generally.

I shall not pause here to consider in detail the implications of my findings for discussions about the nature of scientific theories. Clearly, the old idea (beloved by the logical empiricists) of a scientific theory as a formal axiom system – a hypothetico-deductive system – is inadequate. At a minimum, it needs a major supplement. A popular alternative today to the traditional view is the so-called 'semantic' view of theories, where one thinks less of all-embracing systems and more of families of limited models, which might or might not be applicable in certain situations. I am not convinced that a naturalist necessarily must abandon ideal pictures or the belief that somehow the hypothetico-deductive picture functions as an ideal. Nevertheless, as a naturalist, I do find this alternative position attractive, because it does seem to me to describe truly the way that much science actually functions – and this applies particularly to evolutionary biology. But, even if one embraces the semantic view, let us not forget that this view is itself generally presented in no less a linguistic and formal manner than the standard view. So it too has got to be extended to encompass metaphors and pictures. Perhaps the extension can be done more readily on the semantic view than on the traditional view. I am open to argument.

My real concern now, however, is with the light that my analysis and findings throw on knowledge itself – our relationship to the external world, especially as mediated through science (which I will flatly and provocatively say is our highest form of knowledge). And as I turn to this, first let me go again to history, and let me say more about the context in which Wright's picture appeared. I have said that it is quite probable that he got his picture under the immediate stimulus of a kindred metaphor of Bergson. But this is only part of the story. Wright was looking for, or primed for, such a metaphor, for his whole approach to evolution came from and was shaped by a particular tradition. Wright was (as I have said) unambiguously Darwinian, thinking natural selection a significant factor in evolutionary change. Far more so, however, was he a follower of Herbert Spencer, Darwin's contemporary and fellow Englishman and evolutionist.

This will seem amazing, for Spencer's reputation today is as low as Darwin's is high. But in his time, Spencer was *the* authority, and nowhere more so than in America. Significant for us is the line of influence to Wright, through his first biology teacher. This was the woman who introduced him to evolutionary thought, Wilhelmine Key, a student of C.O. Whitman, one of the most ardent of American biological Spencerians. Even more significant is the yet stronger connection through Wright's graduate school experience at Harvard, for his teacher there was L.J. Henderson, author of the *Fitness of the Environment.*

Many today think that Spencer's main contribution to evolutionary thought was a stern version of 'social Darwinism,' a particularly vicious form of *laissez-faire* economics, where the rich succeed and the weakest go to the wall. Far more influential, however, was Spencer's 'dynamic equilibrium,' a kind of progressive force upwards, from simplicity to complexity, from the valueless to the valued, marked by stages of equilibrium or balance which eventually prove unstable (or are dislodged), forcing a shift up to a new plateau (Pittenger 1993). This view was adopted in its entirety by Henderson, and passed straight on to his pupil – who obviously translated it directly into populational genetical terms and who visualized it exactly in his landscape metaphor.[6]

Parenthetically, once the Spencerian background is made public, one can see precisely why – a matter of puzzlement to many – Wright called his theory the shifting *balance* theory. It is a balance or equilibrium between forces promoting genetic homogeneity and genetic heterogeneity – in themselves very Spencerian notions. It also explains why, despite his thinking having become far more selectionist in the 1940s, Wright always maintained his theory was unchanged. With respect to the crucial Spencerian notion of balance, it always was. What changed was the much-less-significant (for Wright) item of the forces promoting such balance.

Completing the historical background and locating fully the setting of Wright's diagram/metaphor, let me make two final points. First, reconfirming what has been said already, Wright's work does not simply look backwards. It looks forwards also, through the very great influence he had on his fellow evolutionists – even down to this day, where it is the shared underpinning of the thought of people who *prima facie* take very different positions. Fellow Harvard faculty members Stephen Jay Gould and Edward O. Wilson have been at ongoing loggerheads, with Wilson expressing contempt for Gould's palaeontologically inspired theory of 'punctuated equilibria' and Gould being no less critical of Wilson's 'sociobiology.' Yet Gould's punctuated equilibria theory (think of the name!) stands right in the tradition, both through direct debts to the Wrightian inspired synthetic theory and through more indirect debts thanks to its use of the notion of stability or homeostasis, a pet idea of W.B. Cannon, a Harvard (Spencerian) buddy of Henderson. Wilson's sociobiology has been noted already as having come straight from Wrightian aspects of the synthetic theory; although, it does also owe much to another of Henderson's Harvard chums, the entomologist William Morton Wheeler. Wilson, incidentally, openly admires Spencer.

My second point, also picking up on something said earlier, is that the American tradition is not the British tradition. In that country, Charles Darwin was the icon and font of inspiration for evolutionists. And for Darwin – and especially for his ardent followers like Wright's rival, Fisher – the key metaphor had little to do with progress and balance. It was rather that of adaptation – seeing organisms as if they were artifacts, objects of design. It was this that was addressed and highlighted by British natural theology, notably by the author of Darwin's undergraduate reading, William Paley. It was this that was tackled and explained by Darwin's key mechanism of natural selection. It was this that convinced Fisher that selection, working in large groups, could be effective. And it is this that has come right down to the present and inspires and informs the work of leading British evolutionists today: William Hamilton (1964) and Geoffry Parker (1978), to take two major examples. For them, adaptive landscapes are really very small beer.

I do not want to exaggerate. I have noted Wright's Darwinian debts, even as I have noted also that selection was ever for him a secondary mechanism to his dynamic equilibrium view. I have noted also that some British took up Wright's landscapes, although the greatest enthusiast for the landscapes and drift was Julian Huxley, a man who

– for all of his trumpeting of Darwinism – had a very tenuous relationship to the British tradition. He was ever a vitalist, an enthusiast for Bergson (in turn much influenced by Spencer) and (later) the very non-Darwinian Teilhard de Chardin.[7]

My point now is simply that there were/are these two traditions, and Wright (thanks in large part to his landscapes) is a central figure in one and not the other.

8. INTERNAL REALISM

With Wright's work now firmly located in context, what can we or would we want to say philosophically? At one level – and this does seem fairly definite – we are being pointed towards a view of science (and knowledge in general) which takes metaphors, including visual metaphors, very seriously. They inform and structure our thinking. Yet, at the same time and just as crucially, one must accept that no one metaphor seems to be crucial. One can be a good evolutionist and yet deny (or, more likely, ignore) the Wrightian landscape.

In line with a powerful trend in modern history and sociology of science, therefore, one does seem to be pushed to some sort of 'constructivism,' where science is seen as a construct resting on and emerging from the culture of its day and place. And, one might add – although not in the context of my example – precisely such a philosophy has been endorsed by students of biological illustration:

> Scientists intend their pictorial representations, like their verbal expressions, to illuminate reality. Nevertheless, commentators of scientific activity should not give interpretative primacy to the issue of correspondence between representations and nature. Instead we should center our sights on interventions within a nature and society that scientists are continually helping to construct. The multiple references built into diagrams deserve attention because they point to many of the resources mobilized in such constructions. (Taylor and Blum 1991, p. 291)

However, although all of this is fair enough, our example surely gives no warrant for pushing constructivism all the way to rabid subjectivism, where science is seen to be no more than a creation of society (taken as a whole or through individual members). Wrightian inspired evolutionary biology is more than a mere fiction, where anything goes. Reality may be mediated through Wright's picture; but his picture succeeded and was used enthusiastically by others precisely because it did help to make sense of reality – both as known then and as new discoveries came in down through the years. It provided the basis for a fruitful 'paradigm' or ongoing 'research program,' to use the language of the philosophers (Kuhn 1962 and Lakatos 1970 respectively).

We seem therefore to be pushed towards a middle position, one somewhere between the extreme objectivism of the traditional philosopher of science (like the Popper of science as 'knowledge without a knower') and the extreme subjectivism of the constructivist, who sees everything as mere psychological or sociological whim. I cannot, given what has gone before, pretend now to offer any logical argument for what this middle position must be. But, as a naturalist (and a Popperian!), I am allowed to make bold conjectures, and in this spirit I nominate the ontology/epistemology of Hilary Putnam, something he labels 'internal realism.' Recognizing that there are as many versions of realism as there are realists, he writes as follows:

> ... one of these perspectives is the perspective of metaphysical realism. On this perspective, the world consists of some fixed totality of mind-independent objects. There is exactly one true and complete description of 'the way the world is.' Truth involves some sort of correspondence relation between words or thought-signs and external things and sets of things. I shall call this perspective the *externalist* perspective, because its favorite point of view is a God's Eye point of view.
>
> The perspective I shall defend has no unambiguous name. It is a late arrival in the history of philosophy, and even today it keeps being confused with other points of view of a quite different sort. I shall refer to it as the *internalist* perspective, because it is characteristic of this view to hold that *what objects does the world consist of?* is a question that it only makes sense to ask *within* a theory or description. Many 'internalists' philosophers, though not all, hold further that there is more than one 'true' theory or description of the world. 'Truth,' in an internalist view is some sort of (idealized) rational acceptability – some sort of ideal coherence of our beliefs with each other and with our experiences *as those experiences are themselves represented in our belief system* – and not correspondence with mind-independent 'states of affairs.' There is no God's Eye point of view that we can know or usefully imagine; there are only various points of view of actual persons reflecting various interests and purposes that their descriptions and theories subserve. (Putnam 1981, pp. 49–50)

The talk is of coherence. Yet, one is not precluded from the kind of correspondence demanded by the semantic view of theories:

In an internalist view also, signs do not intrinsically correspond to objects, independently of how those signs are employed and by whom. But a sign that is actually employed in a particular way by a particular community of users can correspond to particular objects *within the conceptual scheme of those users.* 'Objects' to not exist independently of conceptual schemes. We cut up the world into objects when we introduce one or another scheme of description. Since objects *and* the signs are alike *internal* to the scheme of description, it is possible to say what matches what. (Putnam 1981, p. 52)

But, we certainly do not and cannot have the correspondence of the traditional objectivist, where 'snow is white' can be slapped onto an independently existing white snow. (Philosophically informed readers will of course recognize the 'snow is white' example as that which Alfred Tarski used to illustrate his correspondence theory of truth. Expectedly, this is a theory much favoured by Popper.)

As it happens, I have argued elsewhere for internal realism, using modern evolutionary biology as my foundation (Ruse 1986). In other words, I have argued for the position on naturalistic grounds – although Putnam himself seems not to be a naturalist and denies the pertinence of evolutionary biology (see Putnam 1982). Here I am happy simply to endorse such realism, pointing merely to the fact that it does seem to be an epistemology/ontology that welcomes my discussion of Sewall Wright's adaptive landscape picture/metaphor. One has the world as mediated through a human creation – the metaphor of a landscape – and one cannot escape from this mediation without a loss of content. Yet, at the same time, one is constrained and stimulated by the empirical discoveries one makes through the creation. There is no God's Eye View, but there is a lot more than mere feeling or intuition.

What I will note, here, now starting to bring my discussion to a close, is that my analysis of Wright's work meshes exactly with some of the most exciting recent work on metaphor, and that (on grounds independent of my own) these thinkers have themselves been pointed towards internal realism. George Lakoff and Mark Johnson (1980) argue that metaphors are essential, uneliminable parts of our thought, themselves in some sense creating reality:

New metaphors, like conventional metaphors, can have the power to define reality. They do this through a coherent network of entailments that highlight some features of reality and hide others. The acceptance of the metaphor, which forces us to focus *only* on those aspects of our experience that it highlights, leads us to view the entailments of the metaphor as being *true*. (Lakoff and Johnson 1980, p. 157)

There is a sense of correspondence. '*We understand a statement as being true in a given situation when our understanding of the statement fits our understanding of the situation closely enough for our purposes*' (1980, p. 179, their italics). But, in an equally crucial sense, because truth is relative to understanding, there can be no absolute, viewer-independent knowledge. We have to work from within a culture; although, this certainly does not mean that all standards are jettisoned and that 'anything goes':

We have seen that truth is relative to understanding, which means that there is no absolute standpoint from which to obtain absolute objective truths about the world. This does not mean that there are no truths; it means only that truth is relative to our conceptual system, which is grounded in, and constantly tested by, our experiences and those of other members of our culture in our daily interactions with other people and with our physical and cultural environments. (1980, p. 193)

In later writings, Lakoff and Johnson (Lakoff 1987; Johnson 1987) tie in their 'experientialist position' to Putnam's internal realism, arguing that the two are the same thing by different names. Recognizing that we are working still in the realm of conjecture rather than proof, this does neatly parallel the way in which my discussion of Wright's work has pointed me to the same ends. And the connection is made yet stronger, giving Lakoff and Johnson's discussion particular immediacy, as one learns that a key plank in their argument for the significance of metaphor is the existence of basic 'orientational' metaphors, rooted in personal bodily experience, that structure all of our thinking:

These spatial orientations arise from the fact that we have bodies of the sort we have and that they function as they do in our physical environment. Orientational metaphors give a concept a spatial orientation; for example, HAPPY IS UP. The fact that the concept HAPPY is oriented UP leads to English expressions like 'I'm feeling *up* today.' (Lakoff and Johnson 1980, p. 14)

Obviously Wright's diagram/metaphor fits right into this thinking, given its stress on 'up/down' (an example highlighted by Lakoff and Johnson) and 'balance,' something just as crucial to us as upright vertebrates. Not only does it fit, it gives just what the naturalist craves, namely an unexpected explanation of the hitherto obscure. If you think for a moment, there is something very odd about Wright's picture, namely the fact that he paints a landscape with the need for genes to climb up mountains. Much more obvious would have been a landscape stressing valleys, where genes have a natural tendency (thanks to gravity) to roll *down*, unless disturbed otherwise. (Interestingly, the English evolutionary geneticist C.H. Waddington [1956] did produce pictures of this nature, in the context of a theory of gene interaction.)

Apart from the more obvious progressionist implications – something certainly seized on by the likes of Dobzhansky and Stebbins (although more recently deplored by Gould) – it seems plausible to suggest that Wright's thinking, having genes defy gravity, was an aspect of the general structural metaphorical thought of human beings, stressed by Lakoff and Johnson. Putting the matter bluntly, because we are upright mammals, we do tend to think in vertical terms, and (for all the obvious reasons) stress the upwards direction as the positive/healthy/valued orientation. Wright was no less human than the rest of us, and so his thinking came out the way that it did.

9. CONCLUSION

As a naturalist committed to evolutionary biology, and as one who has – as I have said – already argued elsewhere for internal realism on biological grounds, I am readily sympathetic to a philosophy which ultimately locates Wright's visual thought in his personal bodily experiences. But, I am much aware that I have long since ceased to prove anything, and am trying simply to fit my example into a pattern of philosophical thought that I find congenial. Yet, the fit is neat and suggestive. Hence, for this reason I commend it to you. Wright's adaptive landscapes have played a crucial role in evolutionary thought in this century. In themselves, they offer much of historical and philosophical interest. My feeling is that they point to matters and conclusions of much broader epistemological and ontological significance.

NOTES

1 R.W. Merriam, SUNY at Stony Brook.

2 To my eternal credit, although I may not have talked about pictures, I have always acknowledged their significance by using them. This began in a minor way in my first book, *The Philosophy of Biology* (1973), and reached a peak in *Darwinism Defended* (1982).

3 I expound my naturalism as a general system in Ruse 1986, and am now writing a book on the concept of progress in evolutionary biology in which I try to show how one does a naturalist philosophy of science. Methodologically and metaphysically I owe much to my long personal and philosophical friendship with David Hull, although we differ widely on many actual issues. See Hull 1989 and Ruse 1989.

4 I have this on the authority of G.L. Stebbins, who was in the audience (interview, May 1988).

5 Provine (1986) deals with this point in some detail.

6 I was led to the Spencerian influence on Wright's thought by a number of letters which he wrote to his brother Quincy, around 1915. These are now in the Quincy Wright Papers, at the University of Chicago. I am as obliged to Will Provine for telling me of them as I am shocked by Provine's refusal to see the influence of Spencer or anyone else of a philosophical mind-frame on Wright's thinking.

7 All of these points, including those in subsequent paragraphs, are dealt with in my forthcoming book, *Monad to Man: The Concept of Progress in Evolutionary Biology.*

Bibliography

Accum, F. 1807. *System of Theoretical and Practical Chemistry*. 2d ed. 2 vols. London: G. Kearsley.
– 1815. *A Practical Treatise on Gas-Lights*. London: R. Ackermann.
– 1828. *Chemical Re-agents, or Tests*. 2d ed. Ed. W. Maugham. London: Charles Tilt.
Achinstein, P. 1968. *Concepts of Science: A Philosophical Analysis*. Baltimore: Johns Hopkins University Press.
Ackerman, James. 1985a. 'The Involvement of Artists in Renaissance Science.' In Shirley and Hoeniger 1985, pp. 94–129.
– 1985b. 'Early Renaissance "Naturalism" and Scientific Illustration.' In Ellenius 1985, pp. 1–17.
Adam, Charles, and Paul Tannery, eds. 1964–74. *Oeuvres de Descartes*. Paris: J. Vrin.
Addington, L.R. 1986. *Lithic Illustration: Drawing Flaked Stone Artifacts for Publication*. Chicago: University of Chicago Press.
Adkins, L., and R.A. Adkins. 1989. *Archaeological Illustration*. Cambridge: Cambridge University Press.
Alberti, Leon Battista. 1486. *De re aedificatoria*. Florence. English ed., 1988. *On the Art of Building in Ten Books*. Trans. J. Rykwert, N. Leach, and R. Tavernor. Cambridge, Mass.: MIT Press.
Alpers, Svetlana. 1983. *The Art of Describing: Dutch Art in the Seventeenth Century*. Chicago: University of Chicago Press.
Anderson, Frank J. 1977. *An Illustrated History of the Herbals*. New York: Columbia University Press.

Andrews, P., and C. Stringer. 1989. *Human Evolution. An Illustrated Guide.* Cambridge: Cambridge University Press.

Andrews, T. 1889. *Scientific Papers.* London: Macmillan.

Apianus, Petrus. 1524. *Cosmographicus liber.* Landshutae: D.I. Weyssenburgers.

– 1534. *Instrumentum primi mobilis.* Nuremberg: I. Petrius.

– 1540. *Astronomicum caesareum.* Ingolstadt: P. Apian.

Arber, Agnes. 1953. 'From Medieval Herbalism to the Birth of Modern Botany.' In Underwood 1953, vol. 1, pp. 317–36.

– 1986. *Herbals: Their Origin and Evolution, a Chapter in the History of Botany, 1470–1670.* Cambridge: Cambridge University Press. Reprint of 2d ed. of 1938, with introd. and annotations by William T. Stearn.

Ardrey, R. 1961. *African Genesis: A Personal Investigation into the Animal Origins and Nature of Man.* London: Collins.

– 1976. *The Hunting Hypothesis: A Personal Conclusion concerning the Evolutionary Nature of Man.* New York: Atheneum.

Arnheim, Rudolf. 1969. *Visual Thinking.* London: Faber and Faber.

Ashton, E.M., and S. Zuckerman. 1950. 'Some Quantitative Dental Characters of Fossil Anthropoids.' *Philosophical Transactions of the Royal Society* B.234: 485–520.

Ashworth, William B., Jr. 1984. 'Marcus Gheeraerts and the Aesopic Connection in Seventeenth-Century Scientific Illustration.' *Art Journal* 44: 132–8.

– 1985. 'The Persistent Beast: Recurring Images in Early Zoological Illustration.' In Ellenius 1985, pp. 46–66.

– 1987. 'Iconography of a New Physics.' *History and Technology* 4: 267–97.

– 1989. 'Light of Reason, Light of Nature: Catholic and Protestant Metaphors of Scientific Knowledge.' *Science in Context* 3: 89–107.

– 1991. 'The Scientific Revolution: The Problem of Visual Authority.' Paper presented at the History of Science Society Conference on Critical Problems and Research Frontiers, Madison, Wisconsin, 30 October–3 November 1991, pp. 326–48.

Augusta, J., and Z. Burian. 1960. *Prehistoric Man.* Trans. M. Schierl. London: P. Hamlyn.

Baigrie, Brian S. 1992. 'Descartes' Mechanical Cosmology.' In *The Encyclopedia of Cosmology: Historical, Philosophical, and Scientific Foundations of Modern Cosmology.* Ed. Norriss Hetherington. New York: Garland, pp. 164–76.

Baillet, Adrien. 1693. *The Life of Monsieur Des-Cartes.* London: R. Simpson.

Balkwill, F.H. 1893. *The Testimony of the Teeth to Man's Place in Nature; with Other Essays on the Doctrine of Evolution.* London: Kegan Paul, Trench, Trubner & Co.

Banks, R.E.R., et al., eds. 1994. *Sir Joseph Banks: A Global Perspective.* Kew: Royal Botanic Gardens.
Barnes, B. 1974. *Scientific Knowledge and Sociological Theory.* London: Routledge and Kegan Paul.
Barnes, B., and D. Edge, eds. 1982. *Science in Context: Readings in the Sociology of Science.* Cambridge, Mass.: MIT Press.
Barr, John H. 1911. *Kinematics of Machinery: A Brief Treatise on Constrained Motions of Machine Elements.* New York: John Wiley and Sons.
Barrett, P.H. 1960. 'A Transcription of Darwin's First Notebook on "Transmutation of Species."' *Bulletin of the Museum of Comparative Zoology* 122: 247–96.
Barzon, Antonio, ed. 1960. *Tractatus astrarii.* Vatican City: Biblioteca Vaticana.
Basalla, George. 1988. *The Evolution of Technology.* Cambridge: Cambridge University Press.
Bastide, F. 1990. 'The Iconography of Scientific Texts: Principles of Analysis.' In Lynch and Woolgar 1990, pp. 187–229.
Baynes, K., and F. Pugh. 1981. *The Art of the Engineer.* Guildford: Lutterworth.
Bedini, Silvio, and F. Maddison. 1966. *Mechanical Universe: The Astrarium of Giovanni de'Dondi.* Philadelphia: American Philosophical Society.
Beer, Arthur, and Peter Beer, eds. 1975. *Kepler. Four Hundred Years: Proceedings of Conferences Held in Honor of Johannes Kepler.* Vistas in Astronomy 18. Oxford and New York: Pergamon Press.
Beer, Arthur, and K.A. Strand, eds. 1975. *Copernicus Yesterday and Today: Proceedings of the Commemorative Conference Held in Washington in Honor of Nicolaus Copernicus.* Vistas in Astronomy 17. Oxford and New York: Pergamon Press.
Behrensmeyer, A.K. 1975. 'The Taphonomy and Paleoecology of Plio-Pleistocene Vertebrate Assemblages East of Lake Rudolph, Kenya.' *Bulletin of the Museum of Comparative Zoology* 146: 473–578.
Behrensmeyer, A.K., and A.P. Hill. 1980. *Fossils in the Making: Vertebrate Taphonomy and Paleoecology.* Chicago: University of Chicago Press.
Belofsky, Harold. 1991. 'Engineering Drawing – a Universal Language in Two Dialects.' *Technology and Culture* 32: 23–46, and discussion 33: 853–7.
Bennett, James A. 1987. *The Divided Circle: A History of Instruments for Astronomy, Navigation and Surveying.* Oxford: Phaidon Christie's.
Berengario da Carpi, Jacapo. 1521. *Commentaria ... super anatomia mudini.* Bologna: Hieronimus de Benedictis.
– 1522. *Isagogae breves ....* Bologna: Hieronimus de Benedictis. English ed.,

1959. *A Short Introduction to Anatomy (Isagogae breves)*. Trans. Levi R. Lind. Chicago: University of Chicago Press.

Bergson, H. 1911. *Creative Evolution.* New York: Holt and Co.

Berkeley, G. 1962. *The Principles of Human Knowledge.* Ed. G. Warnock. London: Collins.

Binford, L.R. 1981. *Bones: Ancient Men and Modern Myths.* New York: Academic Press.

– 1983. *In Pursuit of the Past: Decoding the Archeological Record.* London: Thames and Hudson.

– 1984. *Faunal Remains from Klasies River Mouth.* New York: Academic Press.

– 1988. 'Fact and Fiction about the *Zinjanthropus* Floor: Data, Arguments, and Interpretations.' *Current Anthropology* 29: 123–35.

Blair, Anne. 1990. 'Tycho Brahe's Critique of Copernicus and the Copernican System.' *Journal of the History of Ideas* 51: 355–77.

Bloor, D. 1976. *Knowledge and Social Imagery.* London: Routledge and Kegan Paul.

Boaz, N.T. 1982. 'American Research on Australopithecines and Early Homo, 1925–1980.' In *A History of American Physical Anthropology, 1930–1980.* Ed. F. Spencer. New York: Academic Press, pp. 239–60.

Bogen, J., and J. Woodward. 1988. 'To Save the Phenomena.' *Philosophical Review* 97: 303–52

Boime, Albert. 1984. 'Van Gogh's *Starry Night*: A History of Matter and a Matter of History.' *Arts Magazine* 59(4): 86–103.

Booker, Peter Jeffrey. 1963. *A History of Engineering Drawing.* London: Chatto and Windus.

Boudreau, Linda Hults. 1978. 'Hans Baldung Grien and Albrecht Dürer: A Problem in Northern Mannerism.' Ph.D. diss., University of North Carolina.

Bowler, P.J. 1989. 'Development and Adaptation: Evolutionary Concepts in British Morphology, 1870–1914.' *British Journal for the History of Science* 22: 283–97.

Bradbury, S., and G.L'E. Turner, eds. 1967. *Historical Aspects of Microscopy.* Cambridge: W. Heffer and Sons, Ltd, for the Royal Microscopic Society.

Brady, R. 1985. 'On the Independence of Systematics.' *Cladistics* 1: 113–26.

Brahe, Tycho. 1573. *De nova et nullius aevi memoria prius visa stella.* Copenhagen: Laurentius Benedictus. English ed., 1623. *Learned: Tico Brahe, His Astronomical Conjecture of the New and Much Admired Star which Appeared in the Year 1572.* London: B. Alsop and T. Fawcet.

– 1598. *Astronomiae instauratae mechanica.* Wandesbeck: Ranzovianus. English ed., 1946. *Tycho Brahe's Description of His Instruments and Scientific Work as*

Given in 'Astronomiae instauratae mechanic' (Wandesbeck 1598). Trans. Hans H. Raeder, Elis Strömgren, and Bengt G. Strömgren. Copenhagen: Kongelige Danske Videnskabernes Selskab.
– 1602. *Progymnasmata*. Ed. Johannes Kepler. Prague.
– 1913–29. *Opera omnia*. Ed. John Dreyer. Copenhagen: Libreria Gylendialiana.
Brain, C.K. 1967. 'Bone Weathering and the Problem of Bone Pseudo-Tools.' *South African Journal of Science* 63: 97–9.
– 1968. 'Who Killed the Swartkrans Ape-Men?' *South African Museums Bulletin* 9: 127–39.
– 1981. *The Hunters or the Hunted? An Introduction to African Cave Taphonomy*. Chicago: University of Chicago Press.
Braithwaite, R. 1953. *Scientific Explanation: A Study of the Function of Theory, Probability and Law in Science*. Cambridge: Cambridge University Press.
Brande, W.T. 1830. *A Manual of Chemistry*. 3d ed. 2 vols. London: John Murray.
Brannigan, A. 1981. *The Social Basis of Scientific Discoveries*. Cambridge: Cambridge University Press.
Brock, W.H. 1993. *The Fontana History of Chemistry*. London: Fontana.
Broom, R. 1925. 'Some Notes on the Taungs Skull.' *Nature* 115: 569–71.
– 1936a. 'A New Fossil Anthropoid Skull from South Africa.' *Nature* 138: 486–8.
– 1936b. 'A New Ancestral Link between Ape and Man.' *Illustrated London News* 189: 476–7.
– 1938a. 'More Discoveries of Australopithecus.' *Nature* 141: 828–9.
– 1938b. 'The Missing Link Is No Longer Missing?' *Illustrated London News* 193: 310–11.
– 1942. 'The Hand of the Ape-Man *Paranthropus robustus*.' *Nature* 149: 513–14.
– 1943. 'An Ankle Bone of the Ape-Man *Paranthropus robustus*.' *Nature* 152: 689–90.
– 1950. *Finding the Missing Link*. 2d ed. London: Watts.
Broom, R., and G.W. Schepers. 1946. 'The South African Fossil Ape-Men, the Australopithecinae.' *Transvaal Museum Memoir* 2: 7–144.
Brown, J.R. 1989. *The Rational and the Social*. London and New York: Routledge.
– 1991. *The Laboratory of the Mind: Thought Experiments in the Natural Sciences*. London and New York: Routledge.
– 1994. *Smoke and Mirrors: How Science Reflects Reality*. London and New York: Routledge.

Buchwald, Jed, ed. 1995. *Theories of Practice. Stories of Practice.* Chicago: University of Chicago Press.

Bud, R., and G.K. Roberts. 1984. *Science versus Practice: Chemistry in Victorian Britain.* Manchester: Manchester University Press.

Bunge, M. 1967. 'Analogy in Quantum Theory: From Insight to Nonsense.' *British Journal for the Philosophy of Science* 18: 265–86.

Bunn, H.T. 1981. 'Archaeological Evidence for Meat-Eating by Plio-Pleistocene Hominids from Koobi Fora and Olduvai Gorge.' *Nature* 291: 574–7.

– 1986. 'Patterns of Skeletal Representation and Hominid Subsistence Activities at Olduvai Gorge, Tanzania, and Koobi Fora, Kenya.' *Journal of Human Evolution* 15: 673–90.

Bunn, H.T., and E. Kroll. 1986. 'Systematic Butchery by Plio-Pleistocene Hominids at Olduvai Gorge, Tanzania.' *Current Anthropology* 27: 431–52.

– 1988. 'Reply to "Fact and Fiction" about the FLK *Zinjanthropus* Floor.' *Current Anthropology* 29: 135–49.

Byars, J. 1988. 'Gazes/Voices/Power: Expanding Psychoanalysis for Feminist Film and Television Theory.' In *Female Spectators: Looking at Film and Television.* Ed. E.D. Pribram. London: Verso, pp. 110–31.

Carter, H.B. 1991. *Sir Joseph Banks, 1743–1820.* London: British Museum (Natural History).

Caspar, Max. 1959. *Kepler.* Trans. and ed. C. Doris Hellman. London and New York: Abelard-Schuman.

Chaptal, J.A. 1796. *Elemens de chymie.* 3d ed. Paris: Deterville.

Chemical Society (Great Britain). 1901–14. *Memorial Lectures Delivered before the Chemical Society 1893/1900–1901/13.* 2 vols. London: Gurney and Jackson.

Chemistry, Royal Society of Great Britain. 1984. *Oxygen and the Conversion of Future Feedstocks: 3rd BOC Priestley Conference.* London: Roy. Soc. Chem.

Clapham, F.M. 1976. *The Rise of Man.* London: Sampson Low.

Clarke, Desmond M. 1979. 'Physics and Metaphysics in Descartes' Principles.' *Studies in History and Philosophy of Science* 10(2): 89–122.

– 1982. *Descartes' Philosophy of Science.* University Park, Pa.: Pennsylvania State University Press.

Clusius [Charles de l'Écluse]. 1576. *Rariorum aliquot stirpium per Hispanias observatarum historia.* Antwerp: Plantin.

– 1601. *Rariorum plantarum historia.* Antwerp: Plantin.

Clutton-Brock, J. Clutton, and C. Grigson, eds. 1983. *Animals and Archaeology.* Vol. 1. Oxford: British Archaeology Report.

Cohen, I. Bernard. 1980. *From Leonardo to Lavoisier, 1450–1800.* Album of Science Series. New York: Charles Scribner's Sons.

Cole, F.J. 1953. 'The History of Albrect Dürer's Rhinoceros in Zoological Literature.' In Underwood 1953, vol. 1, pp. 337–56.
Coleman, W. 1971. *Biology in the Nineteenth Century: Problems of Form, Function, and Transformation.* New York: Wiley.
Collins, H. 1985. *Changing Order: Replication and Induction in Scientific Practice.* London: Sage Publications.
Collins, H.M., and T.J. Pinch. 1982. *Frames of Meaning: The Social Construction of Extraordinary Science.* London: Routledge and Kegan Paul.
Collins, James Daniel. 1971. *Descartes' Philosophy of Nature.* Oxford: Basil Blackwell.
Conkey, M.W. 1991. 'Contexts of Action, Contexts for Power: Material Culture and Gender in the Magdalenian.' In *Engendering Archaeology: Women and Prehistory.* Ed. J.M. Gero and M.W. Conkey. Oxford: Basil Blackwell, pp. 57–92.
Constable, John. 1970. *John Constable's Discourses.* Ed. R.B. Beckett. Ipswich: Suffolk Records Society.
Copernicus, Nicolaus. 1543. *De revolutionibus orbium caelestium.* Nuremberg: I. Petrius.
– 1972. *Nicholas Copernicus: Complete Works.* 3 vols. London, Warsaw, and Cracow: Macmillan and Polish Scientific Publishers.
Cornwall, I.W. 1960. *The Making of Man.* London: Phoenix House.
Cottingham, J., ed. 1976. *Descartes' Conversation with Burman.* Oxford: Clarendon Press.
Cottingham, J., R. Stoothoff, and D. Murdoch, eds. 1976. *The Philosophical Writings of Descartes.* 2 vols. Cambridge: Cambridge University Press.
Cottingham, J., R. Stoothoff, D. Murdoch, and A. Kenny, eds. 1991. *The Philosophical Writings of Descartes. Volume III. The Correspondence.* Cambridge: Cambridge University Press.
Cox, A. 1969. 'Geomagnetic Reversals.' *Science* 163: 237–45.
Cox, A., R.R. Doell, and G.B. Dalrymple. 1963. 'Geomagnetic Polarity Epochs and Pleistocene Geochronometry.' *Nature* 198: 1049–51.
– 1964. 'Reversals of the Earth's Magnetic Field.' *Science* 144: 1537–43.
Crombie, Alistair. 1958. 'Some Aspects of Descartes' Attitude to Hypothesis and Experiment.' *Actes du symposium international des sciences physiques et mathématiques dans la première moitié du XVIIe siècle.* Paris: Hermann & Cie.
– 1967. 'The Mechanistic Hypothesis and the Scientific Study of Vision ...' In Bradbury and Turner 1967, pp. 3–112.
– 1985. 'Science and the Arts in the Renaissance: The Search for Truth and Certainty, Old and New.' In Shirley and Hoeniger 1985, pp. 15–26.

Crookes, W. 1886. 'Chemical Science.' *Report of the British Association for the Advancement of Science*, pp. 558–76.

– 1891. 'Electricity in transitu.' *Chemical News* 63: 53–6, 68–70, 77–80, 89–93, 98–100, 112–14.

Crosland, M.P., ed. 1975. *The Emergence of Science in Western Europe*. London: Macmillan.

– 1978. *Historical Studies in the Language of Chemistry*. 2d ed. New York: Dover.

– 1994. *In the Shadow of Lavoisier: The 'Annales de Chimie' and the Establishment of a New Science*. Faringdon: British Society for the History of Science.

Dalton, J. 1808–11. *A New System of Chemical Philosophy*. Manchester: S. Russell

Daniell, J.F. 1839. *An Introduction to the Study of Chemical Philosophy: Being a Preparatory View of the Forces Which Concur to the Production of Chemical Phenomena*. London: J.S. Parker.

Darnell, J., H. Lodish, and D. Baltimore. 1990. *Molecular Cell Biology*. New York: Scientific American Books.

Dart, R.A. 1925. 'Australopithecus Africanus: The Man-Ape of South Africa.' *Nature* 115: 195–9.

– 1948. 'The Makapansgat Proto-Human *Australopithecus promethus*.' *American Journal of Physical Anthropology* n.s. 6 (3): 259–84.

– 1949. 'The Predatory Implemental Technique of *Australopithecus*.' *American Journal of Physical Anthropology* 7: 1–6.

– 1953. 'The Predatory Transition from Ape to Man.' *International Anthropological and Linguistic Review* 1: 201–18.

– 1955. 'The First Australopithecine Fragment from the Makapansgat Pebble Culture Stratum.' *Nature* 176: 170.

– 1957. 'The Makapansgat Australopithecine Osteodontokeratic Culture.' *Proceedings of the Third Pan African Congress on Prehistory, (Livingstone, 1955)*. Ed. Desmond Clark. London: Chatto and Windus, pp. 161–71.

– 1959. 'The Ape-Men Tool-Makers of a Million Years Ago: South African Australopithecus – His Life, Habits and Skills.' *Illustrated London News* 234: 798–801.

Darwin, C. 1859. *On the Origin of Species by Means of Natural Selection; or, The Preservation of Favoured Races in the Struggle for Life*. London: John Murray.

– 1871. *The Descent of Man, and Selection in Relation to Sex*. 2 vols. London: John Murray.

Davy, H. 1839–40. *The Collected Works*. Ed. J. Davy. 9 vols. London: Smith, Elder.

Davy, J. 1836. *Memoirs of the Life of Sir Humphry Davy*. 2 vols. London: Longman, Rees, Orme, Brown, Green & Longman.

Dawkins, W.B. 1874. *Cave Hunting: Researches on the Evidence of Caves Respecting the Early Inhabitants of Europe.* London: Macmillan.
Deforge, Yves. 1981. *Le Graphisme technique: son histoire et son enseignement.* Seyssel: Champs Villon.
de Jong, H.M.E. 1969. *M. Maier's Atalanta Fugiens.* Leiden: E.J. Brill.
Dennis, Michael Aaron. 1989. 'Graphic Understanding: Instruments and Interpretation in Robert Hooke's *Micrographia.*' *Science in Context* 3: 309–64.
de Queiroz, K. 1988. 'Systematics and the Darwinian Revolution.' *Philosophy of Science* 55: 238–59.
De Santillana, Giorgio. 1969 'The Role of Art in the Scientific Renaissance.' In *Critical Problems in the History of Science.* Ed. Marshall Clagett. Madison: University of Wisconsin Press, pp. 33–65.
Descartes, René. 1664. *l'Homme de René Descartes et un traité de la formation du foetus du mesme autheur, avec les remarques de Louys de la Forge.* Paris: Charles Angot.
Desmond, A.J. 1974. 'Central Park's Fragile Dinosaurs.' *Natural History* 83(8): 64–71.
– 1975. *The Hot Blooded Dinosaurs.* London: Blond and Briggs.
– 1979. 'Designing the Dinosaur: Richard Owen's Response to Robert Edmond Grant.' *Isis* 70: 224–34.
Dijksterhuis, E.J. 1986. *The Mechanization of the World Picture: Pythagoras to Newton.* Trans. C. Dikshoorn. Princeton: Princeton University Press.
Dobzhansky, T. 1937. *Genetics and the Origin of Species.* New York: Columbia University Press.
– 1951. *Genetics and the Origin of Species.* 3d ed. New York: Columbia University Press.
Dodoens, Rembert. 1566. *Frumentorum, legumium ... herbarum ... historia.* Antwerp: Plantin.
– 1568. *Florum ... herbarum historia.* Antwerp: Plantin.
Doney, Willis, ed. 1967. *Descartes: A Collection of Critical Essays.* Garden City, N.Y.: Anchor Books.
Donovan, A. 1993. *Antoine Lavoisier: Science, Administration and Revolution.* Oxford: Blackwell
Drake, Stillman. 1975. 'Copernicanism in Bruno, Kepler and Galileo.' In Beer and Strand 1975, pp. 177–90.
– 1989. *History of Free Fall: Aristotle to Galileo.* Toronto: Wall and Thompson.
Dryander, Johannes. 1537. *Anatomiae, hoc est, corporis humani dissectionis.* Marburg: Eucharius Cervicornus.
Duhem, P. 1954. *The Aim and Structure of Physical Theory.* Princeton: Princeton University Press.

Eamon, William. 1980. 'Botanical Empiricism in Late Medieval Technical Writings.' *Res Publica Litterarum* 3: 237–45.

Edgerton, Samuel Y., Jr. 1975. *The Renaissance Rediscovery of Linear Perspective.* New York: Basic Books.

– 1980. 'The Renaissance Artist as Quantifier.' In *The Perception of Pictures.* Ed. Margaret Hagen. New York: Academic Press, vol. 1, pp. 179–213.

– 1985a. 'The Renaissance Development of the Scientific Illustration.' In Shirley and Hoeniger 1985, pp. 168–97.

– 1985b. 'Galileo, Florentine "Disegno" and the "Strange Spotednesse" of the Moon.' *Art Journal* 44: 225–48.

– 1991. *The Heritage of Giotto's Geometry: Art and Science on the Eve of the Scientific Revolution.* Ithaca, N.Y.: Cornell University Press.

Einstein, Albert. 1949a. 'Autobiographical Notes.' In *Albert Einstein: Philosopher-Scientist.* Ed. Paul A. Schilpp. New York: Harper and Row, vol. 1, pp. 3–95.

– 1949b. Letter. In Hadamard 1954, pp. 142–3.

– 1961. *Relativity: The Special and General Theory.* Trans. R. W. Lawson. New York: Crown Reprint.

Einstein, A., and L. Infeld. 1938. *The Evolution of Physics: From Early Concepts to Relativity and Quanta.* New York: Simon and Schuster.

Eisenstein, Elizabeth L. 1969. 'The Advent of Printing and the Problem of the Renaissance.' *Past and Present* 45: 19–89.

– 1970. 'The Advent of Printing in Current Historical Literature: Notes and Comments on an Elusive Transformation.' *American Historical Review* 75: 727–43.

– 1979. *The Printing Press as an Agent of Change: Communications and Cultural Transformations in Early Modern Europe.* Cambridge: Cambridge University Press.

– 1983. *The Printing Revolution in Early Modern Europe.* Cambridge: Cambridge University Press.

Elkins, James. 1984. 'Michelangelo and the Human Form: His Knowledge and Use of Anatomy.' *Art History* 7: 176–86.

Ellenius, Allan, ed. 1985. *The Natural Sciences and the Arts: Aspects of Interaction from the Renaissance to the Twentieth Century.* Stockholm: Almqvist and Wiksell International.

Elliott Smith, G.F.S. 1925. 'Reconstructed: Australopithecus and "The Rhodesian Man."' *Illustrated London News* 166: 239–40.

Eriksson, G. 1992. 'Berelius and the Atomic Theory: The Intellectual Background.' In Melhado and Frängsmyr 1992, pp. 56–84.

Estienne, Charles [Carolus Stephanus]. 1545. *De dissectione partium corporis humani libri tres.* Paris: Simon de Colines.

Eustachio, Bartolomeo. 1714. *Tabulae anatomicae.* Rome: Francisco Gonzaga.
Fagan, B.M. 1985. *The Adventure of Archaeology.* Washington, D.C.: National Geographic Society.
Faraday, M. 1827. *Chemical Manipulation.* London: W. Phillips.
Farrar, W.V. 1975. 'Science and the German University System, 1790–1850.' In Crosland 1975, pp. 179–92.
Ferguson, Eugene S. 1977. 'The Mind's Eye: Nonverbal Thought in Technology.' *Science* 197: 827–36.
– 1992. *Engineering and the Mind's Eye.* Cambridge, Mass.: MIT Press.
Feynman, R. 1985. *QED: The Strange Story of Light and Matter.* Princeton: Princeton University Press.
Ficino, Marsilio. 1493. *De sole.* Florence: Antonius Mischominus, cap. XIII.
Field, J.V. 1988a. *Kepler's Geometrical Cosmology.* Chicago: University of Chicago Press.
– 1988b. 'What Is Scientific about a Scientific Instrument?' *Nuncius* 3(2): 3–26.
Fine, A., M. Forbes, and L. Wessels, eds. 1991. *PSA 1990.* vol. 2. East Lansing, Mich.: Philosophy of Science Association.
Fisch, M., and S. Schaffer. 1991. *William Whewell: A Composite Portrait.* Oxford: Clarendon Press.
Fisher, R.A. 1930. *The Genetical Theory of Natural Selection.* Oxford: Oxford University Press.
Fitzpatrick, M. 1984. 'Priestley in Caricature.' In Chemistry, Royal Society, 1984, pp. 347–69.
Foley, R., ed. 1984. *Hominid Evolution and Community Ecology: Prehistoric Human Adaption in Biological Perspective.* London: Academic Press.
Ford, B.J. 1992. *Images of Science: A History of Scientific Illustration.* London: British Library.
Forgan, S. 1986. 'Context, Image and Function: A Preliminary Enquiry into the Architecture of Scientific Societies.' *British Journal for the History of Science* 19: 89–113.
Forge, Louis de la. 1666. *Traité de l'esprit de l'homme.* In *Louis de La Forge: Oeuvres Philosophiques.* Ed. P. Clair. Paris: Presses Universitaires de France, 1974, pp. 65–349.
Foucault, Michel. 1973. *The Order of Things: An Archaeology of the Human Sciences.* New York: Vintage Books.
French, Roger. 1985. 'Berengario da Carpi and the Use of Commentary in Anatomical Teaching.' In *The Medical Renaissance of the Sixteenth Century.* Ed. A. Wear, R. French, and I. Lonie. Cambridge: Cambridge University Press.

Fries, Lorenz. 1519. *Spiegel der Artzny.* Strasbourg: Johannes Grüninger.

Fuchs, Leonhart. 1542. *De historia stirpium (commentariorum tomi vivae imagines).* Basel: Isingrin.

Fürbringer, M. 1888. 'Untersuchungen zur Morphologie und Systematik der Vögel.' *Bijdragen tot de Dierkunde* 15. Amsterdam: Tj Van Holkema.

Gaffney, E.S. 1984. 'Historical Analysis of Theories of Chelonian Relationship.' *Systematic Zoology* 33(3): 283–301.

Galilei, Galileo. 1610. *Sidereus nuncius.* Venice: Baglioni. English ed., 1989. *Sidereus Nuncius or the Sidereal Messenger.* Trans. and introd. by Albert Van Helden. Chicago: University of Chicago Press, 1989.

Galluzzi, Paolo. 1987a. 'Leonardo da Vinci: From the "elementi macchinali" to the Man-Machine.' *History and Technology* 4: 235–65.

– ed. 1987b. *Leonardo da Vinci: Engineer and Architect.* Montreal: Montreal Museum of Fine Arts.

– 1990. 'Leonardo da Vinci on Creativity, Mechanics and the Imitation of Nature.' In *Creativity in the Arts and Science.* Ed. William R. Shea and Antonio Spadafora. Canton, Mass.: Science History Publications.

Gamble, C. 1992. 'Figures of Fun; Theories about Cavemen.' *Archaeological Review from Cambridge* 11(2): 357–72.

Garber, Daniel. 1978. 'Science and Certainty.' In Hooker 1978, pp. 114–51.

– 1992: 'Descartes' Physics.' In *The Cambridge Companion to Descartes.* Ed. J. Cottingham. Cambridge: Cambridge University Press, pp. 286–334.

Gascoigne, J. 1994. *Joseph Banks and the English Enlightenment.* Cambridge: Cambridge University Press.

Gaukroger, Stephen, ed. 1980. *Descartes: Philosophy, Mathematics and Physics.* New Jersey: Barnes and Noble Books.

Gedzelman, Stanley David. 1989. 'Cloud Classification before Luke Howard.' *Bulletin of the American Meteorological Society* 70: 381–95.

– 1990a. 'The Meteorological Odyssey of Vincent van Gogh.' *Leonardo* 23: 107–16.

– 1990b. 'Leonardo da Vinci and the Downburst.' *Bulletin of the American Meteorological Society* 71: 649–55.

– 1991a. 'Atmospheric Optics in Art.' *Applied Optics* 30(24): 3514–22.

– 1991b. 'Weather Forecasts in Art.' *Leonardo* 24: 441–51.

Gettings, F. 1981. *Dictionary of Occult, Hermetic and Alchemical Sigils.* London: Routledge.

Ghiselin, M.T. 1969. *The Triumph of the Darwinian Method.* Berkeley: University of California Press.

Giere, R.N. 1983. 'Testing Theoretical Hypotheses.' In *Testing Scientific*

Theories. Ed. J. Earman. Minnesota Studies in the Philosophy of Science, vol. 9. Minneapolis: University of Minnesota Press, pp. 269–98.
– 1988. *Explaining Science: A Cognitive Approach.* Chicago: University of Chicago Press.
– 1989a. 'What Does Explanatory Coherence Explain?' *Behavioral and Brain Sciences* 12: 475–6.
– 1989b. 'The Units of Analysis in Science Studies.' In *The Cognitive Turn: Sociological and Psychological Perspectives on Science.* Ed. S. Fuller, M. DeMey, T. Shinn, and S. Woolgar. Sociology of the Sciences Yearbook. Dordrecht: D. Reidel, pp. 3–11.
– 1989c. 'Scientific Rationality as Instrumental Rationality.' *Studies in History and Philosophy of Science* 20(3): 377–84.
– 1991. 'Implications of the Cognitive Sciences for the Philosophy of Science.' In Fine, Forbes, and Wessels 1991, vol. 2, pp. 419–30.
– 1994a. 'No Representation without Representation.' Essay review of *Representation in Scientific Practice,* by Michael Lynch and Steve Woolgar. *Biology and Philosophy* 9: 113–20.
– 1994b. 'The Cognitive Structure of Scientific Theories.' *Philosophy of Science* 61: 276–96.
Gifford, D. 1978. 'Ethnoarchaeological Observations of Natural Process Affecting Cultural Materials.' In *Explorations in Ethnoarchaeology.* Ed. R.A. Gould. Albuquerque: University of New Mexico Press, pp. 77–101.
– 1980. 'Ethnoarcheological Contributions to the Taphonomy of Human Sites.' In *Fossils in the Making Vertebrate Taphonomy and Paleoecology.* Ed. A.K. Behrensmeyer and A.P. Hill. Chicago: University of Chicago Press, pp. 93–106.
– 1981. 'Taphonomy and Palaeoecology: A Critical Review of Archaeology's Sister Disciplines.' *Advances in Archaeological Method and Theory* 4: 365–438.
Gifford-Gonzalez, D. 1993. 'You Can Hide, But You Can't Run: Representations of Women's Work in Illustrations of Palaeolithic Life.' *Visual Anthropology Review* 9: 23–41.
Gingerich, Owen. 1971. 'Apianus's *Astronomicum Caesareum* and Its Leipzig Facsimile.' *Journal for the History of Astronomy* 2: 168–77.
– 1973. 'From Copernicus to Kepler: Heliocentrism as Model and as Reality.' *Proceedings of the American Philosophical Society* 117(6): 513–22.
– ed. 1975. *The Nature of Scientific Discovery: A Symposium Commemorating the Five Hundredth Anniversary of the Birth of Nicolaus Copernicus.* Washington: Smithsonian Institution Press.
Glen, W. 1982. *The Road to Jaramillo.* Stanford: Stanford University Press.

Golinski, J. 1992. *Science as Public Culture: Chemistry and Enlightenment in Britain, 1760–1820*. Cambridge: Cambridge University Press.

Gombrich, E.H. 1968. *Art and Illusion: A Study in the Psychology of Pictorial Representation*. 3d ed. London: Phaidon Press.

– 1972. 'The Visual Image.' *Scientific American* 227(3): 82–96.

– 1974. 'Standards of Truth: The Arrested Image and the Moving Eye.' In Mitchell 1974, pp. 181–218.

Gooding, D. 1989. '"Magnetic Curves" and the Magnetic Field: Experimentation and Representation in the History of a Theory.' In Gooding, Pinch, and Schaffer 1989, pp. 183–223.

Gooding, David, Trevor Pinch, and Simon Schaffer, eds. 1989. *The Uses of Experiment: Studies in the Natural Sciences*. Cambridge: Cambridge University Press.

Gould, Stephen Jay. 1989. *Wonderful Life: The Burgess Shale and the Nature of History*. New York: W.W. Norton.

– 1991. *Bully for Brontosaurus: Reflections in Natural History*. New York: W.W. Norton.

Gray, Andrew. 1804. *The Experienced Millwright; or, A Treatise on the Construction of Some of the Most Useful Machines*. Edinburgh: J. Brown.

Greene, J.C. 1959. *The Death of Adam: Evolution and Its Impact on Western Thought*. Ames: Iowa State University Press.

Greg, R.P., and W.G. Lettsom. 1858. *Mineralogy: Manual of the Mineralogy of Great Britain and Ireland*. London: John van Voorst.

Gregory, W.K. 1938. 'Evidence of the Australopithecine Man-Apes on the Origin of Man.' *Science* 88: 615–16.

– 1945. 'Revised Reconstruction of the Skull of *Pleisanthropus Transvaalensis* Broom.' *American Journal of Physical Anthropology* 3: 267–75.

– 1949. 'The Bearing of the Australopithecinae upon the Problem of Man's Place in Nature.' *American Journal of Physical Anthropology* 7: 485–512.

Gruber, Howard E. 1981. 'Darwin's "Tree of Nature" and Other Images of Wide Scope.' In *On Aesthetics in Science*. Ed. Judith Wechsler. Cambridge, Mass.: MIT Press, pp. 121–40.

Guyton de Morveau, L.B., et al. 1787. *Methode de nomenclature chimique*. Paris: Cuchet. Reprint, 1992. Lisbon: A.M. Nunes dos Santos.

Hacking, Ian. 1991. 'Matters of Graphics.' *Science* 252: 979–80.

Hadamard, Jacques. 1954. *An Essay on the Psychology of Invention in the Mathematical Field*. Princeton: Princeton University Press.

Hall, A.R. 1960. *The Scientific Revolution: 1500–1800: The Formation of the Modern Scientific Attitude*. Boston: Beacon Press.

Hall, Bert S. 1976. 'The New Leonardo.' Review of *Leonardo da Vinci: The*

Madrid Codices, ed. L. Reti; and *The Unknown Leonardo*, by L. Reti. *Isis* 67: 463–75.
– 1978. 'Giovanni de'Dondi and Guido da Vigevano: Notes Toward a Typology of Medieval Technological Writings.' *Annals of the New York Academy of Sciences* 314: 127–44.
– 1979a. *The Technological Illustrations of the So-called 'Anonymous of the Hussite Wars': Codex latinus monacensis 197, Part I.* Weisbaden: Reichert.
– 1979b. '*Der meister sol auch kennen schreiben und lesen*: Writings about Technology, ca.1400–ca.1600 A. D., and Their Cultural Implications.' In *Early Technologies: Invited Lectures on the Middle East at the University of Texas, Austin.* Ed. D. Schmandt-Besserat. Malibou: Undena Publications, vol. 3, pp. 47–58.
– 1982a. 'Guido da Vigevano's *Texaurus Regis Franciae*, 1335.' *Scripta* 6: 33–44.
– 1982b. 'Production et diffusion de certains traités de techniques au Moyen Age.' In *Les arts mécaniques au Moyen-Age.* Ed. G. Allard and S. Lusignan. Montréal: Institut d'études médievales, pp. 147–70.
– 1982c. 'Editing Texts in the History of Early Technology.' In *Editing Texts in the History of Science and Medicine: Papers Given at the Seventeenth Annual Conference on Editorial Problems.* Ed. T.H. Levere. New York: Garland, pp. 69–100.
Hall, Bert, and Ian Bates. 1976. 'Leonardo, the Chiaravalle Clock and Epicyclic Gearing: A Reply to Antonio Simoni.' *Antiquarian Horology* 9: 910–17.
Hall, Marie Boas. 1962. *The Scientific Renaissance, 1450–1630.* New York: Harper Torchbook.
Hallyn, Fernand. 1990. *The Poetic Structure of the World: Copernicus and Kepler.* Trans. D. M. Leslie. New York: Zone Books.
Hamilton, W.D. 1964. 'The Genetical Evolution of Social Behaviour I and II.' *Journal of Theoretical Biology* 7: 1–52
Harms, Wolfgang. 1985. 'On Natural History and Emblematics in the 16th Century.' In Ellenius 1985, pp. 67–83.
Harper, W. 1989. 'Conscilience and Natural Kind Reasoning.' In *An Intimate Relation: Studies in the History and Philosophy of Science.* Ed. J. Brown and J. Mittelstrass. Dordrecht: Kluwer Academic Publishers, pp. 115–52.
– 1990. 'Newton's Classic Deductions from Phenomena.' In Fine, Forbes, and Wessels 1991, vol. 2, pp. 183–96.
Harré, Rom. 1970. *The Principles of Scientific Thinking.* Chicago: University of Chicago Press.
Harrison, C.J.O. 1969. 'The Possible Affinities of the Australian Treecreepers of the Genus *Climacteris.*' *Emu* 69: 161–8.

Harte, N., and J. North. 1978. *The World of University College London, 1828–1978.* London: University College.

Harwood, John T. 1989. 'Rhetoric and Graphics in "Micrographia."' In *Robert Hooke: New Studies.* Ed. Michael Hunter and Simon Schaffer. Woodbridge: Boydell Press, pp. 119–47.

Hatfield, Gary. 1990. 'Metaphysics and the New Science.' In Lindberg and Westman 1990, pp. 93–166.

Haüy, R.J. 1793. 'Exposition de la théorie sur la structure des cristaux.' *Annales de chimie* 17: 225–319.

Hempel, C.G. 1965. *Aspects of Scientific Explanation.* New York: Free Press.

Henderson, L.J. 1913. *The Fitness of the Environment, an Inquiry into the Biological Significance of the Properties of Matter.* New York: Macmillan.

Heninger, S.K., Jr. 1977. *The Cosmographical Glass: Renaissance Diagrams of the Universe.* San Marino: Huntington Library.

Hess, H.H. 1962. 'History of Ocean Basins.' In *Petrologic Studies: A Volume in Honor of A.F. Buddington.* Ed. A.E.J. Engel, H.L. James, and B.F. Leonard. Boulder: Geological Society of America, pp. 599–620.

Hesse, M.B. 1966. *Models and Analogies in Science.* Notre Dame: University of Notre Dame Press.

Hill, A. 1976. 'On Carnivore and Weathering Damage to Bone.' *Current Anthropology* 17: 335–6.

Hoenen, Petrus H. J. 1967. 'Descartes's Mechanism.' In Doney 1967, pp. 353–68.

Hoeniger, David F. 1985. 'How Plants and Animals Were Studied in the Mid-Sixteenth Century.' In Shirley and Hoeniger 1985, pp. 130–48.

Hoffmann, R., and V. Torrence. 1993. *Chemistry Imagined: Reflections on Science.* Washington: Smithsonian Institution Press.

Holmes, A. 1930. 'Radioactivity and Earth Movements.' *Transactions of the Geological Society of Glasgow* 18: 559–606.

– 1944. *Principles of Physical Geology.* London: Nelson.

Holton, Gerald. 1986. *The Advancement of Science, and Its Burdens: The Jefferson Lecture and Other Essays.* Cambridge: Cambridge University Press.

Hooker, Michael, ed. 1978. *Descartes: Critical and Interpretive Essays.* Baltimore: Johns Hopkins University Press.

Houghton, W. E. 1957. 'The History of Trades: Its Relation to Seventeenth Century Thought.' In *Roots of Scientific Thought: A Cultural Perspective.* Ed. Philip P. Wiener and Aaron Noland. New York: Basic Books.

Howell, F.C. 1965. *Early Man.* New York: Time-Life Books.

Hull, D. 1989. *The Metaphysics of Evolution.* Albany: SUNY Press.

Hulton, Paul. 1985. 'Realism and Tradition in Ethnological and Natural History Imagery in the Sixteenth Century.' In Ellenius 1985, pp. 18–31.

Hundt, Magnus. 1501. *Antropologium.* Leipzig: Wolfgang Stöcklin.

Hutchinson, G. Evelyn. 1974. 'Attitudes toward Nature in Medieval England: The Alphonso and Bird Psalters.' *Isis* 65: 4–37.

Hutchinson, Keith. 1991. 'Copernicus, Apollo, and Herakles.' In *The Uses of Antiquity: The Scientific Revolution and the Classical Tradition.* Ed. S. Gaukroger. Dordrecht: Kluwer Academic Publishers, pp. 1–23.

– 1993. 'Harmony and Authority: The Political Symbolism of Copernicus's Personal Seal.' In Mazzolini 1993, pp. 115–68.

Huxley, T.H. 1863. *Evidence as to Man's Place in Nature.* London: Williams and Norgate.

– 1868a. 'On the Classification and Distribution of the *Alectoromorphae* and *Heteromorphae.*' *Proceedings of the Zoological Society of London,* pp. 294–319.

– 1868b. [Letter to A. Newton]. *Ibis* 4 (new series): 357–62.

Isaac, G.L. 1976. 'The Activities of Early African Hominids: A Review of Archaeological Evidence from the Time Span Two and a Half to One Million Years Ago.' In *Human Origins: Louis Leakey and the East African Evidence.* Ed. G.L. Isaac and E. McCown. Menlo Park, Calif.: W.A. Benjamin, pp. 483–514.

– 1978a. 'The Food-Sharing Behavior of Protohuman Hominids.' *Scientific American* 238(4): 90–108.

– 1978b. 'Food Sharing and Human Evolution: Archaeological Evidence from the Plio-Pleistocene of East Africa.' *Journal of Anthropological Research* 34: 311–25.

– 1984. 'The Archaeology of Human Origins: Studies of the Lower Pleistocene in East Africa 1971–1981.' *Advances in World Archaeology* 3: 1–87.

Ivins, William M., Jr. 1953. *Prints and Visual Communication.* Cambridge: Routledge and Kegan Paul.

Jardine, W. 1858. *Memoirs of Hugh Edwin Strickland, M.A.* London: John Van Voorst.

Jaynes, J. 1970. 'The Problem of Animate Motion in the Seventeenth Century.' *Journal of the History of Ideas* 31: 219–34.

Johanson, D.C., and M.A. Edey. 1981. *Lucy: The Beginnings of Humankind.* London: Book Club Associates.

Johnson, M. 1987. *The Body in the Mind: The Bodily Basis of Meaning, Imagination, and Reason.* Chicago: University of Chicago Press.

Joule, J. 1850. 'On the Mechanical Equivalent of Heat.' *Philosophical Transactions of the Royal Society (London)* 140: 61–82.

Jung, C.G. 1968. *Psychology and Alchemy*. 2d ed. Trans. R.F.C. Hull. London: Routledge and Kegan Paul.

Kaiser, M. 1991. 'From Rocks to Graphs – the Shaping of Phenomena.' *Synthese* 89: 111–33.

Kane, R. 1842. *Elements of Chemistry*. Dublin: Hodges and Smith.

Kaufmann, T.D. 1993. *The Mastery of Nature: Aspects of Art, Science, and Humanism in the Renaissance*. Princeton: Princeton University Press.

Kaup, J.J. 1854. 'Einige Worte über die systematische Stellung der Familie der Raben, Corvidae.' *Journal für Ornithologie* 2, Jahresversammlung, XLVII–LVII.

Keele, Kenneth, and Carlo Pedretti. 1979–80. *Leonardo da Vinci: Corpus of the Anatomical Studies in the Collection of Her Majesty, the Queen, at Windsor Castle*. 3 vols. London and New York: Johnson Reprint Corporation.

Keith, A. 1925. 'The Taungs Skull.' *Nature* 116: 11.

– 1948. *A New Theory of Human Evolution*. London: Watts and Co.

Kellett, C. 1964. 'Perino del Vaga et les illustrations pour l'Anatomie d'Estienne.' *Aesculapius* 37: 74–9.

Kemp, Martin. 1970. 'A Drawing for the *Fabrica*; and Some Thoughts upon the Vesalius Muscle-Men.' *Medical History* 14: 277–88.

– 1977. 'From "Mimesis" to "Fantasia": The Quattrocento Vocabulary of Creation, Inspiration and Genius in the Visual Arts.' *Viator* 8: 347–98.

– 1989a. 'Geometrical Bodies as Exemplary Forms in Renaissance Space.' In *World Art: Themes of Unity in Diversity: Acts of the XXVIth International Congress of the History of Art*. Ed. Irving Lavin. 3 vols. University Park: Pennsylvania State University Press, vol. 1, pp. 237–42.

– ed. 1989b. *Leonardo on Painting: An Anthology of Writings by Leonardo da Vinci: With a Selection of Documents Relating to His Career as an Artist*. Trans. Martin Kemp and Margaret Walker. New Haven: Yale University Press.

– 1990. 'Taking It on Trust: Form and Meaning in Naturalistic Representation.' *Archives of Natural History* 17: 127–88.

– 1991. '"Intellectual Ornaments": Style, Function and Society in Some Instruments of Art.' In *Interpretation and Cultural History*. Ed. Joan H. Pittock and Andrew Wear. London: Macmillan, pp. 135–52.

– 1992. *The Science of Art: Optical Themes in Western Art from Brunelleschi to Seurat*. Rev. ed. New Haven: Yale University Press.

– 1993. '"The Mark of Truth": Looking and Learning in Some Anatomical Illustrations from the Renaissance and Eighteenth Century.' In *Medicine and the Five Senses*. Ed. William F. Bynum and Roy Porter. Cambridge: Cambridge University Press, pp. 85–121.

Kepler, Johannes. 1596. *Mysterium cosmographicum.* Tubingen: Georgius gruppen bachius.

– 1604. *Ad vitellionem paralipomena.* In *Johannes Kepler: Gesammelte Werke.* Ed. W. Von Dyck and M. Caspar. Vol. 2. Munich: C.H. Beck'sche. Verlagsbuchhandlung, 1938–.

– 1609. *Astronomia nova.* Heidelberg: G. Vogelinus. English ed., 1992. *New Astronomy.* Trans. William Donahue. Cambridge: Cambridge University Press, 1992.

– 1615. *Nova stereometria doliorum vinariorum.* Linz: Johannes Plancus.

– 1618–22. *Epitome astronomiae Copernicae.* Linz: Johannes Plancus.

– 1627. *Tabulae Rudolphinae.* Ulm: Jonas Saurius.

– 1634. *Somnium.* Published posthumously by Ludovic Kepler. Frankfurt: L. Kepler. English ed., 1967. *Kepler's Somnium: The Dream, or Posthumous Work on Lunar Astronomy.* Trans. Edward Rosen. Madison: University of Wisconsin Press, 1967.

– 1938–88. *Johannes Kepler: Gesammelte Werke.* Ed. W. von Dyck, M. Caspar et al. 20 vols. Munich: C.H. Beck'sche Verlagsbuchhandlung.

King-Hele, D.G., ed. 1992. *John Herschel 1792–1871.* London: Royal Society.

Kinzey, W.G., ed. 1987. *The Evolution of Human Behavior: Primate Models.* Albany: State University of New York Press.

Klingender, Francis Donald. 1971. *Animals in Art and Thought to the End of the Middle Ages.* Ed. Evelyn Antal and John Harthan. Cambridge, Mass.: MIT Press.

Knight, D.M. 1977. *Zoological Illustration: An Essay Towards a History of Printed Zoological Pictures.* London: Dawson.

– 1985. 'Scientific Theory and Visual Language.' In Ellenius 1985, pp. 106–24.

– 1986. 'Accomplishment or Dogma: Chemistry in the Introductory Works of Jane Marcet and Samuel Parkes.' *Ambix* 33: 94–8.

– 1992a. *Humphry Davy: Science and Power.* Oxford: Blackwell.

– 1992b. *Ideas in Chemistry: A History of the Science.* New Brunswick, N.J.: Rutgers University Press. 2d ed. 1995.

Knorr-Cetina, K.D. 1981. *The Manufacture of Knowledge: An Essay on the Constructivist and Contextual Nature of Science.* Oxford: Pergamon Press.

Koch, Robert. 1974. *Hans Baldung Grien: Eve, the Serpent and Death.* Ottawa: National Gallery of Canada.

Kornell, Monique. 1989. 'Rosso Fiorentino and the Anatomical Text.' *Burlington Magazine* 131: 842–7.

– 1993. 'Artists and the Study of Anatomy in Sixteenth-Century Italy.' Ph.D. diss., Warburg Institute, University of London.

Kuhn, T.S. 1962. *The Structure of Scientific Revolutions.* Chicago: University of Chicago Press. 2d ed. 1970.

– 1977. *The Essential Tension.* Chicago: University of Chicago Press.

Lakatos, I. 1970. 'Falsification and the Methodology of Scientific Research Programmes.' In *Criticism and the Growth of Knowledge.* Ed. I. Lakatos and A. Musgrave. Cambridge: Cambridge University Press, pp. 91–195.

Lakoff, G. 1987. *Women, Fire, and Dangerous Things: What Categories Reveal about the Mind.* Chicago: University of Chicago Press.

Lakoff, G., and M. Johnson. 1980. *Metaphors We Live By.* Chicago: University of Chicago Press.

Larkin, J.H., and H.A. Simon. 1987. 'Why a Diagram Is (Sometimes) Worth Ten Thousand Words.' *Cognitive Science* 11: 65–99.

Latour, Bruno. 1986. 'Visualization and Cognition: Thinking with Eyes and Hands.' *Knowledge and Society: Studies in the Sociology of Culture, Past and Present* 6: 1–40.

– 1987. *Science in Action: How to Follow Scientists and Engineers through Society.* Cambridge, Mass.: Harvard University Press.

– 1989. *The Pasteurization of France.* Trans. A. Sheridan and J. Law. Cambridge, Mass.: Harvard University Press.

– 1990. 'Drawing Things Together.' In Lynch and Woolgar 1990, pp. 19–68.

Latour, Bruno, and Steve Woolgar. 1979. *Laboratory Life.* Beverly Hills: Sage Publications.

Laurent, A. 1855. *Chemical Method, Notation, Classification and Nomenclature.* Trans. W. Odling. London: Cavendish Society.

Lavoisier, A.L. 1965. *Elements of Chemistry: In a New Systematic Order, Containing All the Modern Discoveries (1790).* Trans. R. Kerr. New York: Dover Publications.

Layton, Edwin T., Jr. 1976. 'American Ideologies of Science and Engineering.' *Technology and Culture* 17: 688–701.

Leakey, L.S.B. 1946. 'A Pre-Historian's Paradise in Africa: Early Stone Age Sites at Olorgesailie.' *Illustrated London News* 209: 382–5.

– 1959. 'A New Fossil Skull from Olduvai.' *Nature* 184: 491–3.

– 1960a. 'Finding the World's Earliest Man.' *National Geographic* 118: 420–35.

– 1960b. 'From the Taungs Skull to "Nutcracker Man": Africa as the Cradle of Mankind and the Primates – Discoveries of the Last Thirty-five Years.' *Illustrated London News* 236: 44.

– 1961. 'Exploring 1,750,000 Years into Man's Past.' *National Geographic* 120: 564–89.
– 1963. 'Very Early East African Hominidae and Their Ecological Setting.' In *African Ecology and Human Evolution.* Ed. F.C. Howell and F. Bourlière. Chicago: Aldine, pp. 448–57.
Leakey, L.S.B., J.F. Evernden, and G.H. Curtis. 1961. 'Age of Bed I, Olduvai Gorge, Tanganyika.' *Nature* 191: 478–9.
Leakey, M.D. 1966. 'A Review of the Oldowan Culture from Olduvai Gorge, Tanzania.' *Nature* 210: 462–6.
– 1979. 'Footprints in the Ashes of Time.' *National Geographic* 155: 446–57.
– 1984. *Disclosing the Past.* Garden City, N.Y.: Doubleday.
Lee, R.B., and I. De Vore, eds. 1968. *Man the Hunter.* Chicago: Aldine.
Le Grand, H. 1988. *Drifting Continents and Shifting Theories: The Modern Revolution in Geology and Scientific Change.* Cambridge: Cambridge University Press.
– 1990. 'Is a Picture Worth a Thousand Experiments?' In *Experimental Inquiries: Historical, Philosophical and Social Studies of Experimentation in Science.* Ed. H. Le Grand. Dordrecht: Kluwer Academic Publishers, pp. 241–70.
Le Gros Clark, W.E. 1947. 'Observations on the Anatomy of the Fossil Australopithecinae.' *Journal of Anatomy* 81: 300–33.
– 1948. 'Observations on the Anatomy of the Fossil Australopithecinae.' *Yearbook of Physical Anthropology* 3: 143–77.
Leibniz, G.W. 1689. 'Discourse on Metaphysics.' In *G.W. Leibniz: Philosophical Essays.* Ed. R. Ariew and D. Garber. Indianapolis: Hackett, 1989.
Lenoble, Robert. 1964. 'The Seventeenth Century Scientific Revolution.' In *The Beginnings of Modern Science.* Ed. R. Taton. New York: Basic Books, pp. 180–99.
Levenson, Jay A., ed. 1991. *Circa 1492: Art in the Age of Exploration.* Washington: National Gallery of Art.
Lewin, R. 1988. *In the Age of Mankind.* Washington, D.C.: Smithsonian Books.
Lewis, G.A. 1866. *Natural History of Birds: Lectures on Ornithology, in Ten Parts, Part I.* Philadelphia: J.A. Bancroft & Co.
Lewontin, R.C., J.A. Moore, B. Wallace, and W.B. Provine, eds. 1981. *Dobzhansky's Genetics of Natural Populations.* New York: Columbia University Press.
Lindberg, David, and Robert S. Westman, eds. 1990. *Reappraisals of the Scientific Revolution.* Cambridge: Cambridge University Press.
Lobelius [Matthias de l'Obel] and Pierre Pena. 1576. *Stirpium adversaria nova.* Antwerp: Plantin.

López Piñero, J.M. 1987. *El Grabado en la Ciencia Hispanica.* Madrid: Consejo Superior de Investigaciones Cientificas.

Lovejoy, A.O. 1936. *The Great Chain of Being: A Study of the History of an Idea.* Cambridge, Mass.: Harvard University Press.

Lubbock, J. 1865. *Prehistoric Times, as Illustrated by Ancient Remains, and the Manners and Customs of Modern Savages.* London: Williams and Norgate.

Lundgren, A. 1992. 'Berzelius, Dalton, and the Chemical Atom.' In Melhado and Frängsmyr 1992, pp. 85–106.

Lutz, C., and J. Collins. 1992. 'The Photograph as an Intersection of Gazes.' *Visual Anthropology Review.*

Lynch, M., and S. Woolgar, eds. 1990. *Representation in Scientific Practice.* Cambridge, Mass: MIT Press.

Mach, E. 1883. *The Science of Mechanics.* 6th ed. Trans. T.J. McCormack. La Salle, Ill.: Open Court, 1960.

MacKenzie, Ann. 1989. 'Descartes on Life and Sense.' *Canadian Journal of Philosophy* 19(2): 163–92.

Mackenzie, C. 1822. *One Thousand Experiments in Chemistry.* New ed. London: Richard Phillips.

Maddison, Francis. 1991. 'Entries on Martin Bylica's Celestial Globe, Astrolabe and Torquetum.' In Levenson 1991, pp. 120–2.

Mahoney, Michael S. 1985. 'Diagrams and Dynamics: Mathematical Perspectives on Edgerton's Thesis.' In Shirley and Hoeniger 1985, pp. 198–220.

Marcet, J. 1825. *Conversations on Chemistry.* 10th ed. London: Longman, Hurst Rees, Orme, Brown and Green.

Marvin, U.B. 1973. *Continental Drift: The Evolution of a Concept.* Washington, D.C.: Smithsonian Institution Press.

Mayr, E. 1959. 'Where Are We?' *Cold Spring Harbor Symposia on Quantitative Biology* 24: 1–14.

– 1969. *Principles of Systematic Zoology.* New York: McGraw-Hill.

– 1982. *The Growth of Biological Thought: Diversity, Evolution and Inheritance.* Cambridge, Mass.: Harvard University Press.

Mazzolini, Renato G., ed. 1993. *Non-Verbal Communication in Science prior to 1900.* Florence: Leo S. Olschki.

McGowan, C. 1982. 'The Wing Musculature of the Brown Kiwi *Apteryx australis mantelli* and Its Bearing on Ratite Affinities.' *Journal of Zoology* 197: 173–219.

Mee, Charles L., Jr. 1988. *Rembrandt's Portrait: A Biography.* New York: Simon and Schuster.

Melhado, E., and T. Frängsmyr, eds. 1992. *Enlightenment Science in the Roman-*

tic Era: The Chemistry of Berzelius in Its Cultural Setting. Cambridge: Cambridge University Press.

Mendeléeff, D. 1897. *The Principles of Chemistry.* 2d ed. Trans. G. Kamensky. Ed. T.A. Lawson. London: Longmans, Green.

Merchant, Carolyn. 1980. *The Death of Nature: Women, Ecology and the Scientific Revolution.* San Francisco: Harper and Row.

Meyer, Barbara Hochstetler, and Alice Wilson Glover. 1989. 'Botany and Art in Leonardo's *Leda and the Swan.*' *Leonardo* 22: 75–82.

Miller, Arthur I. 1978. 'Visualization Lost and Regained: The Genesis of the Quantum Theory in the Period 1913–27.' In *On Aesthetics in Science.* Ed. Judith Wechsler. Cambridge, Mass.: MIT Press, pp. 73–102.

– 1984. *Imagery in Scientific Thought: Creating Twentieth Century Physics.* Boston: Birkhauser.

Miller, R.P., and V.R. Miller, eds. 1983. *Descartes: The Principles of Philosophy.* Dordrecht: D. Reidel Publishing Company.

Miller, W.H. 1839. *A Treatise on Crystallography.* Cambridge: J. & J.J. Deighton.

Mitchell, W.J.T., ed. 1974. *The Language of Images.* Chicago: University of Chicago Press.

Morrow, James H., ed. 1981. *Hans Baldung Grien: Prints and Drawings.* Washington: National Gallery of Art.

Moser, S. 1992. 'The Visual Language of Archaeology: A Case Study of the Neanderthals.' *Antiquity* 66: 831–44.

– 1993a. 'Gender Stereotyping in Pictorial Reconstructions of Human Origins.' In *Women in Archaeology: A Feminist Critique.* Ed. H. DuCros and L. Smith. Canberra: Australian National University Press.

– 1993b. 'Picturing the Prehistoric.' *Metascience* 4: 58–67.

– n.d. 'Science and Social Values: Presenting Archaeological Findings in Museum Displays.' In *Archaeology and Issues in Heritage Management.* Ed. A. Clarke and L. Smith (in press).

Moss, Jean D. 1993. *Novelties in the Heavens: Rhetoric and Science in the Copernican Controversy.* Chicago: University of Chicago Press.

Mulvey, L. 1975. 'Visual Pleasure and Narrative Cinema.' *Screen* 16(3): 7–18.

Mumford, Lewis. 1967. *The Myth of the Machine: Technics and Human Development.* New York: Harcourt, Brace and World.

Münster, Sebastian. 1536. *Organum Uranicum.* Basel: J. Frobenius.

Murdoch, John E. 1984. *Antiquity and the Middle Ages.* Album of Science Series. New York: Scribner.

Muspratt, J.S. 1860. *Chemistry Theoretical, Practical and Analytical, as Applied and Relating to the Arts and Manufactures.* 2 vols. Glasgow: W. Mackenzie.

Myers, F.W.H. 1992. *Human Personality and Its Survival of Bodily Death (1919)*. Norwich: Pilgrim Books.

Myers, G. 1990. 'Every Picture Tells a Story: Illustrations in E.O. Wilson's *Sociobiology*.' In Lynch and Woolgar 1990, pp. 231–65.

Nebelsick, Harold P. 1985. *Circles of God: Theology and Science from the Greeks to Copernicus*. Edinburgh: Scottish Academic Press.

Needham, Joseph. 1982. *Explorations in the History of Science and Technology in China*. Shanghai: Shanghai Chinese Classics Publishing House.

Needham, Joseph, et al. 1954–. *Science and Civilisation in China*. 6 vols. Cambridge: Cambridge University Press.

Nelson, G.G., and N.I. Platnick. 1981. *Systematics and Biogeography: Cladistics and Vicariance*. New York: Columbia University Press.

Neugebauer, Otto. 1968. 'On the Planetary Theory of Copernicus.' *Vistas in Astronomy* 18: 89–104.

Newton, I. 1934. *Sir Isaac Newton's Principles of Natural Philosophy and His System of the World*. Trans. A Motte. Revised by F. Cajori. Berkeley: University of California Press.

Nicholson, W. 1796. *The First Principles of Chemistry*. 3d ed. London: Robinson.

Nickon, A., and E.F. Silversmith. 1987. *Organic Chemistry: The Name Game, Modern Coined Terms and Their Origins*. Oxford: Pergamon.

North, John D., ed. 1976. 'Richard of Wallingford, *Tractatus horologii Astronomici*.' In *Richard of Wallingford: An Edition of his Writings*. 3 vols. Oxford: Clarendon Press.

Nye, M.J. 1993. *From Chemical Philosophy to Theoretical Chemistry: Dynamics of Matter and Dynamics of Disciplines, 1800–1950*. Berkeley: University of California Press.

Oakley, K.P. 1949. *Man the Tool-Maker*. London: British Museum of Natural History.

– 1956. 'The Earliest Toolmakers.' *Antiquity* 30: 4–8.

O'Hara, R.J. 1988a. 'Homage to Clio, or, toward an Historical Philosophy for Evolutionary Biology.' *Systematic Zoology* 37(2): 142–55.

– 1988b. 'Diagrammatic Classifications of Birds, 1819–1901: Views of the Natural System in 19th-century British Ornithology.' In *Acta XIX Congressus Internationalis Ornithologici*. Ed. H. Ouellet. 2 vols. Ottawa: National Museum of Natural Sciences, pp. 2746–59.

Olson, S.L. 1987. 'More on the Affinities of the Black-Collared Thrush of Borneo (*Chlamydochaera jefferyi*).' *Journal für Ornithologie* 128: 246–8.

Opdyke, N.D., B.P. Glass, J.D. Hays, and J.H. Foster. 1966. 'Paleomagnetic Study of Antarctic Deep-Sea Cores.' *Science* 154: 349–57.

Pächt, Otto. 1950. 'Early Italian Nature Studies and the Early Calendar Landscape.' *Journal of the Warburg and Courtauld Institutes* 13: 13–47.

Panofsky, Erwin. 1927. *Die Perspektive als 'symbolishe Form.* Vorträge der Bibliothek Warburg. English ed., 1991. *Perspective as Symbolic Form.* Trans. Christopher Wood. New York: Zone Books, 1991.

– 1955. *The Life and Art of Albrecht Dürer.* 4th ed. Princeton: Princeton University Press.

– 1962. 'Artist, Scientist, Genius: Notes on the "Renaissance Dämmerung."' In *The Renaissance: Six Essays.* Ed. Wallace K. Ferguson et al. New York: Harper and Row, pp. 121–82.

Paris, J.A. 1831. *The Life of Sir Humphry Davy.* London: Henry Colburn and Richard Bentley.

Parker, G. 1978. 'Searching for Mates.' In *Behavioural Ecology: An Evolutionary Approach.* Ed. J. Krebs and N. Davies. Sunderland: Sinauer Associates, pp. 214–44.

Parkes, S. 1808. *The Chemical Catechism, with Notes, Illustrations and Experiments.* 3d ed. London: Lackington, Allen.

Parkinson, J. 1801. *The Chemical Pocket-Book.* 2d ed. London: Whittingham.

Pera, Marcello, and William R. Shea, eds. 1992. *Persuading Science: The Art of Scientific Rhetoric.* Canton, Mass.: Science History Publications.

Pfeiffer, J.E. 1969. *The Emergence of Man.* London: Cardinal.

Pickering, A. 1984. *Constructing Quarks: A Sociological History of Particle Physics.* Edinburgh: Edinburgh University Press

– ed. 1992. *Science as Practice and Culture.* Chicago: University of Chicago Press.

Pietro d'Abano. 1496. *Conciliator differentiarum philosophorum et precipue medicorum.* Venice: Bonetus Locatellus.

Pitman, W.C., III, and J.P. Heirtzler. 1966. 'Magnetic Anomalies over the Pacific-Antarctic Ridge.' *Science* 154: 1164–71.

Pittenger, M. 1993. *American Socialists and Evolutionary Thought, 1870–1920.* Madison: University of Wisconsin Press.

Pliny, the Elder. 1855–7. *The Natural History of Pliny.* Trans. John Bostock and H.T. Riley. 6 vols. London: H.G. Bohn.

Poisson, N.J. 1671. *Commentaire ou remarques sur la méthode de M. Descartes.* Paris.

Popper, K.R. 1972. *Objective Knowledge: An Evolutionary Approach.* Oxford: Clarendon Press.

Potts, R. 1983. 'Foraging for Faunal Resources by Early Hominids at Olduvai Gorge, Tanzania.' In Clutton-Brock and Grigson 1983, pp. 51–62.

– 1984a. 'Home Bases and Early Hominids.' *American Scientist* 72: 338–47.
– 1984b. 'Hominid Hunters? Problems of Identifying the Earliest Hunter/Gatherers.' In *Hominid Evolution and Community Ecology*. Ed. R. Foley. New York: Academic Press, pp. 129–59.
– 1987. 'Reconstructions of Early Hominid Socioecology: A Critique of Primate Models.' In Kinzey 1987, pp. 28–47.
– 1988. *Early Hominid Activities at Olduvai*. New York: Aldine De Gruyter.
Potts, R., and P. Shipman. 1981. 'Cutmarks Made by Stone Tools on Bones from Olduvai Gorge, Tanzania.' *Nature* 291: 577–80.
Poulle, Emanuel, ed. 1987. *Johannis de Dondis paduani civis astrarium*. Padua: Edizione 1+1.d
Prager, Frank D. 1973. 'Kepler als Erfinder.' In *Internationales Kepler-Symposium Weil der Stadt 1971*. Ed. Fritz Krafft, Karl Meyer, and Bernard Sticker. Hildesheim: Verlag Dr. H.A. Gerstenberg, pp. 385–405.
Price, Derek J. de Solla. 1964. 'Automata and the Origins of Mechanism and the Mechanistic Philosophy.' *Technology and Culture* 5: 9–23.
Provine, W. 1986. *Sewall Wright and Evolutionary Biology*. Chicago: University of Chicago Press.
Putnam, H. 1981. *Reason, Truth, and History*. Cambridge: Cambridge University Press.
– 1982. 'Why Reason Can't Be Naturalized.' *Synthese* 52: 3–23.
Ramelli, Agostino. 1588. *The Various and Ingenious Machines of Agostino Ramelli (1588)*. Ed. and trans. by M.T. Gnudi and E. Ferguson. Baltimore: Johns Hopkins University Press, 1976.
Randall, Lilian M.C. 1966. *Images in the Margins of Gothic Manuscripts*. Berkeley and Los Angeles: University of California Press.
Reeds, Karen Meier. 1976. 'Renaissance Humanism and Botany.' *Annals of Science* 33: 519–42.
Reichenow, A. 1882. *Die Vögel der zoologischen Gärten*. Leipzig: Verlag von L.U. Kittler.
Reif, W.-E. 1983. 'Hilgendorf's (1863) Dissertation on the Steinheim Planorbids (Gastropoda; Miocene): The Development of a Phylogenetic Research Program for Paleontology.' *Paläontologische Zeitschrift* 57: 7–20.
Rennie, J. 1833. *Ornithological Dictionary of British Birds, by Colonel G. Montagu, F.L.S.: A New Edition*. London: W.S. Orr and W. Smith.
Rensberger, B. 1981. 'Facing the Past.' *Science* 81: 40–51.
Reti, Ladislao. 1980. 'The Engineer.' In *Leonardo the Inventor*. Ed. Ludwig H. Heydenreich et al. New York: McGraw-Hill, pp. 124–85.
Rheticus, Georg Joachim. 1541. *Narratio prima*. Basel: A.P. Gassarus.

Riddle, John M. 1985. *Dioscorides on Pharmacy and Medicine.* Austin: University of Texas Press.

Ritterbush, Philip C. 1985. 'The Organism as Symbol: An Innovation in Art.' In Shirley and Hoeniger 1985, pp. 149–67.

Roberts, C. 1992. 'The Photographic Pioneer.' In King-Hele 1992, pp. 67–76.

Roberts. K., and J. Tomlinson. 1992. *The Fabric of the Body: European Traditions of Anatomical Illustration.* Oxford: Clarendon Press.

Robin, H. 1992. *The Scientific Image: From Cave to Computer.* New York: Harry N. Abrams.

Robinson, J.T. 1954. 'The Genera and Species of the Australopithecinae.' *American Journal of Physical Anthropology* 12: 181–200.

Rodis-Lewis, Geneviève. 1978. 'Limitations of the Mechanical Model in the Cartesian Conception of the Organism.' In Hooker 1978, pp. 152–70.

Rodowick, D.N. 1982. 'The Difficulty of Difference.' *Wide Angle* 5(1): 4–15.

Roe, D. 1980. 'Introduction: Precise Moments in Remote Time.' *World Archaeology* 12: 107–8.

Root-Bernstein, Robert. 1985. 'Visual Thinking: The Art of Imagining Reality.' *Transactions of the American Philosophical Society* 75(6): 50–67.

Rose, Paul L. 1975. 'Universal Harmony in Regiomontanus and Copernicus.' In *Avant, avec, après Copernic: la representation de l'univers et ses conséquences épistémologiques.* Paris: A. Blanchard, pp. 153–8.

Rosen, Edward. 1971. *Three Copernican Treatises.* 3d. ed. New York: Octagon Books.

– 1975. 'Kepler and the Lutheran Attitude towards Copernicanism in the Context of the Struggle between Science and Religion.' In Beer and Beer 1975, pp. 317–38.

Rossi, Paulo. 1970. *Philosophy, Technology, and the Arts in the Early Modern Era.* Trans. Salvator Attanasio. New York: Harper and Row.

Rothermel, H. 1993. 'Images of the Sun: Warren de la Rue, George Biddell Airy and Celestial Photography.' *British Journal for the History of Science* 26: 137–69.

Rudwick, Martin J.S. 1976. 'The Emergence of a Visual Language for Geological Science 1760–1840.' *History of Science* 14: 149–95.

– 1989. 'Encounters with Adam, or at Least the Hyaenas; Nineteenth-Century Visual Representations of the Deep Past.' In *History, Humanity and Evolution: Essays for John C. Greene.* Ed. J.R. Moore. Cambridge: Cambridge University Press, pp. 231–51.

– 1992. *Scenes from Deep Time: Early Pictorial Representations of the Prehistoric World.* Chicago: University of Chicago Press.

Ruse, M. 1973. *The Philosophy of Biology*. London: Hutchinson.

– 1982. *Darwinism Defended: A Guide to the Evolution Controversies*. Reading, Mass.: Addison-Wesley.

– 1986. *Taking Darwin Seriously: A Naturalistic Approach to Philosophy*. Oxford: Blackwell.

– 1989. *The Darwinian Paradigm: Essays on Its History, Philosophy and Religious Implications*. London: Routledge.

Saunders, J.B., and C.D. O'Malley. 1950. *The Illustrations from the Works of Andreas Vesalius of Brussels*. New York: Dover Books.

Schaaf, L.J. 1992. 'The Poetry of Light: Herschel, Art and Photography.' In King-Hele 1992, pp. 67–76, 77–99.

Schaller, G.B., and G.R. Lowther. 1969. 'The Relevance of Carnivore Behavior to the Study of Early Hominids.' *Southwestern Journal of Anthropology* 25: 307–41.

Scheele, W.E. 1957. *Prehistoric Man and the Primates*. Cleveland: World Publishing Company.

Scheffler, I. 1967. *Science and Subjectivity*. Indianapolis: Bobbs-Merrill.

Schiffer, M.B. 1983. 'Toward the Identification of Formation Processes.' *American Antiquity* 48: 675–706.

– 1987. *Formation Processes of the Archaeological Record*. Albuquerque: University of New Mexico Press.

Schorlemmer, C. 1894. *The Rise and Development of Organic Chemistry*. 2d ed. London: Macmillan.

Schultz, Bernard. 1985. *Art and Anatomy in Renaissance Italy*. Ann Arbor, Mich.: UMI Research Press.

Schupbach, William. 1982. *The Paradox of Rembrandt's 'Anatomy of Dr. Tulp.'* London: Welcome Institute for the History of Medicine.

Schuster, John, and Richard Yeo, eds. 1986. *The Politics and Rhetoric of Scientific Method: Historical Studies*. Dordrecht: Kluwer Academic Publishers Group.

Schweber, S. 1985. 'Feynman and the Visualization of Space-time Processes.' *Review of Modern Physics* 58: 449–508.

Shapere, D. 1964. 'The Structure of Scientific Revolutions.' *Philosophical Review* 73: 383–94.

Shapin, S. 1982. 'History of Science and Its Sociological Reconstructions.' *History of Science* 20: 157–211.

Sharpe, R.B. 1891. *A Review of Recent Attempts to Classify Birds*. Budapest: Second International Ornithological Congress.

Shea, William, ed. 1991. *Storia delle scienze: Le scienze fisiche e astronomiche*. Milan: Mondadori.

Shea, William R. 1991a. *The Magic of Numbers and Motion: The Scientific Career of René Descartes.* Canton: Science History Publications.
– 1991b. 'La rivoluzione scientifica.' In Shea 1991, pp. 168–233.
Sheehan, William. 1988. *Planets and Perception: Telescopic Views and Interpretations, 1609–1909.* Tucson: University of Arizona Press.
Shipman, P. 1983. 'Early Hominid Lifestyle: Hunting and Gathering or Foraging and Scavenging?' In Clutton-Brock and Grigson 1983, pp. 31–49.
– 1986. 'Scavenging or Hunting in Early Hominids: Theoretical Framework and Tests.' *American Anthropologist* 88: 27–43.
Shirley, J.W., and F.D. Hoeniger, eds. 1985. *Science and the Arts in the Renaissance.* Washington, D.C.: Folger Shakespeare Library.
Simon, H.A. 1978. 'On the Forms of Mental Representation.' In *Perception and Cognition: Issues in the Foundations of Psychology.* Ed. C.W. Savage. Minnesota Studies in the Philosophy of Science, vol. 9. Minneapolis: University of Minnesota Press, pp. 3–18.
Simpson, G.G. 1944. *Tempo and Mode in Evolution.* New York: Columbia University Press.
Smith, W.G. 1894. *Man the Primeval Savage.* London: Edward Stanford.
Snyder, Joel. 1974. 'Picturing Vision.' In Mitchell 1974, pp. 219–46.
Stafford, Barbara Marie. 1984. *Voyage into Substance: Art, Science, Nature, and the Illustrated Travel Account, 1760–1840.* Cambridge, Mass.: MIT Press.
Stebbins, G.L. 1969. *The Basis of Progressive Evolution.* Chapel Hill: University of North Carolina Press.
Stevens, P.F. 1983. 'Augustin Augier's "Arbre Botanique" (1801), a Remarkable Early Botanical Representation of the Natural System. *Taxon* 32: 203–11.
– 1984. 'Metaphors and Typology in the Development of Botanical Systematics 1690–1960, or the Art of Putting New Wine in Old Bottles.' *Taxon* 33: 169–211.
Stewart, J.A. 1990. *Drifting Continents and Colliding Paradigms: Perspectives on the Geoscience Revolution.* Bloomington: Indiana University Press.
Stokes, Charlotte. 1980. 'The Scientific Methods of Max Ernst: His Use of Scientific Subjects from *La Nature.*' *Art Bulletin* 62(3): 453–65.
Stresemann, E. 1975. *Ornithology from Aristotle to the Present.* Cambridge, Mass.: Harvard University Press.
Strickland, H.E. 1840. 'Observations upon the Affinities and Analogies of Organized Beings.' *Magazine of Natural History* 4 (new series): 219–26.
– 1841. 'On the True Method of Discovering the Natural System in Zoology and Botany.' *Annals and Magazine of Natural History* 6: 184–94.
– 1844. 'Description of a Chart of the Natural Affinities of the Insessorial

Order of Birds.' *Report of the Thirteenth Meeting of the British Association for the Advancement of Science Held at Cork in August 1843, Notices and Abstracts of Communications*, p. 69.

Stringer, C., and C. Gamble. 1993. *In Search of the Neanderthals: Solving the Puzzle of Human Origins.* London: Thames and Hudson.

Suppe, F. 1989. *The Semantic Conception of Theories and Scientific Realism.* Urbana: University of Illinois Press.

Swainson, W. 1834. *A Preliminary Discourse on the Study of Natural History.* London: Longman, Rees, Orme, Brown, Green & Longman and J. Taylor.

– 1835. *A Treatise on the Geography and Classification of Animals.* London: Longman, Rees, Orme, Brown, Green & Longman and J. Taylor.

– 1837. *On the Natural History and Classification of Birds, Vol. II.* London: Longman, Rees, Orme, Brown, Green & Longman and J. Taylor.

Swainson, W., and J. Richardson. 1829–37. *Fauna Boreali-Americana ... Part Second, the Birds.* 4 vols. London: John Murray.

Swerdlow, Noel, and Otto Neugebauer. 1984. *Mathematical Astronomy in Copernicus's 'De Revolutionibus.'* 2 vols. New York: Springer-Verlag.

Tabernaemontanus, Jacobus [Dietrich, Jacob]. 1590. *Eicones plantarum seu stirpium.* Frankfurt: Nicholas Bassaeus.

Taylor, P.J., and A.S. Blum. 1991. 'Ecosystems as Circuits: Diagrams and the Limits of Physical Analogies.' *Biology and Philosophy* 6: 275–94.

Thagard, P. 1991. *Conceptual Revolutions.* Princeton: Princeton University Press.

Thomson, C.W. 1887. *The Voyage of the Challenger: The Atlantic.* London: Macmillan.

Thomson, T. 1831. *System of Chemistry of Inorganic Bodies.* 7th ed. Edinburgh: Blackwood.

– 1835. 'On Calico-Printing.' *Records of General Science* 1: 3–19, 161–73, 321–30.

Thoren, Victor. 1990. *The Lord of Uraniborg: A Biography of Tycho Brahe.* Cambridge: Cambridge University Press.

Topper, David R. 1980. '"To Reason by Means of Images": J.J. Thomson and the Mechanical Picture of Nature.' *Annals of Science* 37: 31–57.

– 1988. 'On a Ghost of Historiography Past.' *Leonardo* 21: 76–8.

– 1990a. 'The Parallel Fallacy: On Comparing Art and Science.' *British Journal of Aesthetics* 30: 311–18.

– 1990b 'Natural Science and Visual Art: Reflections on the Interface.' In *Beyond History of Science: Essays in Honor of Robert E. Schofield.* Ed. Elizabeteh Garber. Bethlehem: Lehigh University Press, pp. 296–310.

Toynbee, J.M.C. 1973. *Animals in Roman Life and Art.* Ithaca, N.Y.: Cornell University Press.

Trigg, G.L. 1975. *Landmark Experiments in Twentieth Century Physics.* New York: Crane, Russak & Co.

Trinkaus, E., and P. Shipman. 1993. *The Neanderthals: Changing the Image of Mankind.* New York: Knopf.

Tufte, Edward R. 1983. *The Visual Display of Quantitative Information.* Cheshire, Conn.: Graphics Press.

– 1990. *Envisioning Information.* Cheshire, Conn: Graphics Press.

Turner, Anthony J. 1987. *Early Scientific Instruments: Europe, 1400–1800.* London: Philip Wilson Publishers.

Turner, E. 1831. *Elements of Chemistry.* London: John Taylor.

Turner, Gerard L'E., ed. 1991. *Storia delle scienze: Gli strumenti.* Turin: Einaudi.

Twyman, Michael. 1970a. *Printing 1770–1970: An Illustrated History of Its Development and Uses in England.* London: Eyre and Spottiswoode.

– 1970b. *Lithography 1800–1850: The Techniques of Drawing on Stone in England and France and Their Application in Works of Topography.* London: Oxford University Press.

Underwood, E.A., ed. 1953. *Science, Medicine and History: Essays on the Evolution of Scientific Thought and Medical Practice Written in Honour of Chas. Singer.* Vol. 1. London: Oxford University Press.

van der Gracht, W., A.J.M. van Waterschoot et al. 1928. *Theory of Continental Drift.* Tulsa, Okla.: American Association of Petroleum Geologists.

Van Helden, Albert, trans., 1989. *Sidereus Nuncius, or, the Sidereal Messenger.* Chicago: University of Chicago Press.

van't Hoff, J.H. 1898. *The Arrangement of Atoms in Space.* 2d ed. Trans. A. Eiloart. London: Longmans, Green.

Verdet, Jean-Paul. 1991. 'L'Astronomia dalle origini a Copernico.' In Shea 1991, pp. 38–109.

Vesalius, Andreas. 1538. *Tabulae sex.* Venice: B. Vitali.

– 1543. *De humani corporis fabrica.* Basel: Johannes Oporinus.

Vigors, N.A. 1824. 'Observations on the Natural Affinities That Connect the Orders and Families of Birds.' *Transactions of the Linnean Society* 14: 395–517.

Vincent, C.W. n.d. *Chemistry .. as Applied to the Arts and Manufacturers, by Writers of Eminence.* London: William Mackenzie.

Vine, F.J., and D.H. Matthews. 1963. 'Magnetic Anomalies over Oceanic Ridges.' *Nature* 199: 947–9.

Vine, F.J., and J.T. Wilson. 1965. 'Magnetic Anomalies over a Young Oceanic Ridge off Vancouver Island.' *Science* 150: 485–9.

Von Gersdorf, Hans. 1517. *Feldtbüch der Wundartzney.* Strasbourg: Scott.

Voss, E.G. 1952. 'The History of Keys and Phylogenetic Trees in Systematic

Biology.' *Journal of the Scientific Laboratories of Denison University* 43: 1–25.

Vrooman, Jack Rochford. 1970. *René Descartes: A Biography*. New York: G.P. Putnam's Sons.

Waddington, C.H. 1956. *Principles of Embryology*. New York: Macmillan.

Waechter, J. 1976. *Man before History*. Oxford: Elsevier, Phaidon.

Wallace, A.R. 1856. 'Attempts at a Natural Arrangement of Birds.' *Annals and Magazine of Natural History*, 2d series 18: 193–216.

Warner, D.J. 1979. *Graceanna Lewis, Scientist and Humanitarian*. Washington, D.C.: Smithsonian Institution Press.

Washburn, S.L. 1957. 'Australopithecines: The Hunters or the Hunted?' *American Anthropologist* 59: 612–14.

Washburn, S.L., and V. Avis. 1958. 'Evolution of Human Behavior.' In *Behavior and Evolution*. Ed. A. Roe and G.G. Simpson. New Haven: Yale University Press, pp. 421–36.

Washburn, S.L., and C. Lancaster. 1968. 'The Evolution of Hunting.' In Lee and De Vore 1968, pp. 293–303.

Weaver, K. F. 1985. 'Stones, Bones, and Early Man: The Search for Our Ancestors.' *National Geographic* 168: 561–623.

Wegener, A. 1915. *Die Entstehung der Kontinente und Ozeane*. Braunschweig: F. Vieweg & Sohn.

– 1922. *The Origin of Continents and Oceans*. Trans. J.G.A. Skerl. London: Methuen.

Welu, James A. 1975. 'Vermeer: His Cartographic Sources.' *Art Bulletin* 57(4): 529–47.

Westman, Robert S. 1975. 'Three Responses to the Copernican Theory: Johannes Praetorius, Tycho Brahe and Michael Maestlin.' In *The Copernican Achievement*. Ed. Robrt S. Westman. Berkeley, Los Angeles, and London: University of California Press, pp. 285–345.

– 1984. 'Nature, Art and Psyche: Jung, Pauli, and the Kepler-Fludd Polemic.' In *Occult and Scientific Mentalities in the Renaissance*. Ed. Brian Vickers. Cambridge: Cambridge University Press, pp. 177–229.

– 1990. 'Proof, Poetics, and Patronage: Copernicus's Preface to *De revolutionibus*.' In Lindberg and Westman 1990, pp. 167–205.

Whewell, William. 1897. *History of the Inductive Sciences, from the Earliest to the Present Time*. 2 vols. New York: D. Appleton and Co.

White, Lynn, Jr. 1947. 'Natural Science and Naturalistic Art in the Middle Ages.' *American Historical Review* 52: 421–35.

– 1978. 'Medical Astrologers and Late Medieval Technology.' In *Medieval Religion and Technology*. Berkeley and Los Angeles: University of California Press, pp. 297–316.

Whitney, Charles A. 1986. 'The Skies of Vincent Van Gogh.' *Art History* 9: 351–62.

Whitney, Elspeth. 1990. *Paradise Restored: The Mechanical Arts from Antiquity through the Thirteenth Century.* Philadephia: American Philosophical Society.

Wickersheimer, Ernest. 1913. 'L'Anatomie de Guido de Vigevano, médecin de la reine Jeanne de Bourgogne (1345).' *Sudhoffs Archiv für Geschichte der Medizin* 7: 1–25.

– 1926. *Anatomies de Mondino dei Luzzi et de Guido de Vigevano.* Documents scientifiques du XVe siècle, t. 3. Paris: Droz.

Williams, G.C. 1966. *Adaptation and Natural Selection: A Critique of Current Evolutionary Thought.* Princeton: Princeton University Press.

Wilson, Edward O. 1975. *Sociobiology: The New Synthesis.* Cambridge, Mass.: Harvard University Press.

Wilson, H.F., and M.H. Doner. 1937. *The Historical Development of Insect Classification.* Privately printed.

Winckler, Mary G., and Albert Van Helden. 1992. 'Representing the Heavens: Galileo and Visual Astronomy.' *Isis* 83: 195–217.

Winsor, M.P. 1976. *Starfish, Jellyfish, and the Order of Life: Issues in Nineteenth Century Science.* New Haven: Yale University Press.

Wollaston, W.H. 1813. 'The Bakerian Lecture: On the Elementary Particles of Certain Crystals.' *Philosophical Transactions of the Royal Society* 103: 51–63.

Wood, B. 1977. *The Evolution of Early Man.* Melbourne: Cassell.

Woodward, David, ed. 1987. *Art and Cartography: Six Historical Essays.* Chicago: University of Chicago Press.

Woodward, J. 1989. 'Data and Phenomena.' *Synthese* 79: 393–472.

Woolgar, S., ed. 1988. *Knowledge and Reflexivity: New Frontiers in the Sociology of Knowledge.* London: Sage.

Wright, S. 1931. 'Evolution in Mendelian Populations.' *Genetics* 16: 97–159.

– 1932. 'The Roles of Mutation, Inbreeding, Crossbreeding and Selection in Evolution.' *Proceedings of the Sixth International Congress of Genetics* 1: 356–66.

Yates, Frances A. 1969. *Giordano Bruno and the Hermetic Tradition.* New York: Vintage Books.

Zeki, Semir. 1992. 'The Visual Image in Mind and Brain.' *Scientific American.* 267(3): 68–76.

Zeller, Suzanne. 1987. *Inventing Canada : Early Victorian Science and the Idea of a Transcontinental Nation.* Toronto : University of Toronto Press.

Zimmer, E. 1987. *Deutsche und Niederländische astronomische Instrumente des 11. – 18. Jahrhunderts.* Munich: Sotheby's Publications.

Notes on Contributors

Brian S. Baigrie is associate professor of history and philosophy of science at the Institute for the History and Philosophy of Science and Technology, University of Toronto, Canada. He is the author of numerous papers on the Scientific Revolution of the seventeenth century. His current research is divided between two ongoing projects – a definitive Descartes biographical dictionary and a manuscript on conceptions of artifice and nature in Descartes's natural philosophy.

James Robert Brown is professor of philosophy at the University of Toronto, Canada. His interests include philosophy of mathematics, foundations of physics, realism, thought experiments, and methodological issues in the natural sciences. He is the author of *The Rational and the Social* (1989), *The Laboratory of the Mind: Thought Experiments in the Natural Sciences* (1991), and *Smoke and Mirrors: How Science Reflects Reality* (1994).

Ronald N. Giere is professor of philosophy and director of the Center for Philosophy of Science at the University of Minnesota (U.S.A.). In addition to many papers in the philosophy of science, he is the author of an elementary textbook, *Understanding Scientific Reasoning* (3d edition, 1991), and of *Explaining Science: A Cognitive Approach* (1988). His current research focuses on models of representation in science.

Bert S. Hall is associate professor of history of technology at the Institute for the History and Philosophy of Science and Technology, University of Toronto,

Canada. He has written extensively on late medieval and early modern technology, especially on the manuscript sources.

Martin Kemp is British Academy Wolfson Research Professor and professor of the history of art, University of Oxford. His research has concentrated on Renaissance art and more broadly on the interface of the sciences and the visual arts. As the author of *Leonardo da Vinci: The Marvellous Works of Nature and Man* (1981 and 1993) and *The Science of Art: Optical Themes in Western Art from Brunelleschi to Seurat* (1990 and 1992), he is currently exploring themes in the natural and human sciences.

David Knight has taught history of science at Durham University in England since 1964, after reading chemistry at Oxford. He is now professor of the history and philosophy of science. In 1993 he was president of the History of Science Section of the B.A.A.S.; during 1981–7 he was editor of the *British Journal for the History of Science.* His most recent books are *Ideas in Chemistry: A History of the Science* (1992); and *Humphry Davy: Science and Power* (1992).

Stephanie Moser is research fellow in archaeology at the University of Southampton. She has published several articles on visual imagery, including 'The Visual Language of Archaeology: A Case Study of the Neanderthals' (*Antiquity* 1992); 'Visions of the Australian Pleistocene: Prehistoric Life at Lake Mungo and Kutikina' (*Australian Archaeology* 1992); and 'Gender Stereotyping in Pictorial Reconstructions of Human Origins' (*Women in Archaeology: A Feminist Critique,* ed. H. DuCros and L. Smith, 1993). She is currently writing a book on the history of pictorial representation in archaeology.

Robert J. O'Hara is a postdoctoral fellow at the Center for Critical Inquiry in the Liberal Arts and adjunct assistant professor in the Department of Biology at the University of North Carolina at Greensboro, U.S.A. He has held postdoctoral appointments at the Smithsonian Institution and at the University of Wisconsin (Madison). He is author of 'Homage to Clio, or, toward an Historical Philosophy for Evolutionary Biology' (*Systematic Zoology* 1988); 'Telling the Tree: Narrative Representation and the Study of Evolutionary History' (*Biology and Philosophy* 1992); and 'Systematic Generalization, Historical Fate, and the Species Problem' (forthcoming in *Systematic Biology*). Professor O'Hara's research centres on the history, theory, and practice of systematics and the historical sciences.

Michael Ruse is professor of philosophy and zoology at the University of Guelph, Canada. He is founder and editor of the journal *Biology and Philosophy*, as well as editor of a new series of books on the philosophy of biology, to be published by Cambridge University Press. During 1992–3 he was chair (and now past chair) of the History and Philosophy Section of the American Association for the Advancement of Science. His books include: *The Darwinian Revolution: Science Red in Tooth and Claw* (1979), *Taking Darwin Seriously: A Naturalistic Approach to Philosophy* (1986), *Homosexuality: A Philosophical Inquiry* (1988), *Philosophy of Biology Today* (1988), *The Darwinian Paradigm* (1989), and *Evolutionary Naturalism* (1993).

David Topper is professor of history at the University of Winnipeg, Canada. He is a coeditor of the journal *Leonardo*. He has also won two awards for excellence in teaching. The paper in this volume is the third of a trilogy on the relationship between science and art. The first, 'Natural Science and Visual Art: Reflections on the Interface' (*Beyond History of Science*, ed. E. Garber, 1990), argued for the relative autonomy of art and science. The second, 'The Parallel Fallacy: On Comparing Arts and Science' (*British Journal of Aesthetics* 1990), clarified some aspects of art and science that should and shouldn't be compared. In general, his research has involved issues in the relationship among art, science, and visual perception.

Index